Perspectives in plant cell recognition presents a review of recent advances in understanding the cellular, molecular and genetic mechanisms governing cell-cell interactions in plants. In the case of the interaction between different cells of the same plant, most progress has been made in the study of gametes during sexual reproduction, and the volume begins by considering this topic. Exciting progress in the study of associations between somatic cells crucial to coordinated tissue development is also reported. Interactions between plant cells and cells of other organisms are then represented by consideration of plant pathogenesis and examples of mutual symbiosis; the study of both of these areas has yielded significant information about this category of interaction. In particular, the *Rhizobium/* legume symbiosis has been studied extensively and the genes controlling the specificity of the interaction and involved in creating a harmonious mutualism have been cloned and their products identified.

SOCIETY FOR EXPERIMENTAL BIOLOGY
SEMINAR SERIES: *48*

PERSPECTIVES IN PLANT CELL RECOGNITION

SOCIETY FOR EXPERIMENTAL BIOLOGY SEMINAR SERIES

A series of multi-author volumes developed from seminars held by the Society for Experimental Biology. Each volume serves not only as an introductory review of a specific topic, but also introduces the reader to experimental evidence to support the theories and principles discussed, and points the way to new research.

PERSPECTIVES IN PLANT CELL RECOGNITION

Edited by

J. A. Callow

School of Biological Sciences, University of Birmingham

J. R. Green

School of Biological Sciences, University of Birmingham

CAMBRIDGE
UNIVERSITY PRESS

Published by the Press Syndicate of the University of Cambridge
The Pitt Building, Trumpington Street, Cambridge CB2 1RP
40 West 20th Street, New York, NY 10011-4211, USA
10 Stamford Road, Oakleigh, Victoria 3166, Australia

First published 1992

Printed in Great Britain at the University Press, Cambridge

A catalogue record for this book is available from the British Library

Library of Congress cataloguing in publication data
Perspectives in plant cell recognition / edited by J.A. Callow, J.R. Green.
   p.   cm. – (Society for Experimental Biology seminar series : 48)
Papers presented at a symposium held in Apr. 1991 at Birmingham University.
Includes index.
ISBN 0 521 40445 2 (hc)
1. Plant cell recognition – Congresses.   I. Callow, J.A.   II. Green, J.R.
III. Series: Seminar series (Society for Experimental Biology) : 48.
QK725.P416   1992
581.87 – dc20   91–40252 CIP

ISBN 0 521 40445 2 hardback

Contents

Contributors

R. BAKHUIZEN
RUL/TNO Centre for Phytotechnology, Nonnensteeg 3, 2311VJ Leiden,
The Netherlands
M.-A. BARNY
John Innes Institute, John Innes Centre for Plant Science Research,
Colney Lane, Norwich NR4 7UH, UK
P. BONFANTE-FASOLO
Dipartimento di Biologia Vegetale dell'Università di Torino, Italy
D. BOWLES
Centre for Plant Biochemistry and Biotechnology, University of Leeds,
Leeds LS2 9JT, UK
N. J. BREWIN
John Innes Institute, John Innes Centre for Plant Science Research,
Colney Lane, Norwich NR4 7UH, UK
J. A. CALLOW
School of Biological Sciences, University of Birmingham, Birmingham
B15 2TT, UK
H. C. J. CANTER CREMERS
Department of Plant Molecular Biology, Nonnensteeg 3, 2311VJ Leiden,
The Netherlands
A. CHABOUD
Reconnaissance cellulaire et Amélioration des Plantes, Université de
Lyons 1, ICBMC/LA INRA.23 879, Bâtiment 741, 5ème étage, 43
Boulevard du 11 Novembre 1918, 69622 Villeurbanne Cedex, France
T. M. CUBO
John Innes Institute, John Innes Centre for Plant Science Research,
Colney Lane, Norwich NR4 7UH, UK
Present address: Departimento Microbiologica, Universidad de Sevilla,
4180 Sevilla, Spain.
A. DAVIES
John Innes Institute, John Innes Centre for Plant Science Research,
Colney Lane, Norwich NR4 7UH, UK

B. S. DE PATER
RUL/TNO Centre for Phytotechnology, Nonnensteeg 3, 2311VJ Leiden, The Netherlands

C. L. DIAZ
RUL/TNO Centre for Phytotechnology, Nonnensteeg 3, 2311VJ Leiden, The Netherlands

C. DIGONNET
Reconnaissance cellulaire et Amélioration des Plantes, Université de Lyons 1, ICBMC/LA INRA.23 879, Bâtiment 741, 5ème étage, 43 Boulevard du 11 Novembre 1918, 69622 Villeurbanne Cedex, France

J. A. DOWNIE
John Innes Institute, John Innes Centre for Plant Science Research, Colney Lane, Norwich NR4 7UH, UK

C. DUMAS
Reconnaissance cellulaire et Amélioration des Plantes, Université de Lyons 1, ICBMC/LA INRA.23 879, Bâtiment 741, 5ème étage, 43 Boulevard du 11 Novembre 1918, 69622 Villeurbanne Cedex, France

A. ECONOMOU
John Innes Institute, John Innes Centre for Plant Science Research, Colney Lane, Norwich NR4 7UH, UK

J. L. FIRMIN
John Innes Institute, John Innes Centre for Plant Science Research, Colney Lane, Norwich NR4 7UH, UK

F. C. H. FRANKLIN
School of Biological Sciences, University of Birmingham, Birmingham B15 2TT, UK

V. E. FRANKLIN-TONG
School of Biological Sciences, University of Birmingham, Birmingham B15 2TT, UK

G. W. GOODAY
Department of Molecular and Cell Biology, Marischal College, University of Aberdeen, Aberdeen AB9 1AS, UK

N. A. R. GOW
Department of Molecular and Cell Biology, Marischal College, University of Aberdeen, Aberdeen AB9 1AS, UK

J. R. GREEN
School of Biological Sciences, University of Birmingham, Birmingham B15 2TT, UK

A. GUNDER
John Innes Institute, John Innes Centre for Plant Science Research, Colney Lane, Norwich NR4 7UH, UK

R. M. HACKETT
School of Biological Sciences, University of Birmingham, Birmingham
B15 2TT, UK
A. W. B. JOHNSTON
John Innes Institute, John Innes Centre for Plant Science Research,
Colney Lane, Norwich NR4 7UH, UK
Present address: Department of Biological Sciences, University of East
Anglia, Norwich NR4 7UH, UK
E. I. KANNENBERG
John Innes Institute, John Innes Centre for Plant Science Research,
Colney Lane, Norwich NR4 7UH, UK
L. J. KELLOCK
Institute of Cell and Molecular Biology, University of Edinburgh,
Edinburgh EH9 3JH, UK
J. W. KIJNE
Department of Plant Molecular Biology and RUL/TNO Centre for
Phytotechnology, Nonnensteeg 3, 2311VJ Leiden, The Netherlands
H. KNIGHT
Institute of Cell and Molecular Biology, University of Edinburgh,
Edinburgh EH9 3JH, UK
B. J. J. LUGTENBERG
Department of Plant Molecular Biology and RUL/TNO Centre for
Phytotechnology, Nonnensteeg 3, 2311VJ Leiden, The Netherlands
A. J. MACKIE
School of Biological Sciences, University of Birmingham, Birmingham
B15 2TT, UK
C. MARIE
John Innes Institute, John Innes Centre for Plant Science Research,
Colney Lane, Norwich NR4 7UH, UK
A. MAVRIDOU
John Innes Institute, John Innes Centre for Plant Science Research,
Colney Lane, Norwich NR4 7UH, UK
B. M. MORRIS
Department of Molecular and Cell Biology, Marischal College, Univer-
sity of Aberdeen, Aberdeen AB9 1AS, UK
R. I. PENNELL
Department of Biology, University College London, Gower Street,
London WC1E 6BT, UK
R. PERETTO
Dipartimento di Biologia Vegetale dell'Università di Torino, Italy
R. PEREZ
Reconnaissance cellulaire et Amélioration des Plantes, Université de

Lyons 1, ICBMC/LA INRA.23 879, Bâtiment 741, 5ème étage, 43 Boulevard du 11 Novembre 1918, 69622 Villeurbanne Cedex, France

S. PEROTTO
John Innes Institute, John Innes Centre for Plant Science Research, Colney Lane, Norwich NR4 7UH, UK

A. L. RAE
John Innes Institute, John Innes Centre for Plant Science Research, Colney Lane, Norwich NR4 7UH, UK

E. A. RATHBUN
John Innes Institute, John Innes Centre for Plant Science Research, Colney Lane, Norwich NR4 7UH, UK

N. D. READ
Institute of Cell and Molecular Biology, University of Edinburgh, Edinburgh EH9 3JH, UK

B. REID
Department of Molecular and Cell Biology, Marischal College, University of Aberdeen, Aberdeen AB9 1AS, UK

J. P. RIDE
School of Biological Sciences, University of Birmingham, Birmingham B15 2TT, UK

R. RIVILLA
John Innes Institute, John Innes Centre for Plant Science Research, Colney Lane, Norwich NR4 7UH, UK

A. M. ROBERTS
School of Biological Sciences, University of Birmingham, Birmingham B15 2TT, UK

A.-K. SCHEU
John Innes Institute, John Innes Centre for Plant Science Research, Colney Lane, Norwich NR4 7UH, UK
Present address: Centro Nacional de Biotechnologia, Serrano 115, 28006 Madrid, Spain

G. SMIT
Department of Plant Molecular Biology, Nonnensteeg 3, 2311VJ Leiden, The Netherlands

H. P. SPAINK
Department of Plant Molecular Biology, Nonnensteeg 3, 2311VJ Leiden, The Netherlands

C. J. STAFFORD
School of Biological Sciences, University of Birmingham, Birmingham B15 2TT, UK

J. M. SUTTON
John Innes Institute, John Innes Centre for Plant Science Research, Colney Lane, Norwich NR4 7UH, UK

S. SWART
Department of Plant Molecular Biology, Nonnensteeg 3, 2311VJ Leiden, The Netherlands

A. J. TREWAVAS
Institute of Cell and Molecular Biology, University of Edinburgh, Edinburgh EH9 3JH, UK

A. A. N. VAN BRUSSEL
Department of Plant Molecular Biology, Nonnensteeg 3, 2311VJ Leiden, The Netherlands

H. VAN DEN ENDE
Department of Molecular Cell Biology, University of Amsterdam, Kruislaan 318, 1098 SM Amsterdam, The Netherlands

C. A. WIJFFELMAN
Department of Plant Molecular Biology, Nonnensteeg 3, 2311VJ Leiden, The Netherlands

K. E. WILSON
John Innes Institute, John Innes Centre for Plant Science Research, Colney Lane, Norwich NR4 7UH, UK

Preface

This volume is a collection of papers presented at the 'Perspectives in Cell Recognition' Seminar Series symposium during the annual SEB conference held at Birmingham University, UK, in April 1991. The basic purpose of the sessions was to bring together biologists working on diverse aspects of recognition in plants to discuss recent progress in our understanding of 'self' and 'non-self' interactions between cells. The two major areas of biological interactions covered were those involving dissimilar cells of the same organism associating as gametes in sexual reproduction, and those involving cells of different or 'foreign' organisms, associating in either pathogenesis or mutualistic symbioses. It is in these areas that greatest progress has been made in understanding the cellular, molecular and genetic mechanisms involved and in some cases, notably in the *Rhizobium*–legume symbiosis, the actual genes that control specificity and which are involved in creating a harmonious mutualism have been cloned and their products characterised. Also included are contributions on the exciting progress that is being made in characterising the surface glycoproteins of higher plants that are involved in associations between somatic cells that are crucial to coordinated tissue development, and aspects of cell–cell communication involved in the systemic responses of plants to wounding or pathogenic stimuli.

It was, of course, tempting to include contributions on many other aspects of recognition, such as work on plant hormone receptors and environmental cues, and the intracellular signalling responses associated with recognition. However, it was felt that this would inevitably lead to a more superficial treatment of the cellular interactions in the limited time available, and we were aware of other recent symposia volumes that cover these additional topics.

We hope this book will be of value to advanced undergraduates and research students seeking a broad introduction to plant recognition systems but also trust that the specialist senior scientist will find the juxtaposition of articles on diverse systems illuminating.

Finally, we thank the SEB, ICI and Schering Agrochemicals for financial support without which the symposium could not have taken place. Thanks also to Dr John Anstee, convenor of the SEB Cell Biology section and Dr Julian Dow, secretary of the Cell Signalling Group, under whose auspices the symposium was held. Last but not least, thanks to all contributors for the promptness in delivering their manuscripts.

J. A. Callow and J. R. Green
Birmingham, September 1990

H. VAN DEN ENDE

Sexual signalling in *Chlamydomonas*

Introduction

In the unicellular green alga *Chlamydomonas* the two flagella are used not only for locomotion but also for sexual cell–cell interactions. In a compatible combination of mating-type plus and minus (mt^+ and mt^-) partners, the cells adhere to each other by their flagella. This results not only in a close proximity of the cell bodies that are going to fuse with each other, but also in the generation of a signal, telling the cells to prepare for fusion. This implies at least the total or partial hydrolysis of the cell wall, and the activation of a specialised zone at the anterior part of the cell surface. Thus, *Chlamydomonas* is an example in which signalling is generated by physical cell-to-cell contact. This allows us to investigate the nature and behaviour of the surface receptors involved in this process and the mechanism of signal transduction over the surface membrane and inside the cells. Despite the clarity in which these processes in this simple eukaryotic system are exhibited, and the considerable progress that has been made in a number of laboratories, several enigmas remain, which I address in this chapter.

Strategies of gametic approach

Most of the research concerning sexual interactions has been carried out with the heterothallic *Chlamydomonas eugametos* and *Chlamydomonas reinhardtii*. In these species, cells within one clone all have the same mating type. So when mt^+ and mt^- clonal populations are mixed, the cells exhibit sexual conjugation when they are mating-competent. Generally, mating competence is induced by nitrogen deficiency and the transition of vegetative cells to gametes requires differential gene expression (Bulté & Bennoun, 1990; Treier *et al.*, 1989) and cell division. That is the reason why optimal mating competence in a batch culture is observed at the end of the log phase of growth (Tomson *et al.*, 1985).

Society for Experimental Biology Seminar Series 48: *Perspectives in Plant Cell Recognition*, ed. J. A. Callow & J. R. Green. © Cambridge University Press 1992, pp. 1–17.

In a mixture of mating-competent mt^+ and mt^- cells the zygote yield is dependent on the number of successful collisions between the cells. In principle, this number is proportional to the square of the cell density, while in the case of a 1:1 mixture of opposite mating types, only one out of two collisions can be effective. It is easy to understand, therefore, that the rate of zygote formation is directly dependent on the cell density (Fig. 1). In such species, a strategy to promote physical contact between cells and opposite mating type in a dilute suspension might be advantageous. Such a strategy apparently is mass agglutination. When cells of opposite mating type are mixed, one observes the rapid formation of a number of clumps of cells which agglutinate by means of their flagella. Although, with time, more and more cells become engaged in agglutination, the number of clumps does not increase; they just become larger. This is because cells engaged in sexual contacts have flagella that are more adhesive than are single swimming cells (Demets *et al.*, 1988; Tomson *et al.*, 1990*a*). Owing to this self-enhanced adhesiveness, cells are trapped in clumps of agglutinating cells and thus have a greater chance of adhering to potential partners than those that are still swimming freely. Neverthe-

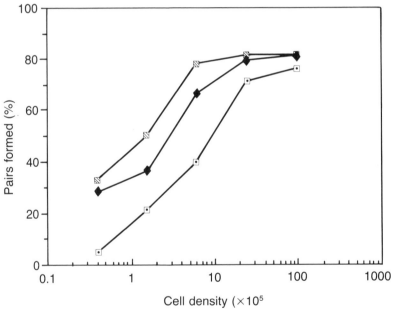

Fig. 1. Rate of cell fusion (expressed as the percentage of cells paired per unit time) in *Chlamydomonas eugametos* in relation to the cell density. Equal amounts of mt^+ and mt^- gametes were mixed and the percentage of pairs formed determined after 30 (□), 60 (◆) and 120 min (■). (From Tomson *et al.*, 1986, with permission.)

less, there remains a strong dependency of the rate of zygote formation on the cell density (because the average size of the clumps seems to increase with increasing cell density), and only at high cell densities (in the order of magnitude of 2×10^6 cells ml^{-1}) is optimal zygote production approaching 100% reached (Tomson *et al.*, 1986).

Massive agglutination is not observed in nitrogen-starved populations of the homothallic *C. monoica*, even while zygote production may be quite substantial in late log phase cell cultures. In this species the two mating types occur within one clone, due to continuous mating-type switching (see e.g. VanWinkle-Swift & Thuerauf, 1991), and one would expect that a direct correlation between cell density and zygote yield would be evident. However, when the zygote yield is determined with varying (initial) cell densities, a complex relationship appears, in which zygote formation is clearly promoted by low cell densities (approximately 10^5 cells ml^{-1}; Fig. 2). Zygotes are not formed in conditions where cell

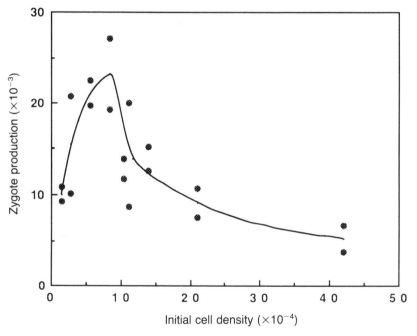

Fig. 2. Zygote formation by *Chlamydomonas monoica* in batches with different initial cell densities. The cells were derived from a continuous culture with Bold's basal medium (cf. VanWinkle-Swift & Thuerauf, 1991) in which the nitrate content was reduced to 30% of the standard value; they were incubated in non-agitated titre wells 0.5 cm^3 in size. The zygote production was determined after 7 days.

divisions are absent, e.g. during nutrient deprival. A significant feature, moreover, is that zygote formation preferably takes place in non-agitated cell suspensions. This can be explained by assuming that sexual interaction between opposite mating-type cells does not occur between cells that swim freely in suspension, but takes place predominantly between cells newly born out of each mother cell. A preliminary model describing the relationship between zygote production and increase of cell number was developed (with Dr R. Lingeman, University of Amsterdam) which fits the data (Fig. 3). This model is a modification of the general formula describing exponential growth; it introduces a correction describing a fraction of daughter cells that forms zygotes by sexual fusion:

$$dC/dt = (1-2x)kC$$

$$dZ/dt = xkC$$

in which k is the growth rate, C is the cell density, Z the number of zygotes and $2x$ is the cell fraction yielding zygotes. Such a correction

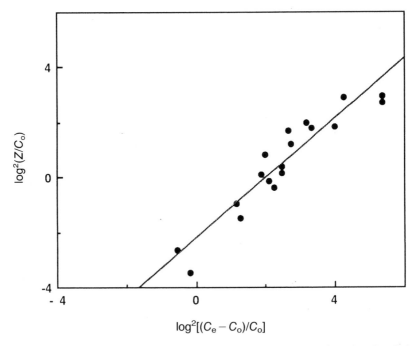

Fig. 3. Double logarithmic plot of the data in Fig. 2, fitting $y = -2.22 + 1.08x$, with confidence level $R = 0.94$. C_o, initial cell density; C_e, final cell density; Z, total number of zygotes formed.

results in a relationship showing a linear dependence between the zygote yield and the increase of cell number in a population:

$$Z_e = \frac{x}{1-2x}(C_e - C_o)$$

The correspondence with the experimental data suggests that the zygote production in *C. monoica* is dependent on the number of cell divisions rather than on the cell density. From the data shown in Fig. 3 it can be inferred that approximately 30–50% of the cells within one mother cell become involved in zygote formation. This hypothesis has one important implication, namely that a given mother cell gives rise to daughter cells of different mating type, even though they are morphologically identical. In this respect *C. monoica* resembles the yeast *Schizosaccharomyces pombe*. In this species, cell division results in two identical daughter cells, but only one of the sister cells produces progeny with interconverted mating type (cf. Klar, 1987). Such developmental asymmetry of daughter cells might also occur in *C. monoica*.

The sexual receptor complex

Sexual adhesiveness of the flagella in *C. reinhardtii* and *C. eugametos*, the two best studied and heterothallic species, is brought about by the presence of extracellular molecules associated with the membrane of the flagella. They have been extensively described and characterised as hydroxyproline-rich glycoproteins with molecular masses in the order of magnitude of 10^6 Da (for a review, see van den Ende *et al.*, 1990). The molecules consist of a shaft with some species-characteristic features, such as a number of flexible joints, a globular domain at one end and a hook-like domain at the other end, by which the molecule is probably attached to the membrane. The carbohydrate part, approximately 50% of the molecular mass, consists of sulphated oligosaccharides which are all *O*-glycosylated and attached to serine, threonine and hydroxyproline residues. This is evidenced by the fact that, after cells are labelled with [^{35}S]sulphate, the agglutinins carry label which can be released completely by mild acid hydrolysis and β-elimination (R. Versluis, unpublished results). Presumably, these oligosaccharides confer the rod-like appearance to the agglutinins (Jentoft, 1990), but they may also carry recognition determinants, since the agglutinins are inactivated by specific glycosidases (Samson *et al.*, 1987).

Since the agglutinins are the only components that have been established to be involved in flagellar adhesion, it is attractive to assume that they interact with each other and not with other proteins at the flagellar surface. A monoclonal antibody (MAb), specifically directed against the mt^- agglutinin in *C. eugametos* (MAb 66.3), and also the Fab fragments derived from it, block flagellar adhesion, which suggests that the mt^- agglutinin interacts with the mt^+ agglutinin and that no other type of interaction is involved in flagellar adhesion (Homan *et al.*, 1988). This view is strengthened by the fact that a mixture of suspended charcoal particles adsorbed with either agglutinin, forms large aggregates when mixed. Another piece of evidence is that in cells carefully fixed with a low concentration of glutaraldehyde, so that the sex-specific adhesiveness is conserved, the interaction is inhibited by solubilised agglutinins in a competitive way. This raises the question, however, as to why this type of result has never been obtained in living cells. I favour the explanation that this is due to the strongly dynamic character of the flagellar membrane during mating. Sexual interaction is accompanied by an increased turnover of agglutinin receptors (Snell & Moore, 1980; Pijst *et al.*, 1984a; Tomson *et al.*, 1990b), due to their inactivation on the flagellar surface. They are apparently replaced by molecules derived from the cell body, because, when mt^- gametes of *C. reinhardtii* interact with a non-fusing mt^+ mutant, the amount of cell body agglutinin initially drops sharply and then recovers to its original level.

There has been considerable discussion about the subcellular compartment from which the agglutinins might be derived. In *C. eugametos* the cell body-associated agglutinins are located predominantly on the plasma membrane. This is based on experiments that involve labelling the cellular surface with agglutinin-specific antibodies after removal of the cell walls (Musgrave *et al.*, 1986). However, several arguments can be used against the view that the plasma membrane is a mobilisable pool of agglutinins. Musgrave *et al.* (1986) demonstrated that there is no free exchange of membrane glycoproteins between the flagellar cell surface and the cell body plasma membrane, while Goodenough (1989), using quick-freeze, deep-etch electron microscopy, noted that agglutinins are not detectable on the outer surface of the cell body plasma membrane. Hunnicutt *et al.* (1990) have confirmed that in *C. reinhardtii*, as in *C. eugametos*, only a small fraction of the total cellular agglutinins is present on the flagella, whereas the majority of the molecules is on the external surface of the cell body plasma membrane but in a non-functional form. This is apparent from the fact that naked gametes do not exhibit any adhesiveness with partner cells. The agglutinins become mobilised to the flagellar surface during sexual interaction and then become activated by an unknown mechanism. Nevertheless, when deflagellated cells were extracted using ammonium sulphate, agglutinin molecules were isolated that were indistinguishable in size and function from active agglutinins.

This was determined using the dried-spot assay, in which samples are dried on a glass slide, after which a suspension of living cells is applied. The cells adhere to the spot when it contains active agglutinin. This fraction was completely destroyed by treating deflagellated cells with trypsin, indicating that they were derived from the outer surface of the cell body plasma membrane. The problem is to determine not only the mechanism of the activation upon transfer from the cell body to the flagellar surface, but also why these molecules are inactive *in situ* on the plasma membrane but become active after extraction. One possibility is that at the plasma membrane they are in an inactive conformation, due to an interaction (non-covalent) with other membrane proteins, which is changed by the extraction procedure or by the adhesion onto the glass surface.

The situation in *C. eugametos* is not much clearer. While the major fraction of total cellular agglutinin is present in the cell body plasma membrane, it is presumably in an active form, since partial removal of the cell wall makes the cell bodies strongly adhesive (F. Schuring, unpublished results). As mentioned above, Musgrave *et al.* (1986) found no evidence for transport of this fraction to the flagellar membrane during mating. Rather, Tomson *et al.* (1990*b*) reported evidence for the existence of two distinct agglutinin pools in the cell body, one of which, located internally, becomes connected with the flagellar membrane during mating and, as in *C. reinhardtii*, is rapidly depleted during mating. However, ultrastructural evidence for its location or for a specific pathway to the flagellar membrane (e.g. in the form of membrane flow) is absent. Also in this species an artificial modification of agglutinin molecules has been reported, namely when inactive agglutinins from darkened light-requiring cells are extracted in reducing conditions using sodium dodecyl sulphate (Kooijman *et al.*, 1988). Thus, in the recent literature about the agglutinins, there is very little understanding of their molecular structure in relation to their mode of action.

Since the agglutinins are extracellular membrane proteins, they must be associated with an integral component of the flagellar membrane. Such integral membrane proteins must also play a role in transmitting the sexual signal generated by sexual adhesion to the cytosol. Evidence for such an anchoring protein came from Homan *et al.* (1982), who showed that the agglutinins could be removed from flagellar membrane vesicles by mild sonication. Re-binding to the vesicles was prevented by trypsinising the vesicles. Kooijman *et al.* (1989) showed that a wheat germ agglutinin (WGA)-binding protein is associated with the agglutinin, because the two proteins co-migrate in the plane of the flagellar membrane (see below). Kalshoven *et al.* (1990) showed by affinity chromatography and Western blotting using MAb 66.3 that this molecule is part of a 120 kDa fraction of flagellar membrane proteins. Part of this fraction has a hydrophobic nature and thus might be embedded in the flagellar membrane itself. From recent experiments, however, it appears that the WGA-binding, agglutinin-associated protein at least is probably an extrinsic protein. Even while the nature of the agglutinin anchor has not been revealed, it is apparent that the agglutinin occurs as a complex, consisting of at least two components in the flagellar membrane (Fig. 4). It is interesting that one of the four 120 kDa proteins now resolved cross-reacts with an antibody directed to the β-subunit of integrins (Marcantonio & Hynes, 1988), and shows a shift in electrophoretic mobility in the presence or absence of reducing agents comparable to other integrins

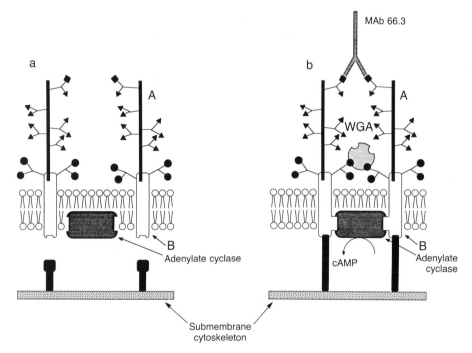

Fig. 4. A model for the mating-receptor complex in the flagellar membrane of *Chlamydomonas eugametos mt⁻* gametes in the (a) non-activated and (b) activated state. The agglutinin (A) is bound to the extracellular domain of an intrinsic protein (B). In both A and B oligosaccharide side-chains are presented to carry epitopes for MAb 66.3 and WGA, respectively. MAb 66.3 and WGA cross-link the complex via A and B, respectively. Cross-linking results in the production of cAMP and the binding of the intracellular domain of B to a submembrane cytoskeleton (from Kalshoven *et al.*, 1990, with permission).

(Hynes, 1987). The integrins generally consist of two non-covalently linked subunits, but so far there is no evidence that this putative integrin molecule is directly associated with the *mt*⁻ agglutinin.

Sexual receptor redistribution

The visualisation of the agglutinin receptor as a multi-protein complex anchored to the flagellar membrane by one or more integral proteins that have a cytosolic domain is appealing because it appears to operate in a fashion similar to that of animal peptide hormone receptors, which are also protein complexes exposed to both the outside and the inside of the cells. These receptors exhibit a rearrangement in the plane of the membrane as part of the signal generation (see e.g. Ellis *et al.*, 1986; Johnson *et al.*, 1988). Likewise, the agglutinin receptor is mobile in the plane of the flagellar membrane. Treatment of intact gametes with MAb 66.3 or WGA causes a redistribution which ultimately results in their accumulation at the flagellar tips. A similar process is exhibited during sexual interaction between gametes of opposite mating type (Homan *et al.*, 1987; Musgrave & van den Ende, 1987). The mechanism of this transport is unknown, but, since it is inhibited by colchicine (Homan *et al.*, 1988), it may well be tubulin-mediated (see also Bloodgood, 1990). It is probable that cross-linking of the receptors by multivalent antibodies or WGA is essential for tipward distribution of the receptor complex, because an Fab fragment of MAb 66.3 fails to induce the redistribution. It does not tell us, however, whether cross-linking is part of the mechanism of movement or part of a signalling mechanism switching the motility machinery on. Bloodgood & Salomonsky (1990) have shown that in *C. reinhardtii* the redistribution of flagellar membrane proteins requires micromolar levels of free calcium in the medium. Redistribution is inhibited by calmodulin antagonists such as trifluoroperazine, the calcium channel blockers diltiazem and D-600, and local anaesthetics, such as lidocaine and procaine. The action of these agents can be interpreted in terms of a requirement for calcium in the signalling mechanism, associated with flagellar glycoprotein redistribution. Also, glycoprotein redistribution is inhibited reversibly by the protein kinase inhibitors H-7, H-8 and staurosporine. Indeed, these workers found in the membrane-matrix fraction obtained by extracting flagella with low concentrations of Nonidet P-40 both Ca^{2+}-dependent kinase and phosphatase activities (Bloodgood & Salomonsky, 1991). A small group of polypeptides in the 26–58 kDa range exhibited a dramatic increase in phosphorylation in the presence of 20 μM free calcium. The protein kinase inhibitors H-7 and H-8 inhibited flagellar glycoprotein redistribution in intact cells as well as protein phosphorylation in the flagellar fractions extracted with detergent. These observations suggest that a signalling pathway is operational in the flagellum, involving calcium influx induced by glycoprotein cross-linking. This is followed by Ca^{2+}-activated protein phosphorylation by which surface mobility is initiated. Protein phosphorylation might directly or indirectly activate an ATPase-containing cross-bridge located between the outer doublet microtubules and the cytoplasmic domains of the transmembrane flagellar glycoproteins. Connections between the flagellar membrane and the axoneme have been described extensively by Dentler (1990).

Whatever may be the mechanism underlying membrane transport, receptor clustering and tipping is particularly interesting for our understanding of sexual signalling in *Chlamydomonas* in that it causes a signal by which mating-structure activation is induced (Goodenough, 1989; van den Ende *et al.*, 1990). Homan *et al.* (1988) demonstrated that treating single *mt⁻* cells with MAb 66.3 induced all the mating reactions, including mating-structure activation. The monovalent Fab fragment was not effective. Kooijman *et al.* (1989) showed that WGA has the same action. In cells having flagellar agglutinins in a non-functional state (as in light-sensitive strains, kept in the dark), WGA still evoked some, but not all, of the mating responses, suggesting that the integrity of the receptor complex is required for complete signalling, although redistribution still occurred.

The issue of glycoprotein redistribution in the flagellar membrane is further complicated by the intriguing phenomenon that flagellar adhesiveness increases five to tenfold as a response to cell–cell adhesion (Demets *et al.*, 1988). Two processes might be involved. One is that this rise is the result of a higher agglutinin concentration at the flagellar surface, which indeed has been found by Goodenough (1989) in *C. reinhardtii*. In *C. eugametos*, in contrast, the increase in adhesiveness occurs without concomitant increase in agglutinin concentration, so it has been postulated that the effect is due to the clustering of agglutinins into more adhesive sites (Tomson *et al.*, 1990b). An interesting aspect is that this phenomenon is inhibited by a number of tubulin-directed drugs (colchicin, vinblastine, N-phenylcarbaminic acid), which again suggests that the behaviour of glycoproteins in the flagellar membrane is directly or indirectly tubulin-mediated.

Sexual signal transduction in *Chlamydomonas*

A role for cyclic AMP (cAMP) in the internal signal transduction pathway is clearly implicated by the fact that there is a transient, but dramatic, increase in intracellular cAMP level as an immediate consequence of sexual adhesion between gametes of opposite mating type in both *C. eugametos* and *C. reinhardtii* (Pijst *et al.*, 1984b; Pasquale &

Goodenough, 1987; Kooijman *et al.*, 1990). A similar increase in cAMP content can be elicited by presenting single mating-type cells with isolated flagella of the other mating type, which illustrates that it is a direct consequence of flagellar interaction. Presenting gametes with dibutyryl-cAMP, even if they are without flagella (Pasquale & Goodenough, 1987) induces the mating responses, in particular mating-structure activation. A difference between both species is that *C. eugametos* responds to dibutyryl-cAMP only in the dark, presumably as the consequence of a light-activated phosphodiesterase. It is of interest to question what is the specific role of cAMP in the process of sexual signalling. Goodenough (1989) observed that in *C. reinhardtii* a transport of agglutinins from the cell body to the flagellar membrane is stimulated by cAMP. A similar phenomenon might occur in *C. eugametos*, where it was shown that dibutyryl-cAMP induced an increase of flagellar adhesiveness, particularly in the dark. Since an increase in adhesiveness is also found during interaction of opposite mating-type cells, a primary effect of cAMP might be to stimulate the flagella-bound transport of agglutinins. However, this can only be part of the action because cAMP clearly has an inductive effect on mating-structure activation. This may be through a coupling with other signalling systems, particularly the Ca^{2+}/phosphoinositide cycle which in *C. eugametos* is also well developed and is clearly involved in sexual signalling. The recent evidence can be summarised as follows. Phosphatidylinositol 4-phosphate (PtdIns(4)P) and phosphatidylinositol 4,5-bisphosphate (PtdIns(4,5)P_2) are present in *C. eugametos* (Irvine *et al.*, 1989), constituting approximately 0.4% of the total phospholipids in *C. eugametos* (Brederoo *et al.*, 1991). These phosphoinositide lipids are concentrated in the plasma membrane of the cell body. Since PtdIns(4,5)P_2 is the precursor of inositol 1,4,5-trisphosphate (Ins(1,4,5)P_3) and is turning over rapidly in *C. eugametos*, one should expect the presence of at least Ins(1,4,5)P_3, which has been confirmed. The basal concentration in single mating-type cells is about 0.01 pmol per 10^6 cells. Most relevant, however, for our understanding of a role of the Ins(1,4,5)P_3/Ca^{2+} system is the recent result that sexual interaction leads to in a tenfold transient increase of Ins(1,4,5)P_3, with a maximal concentration at approximately 15 min after mixing of the cells, when the rate of cell fusion is also maximal. Furthermore, a treatment of *C. eugametos* cells with 5–12% (v/v) ethanol results in a dramatic rise in the intracellular Ins(1,4,5)P_3 concentration (F. Schuring & A. Musgrave, unpublished results). This is significant in view of the fact that alcohols activate a G-protein-associated phospholipase C complex in animals (Hoek & Rubin, 1990). These results indicate clearly that Ins(1,4,5)P_3/Ca^{2+} metabolism is involved in sexual signalling *in vivo*.

It is well known that the Ins(1,4,5)P_3/Ca^{2+} signal transduction system is operative in a large variety of cell types. In this system at least two second messengers are produced, diacylglycerol (DAG) and Ins(1,4,5)P_3. They are generated by a signal transduction process comprising a receptor, a coupling G-protein and a phospholipase. DAG acts by stimulating protein kinase C, whereas Ins(1,4,5)P_3 releases calcium from internal stores (Berridge & Irvine, 1989). While in *C. eugametos* no evidence has been obtained for the presence of a protein kinase C-type enzyme, there is some evidence from earlier literature that Ca^{2+} is released from internal reservoirs during mating (Bloodgood & Levin, 1983; Kaska *et al.*, 1985). In addition, Schuring *et al.* (1990) recently found that A23187 in combination with other agonists induces mating structure activation, thereby confirming an earlier report (Claes, 1980). So at present, it seems that a rise in internal free Ca^{2+} may be responsible for mating-structure activation, although the precise nature of the mobilised Ca^{2+} pools, as well as the mode of action of the increased Ca^{2+} levels, remains elusive. Another problem is the spatial relationship. It appears that the majority of phosphoinositide lipids are present in the cell body plasma membrane, and not in the flagellar membrane, where the primary membrane transduction takes place. The fact that an increase of Ins(1,4,5)P_3 occurs only some 10 min after the start of sexual interaction, while the transient cAMP rise is observed within seconds, suggests that the change in phosphoinositide metabolism is causally coupled to the cAMP metabolism. An obvious experiment to do would be to see if treating cells with dibutyryl-cAMP gives rise to an increase in the level of Ins(1,4,5)P_3. Another possibility is that the production of cAMP, inositol phospholipid turnover and Ca^{2+} mobilisation are integrated in a single receptor cascade system. Unravelling the role and interplay of these mechanisms in *Chlamydomonas* is now quite feasible and this is unique for a plant system.

The sexual response

Essentially, sexual interaction elicits three responses: (1) a rapid increase in adhesiveness of the flagella; (2) redistribution of agglutinin receptors to the flagellar tips; (3) mating-structure activation. As mentioned above, this implies the removal of the cell wall and the formation of a papilla or fertilisation tubule by the outgrowth of a specialised zone of the plasma membrane between the flagellar bases. Work by Detmers *et al.* (1983) clearly suggests that in *C. reinhardtii* the formation of this structure is microfilament-driven. In *C. eugametos*, in contrast, activation of the mating structure leads to a very small papilla which just penetrates the cell wall in both mating types (Brown *et al.*, 1968; Mesland, 1976). This involves the formation of a hole in the cell wall that, despite its small diameter, can be visualised with fluorescent antibodies raised against degradation products of the cell wall. Recent work by Schuring *et al.* (1990) has shown that in exceptional cases, for example when wild-type mt^+ cells are mixed with non-fusing mutant mt^- cells, the papilla is blown up into a stable balloon of considerable size and is easily observable by low-power light microscopy. Such balloon-like mating structures are also formed when the cells are presented with compounds that increase intracellular free Ca^{2+}, such as ethanol, trifluoroperazine together with A23187, or ethanol with $Ins(1,4,5)P_3$. Mutants that exhibit sexual interactions but are unable to produce a hole in the cell wall (Schuring *et al.*, 1991) do not react. Similarly, wild-type gametes treated with 5 mM cysteine, which prevents wall lysis during sexual interaction (Samson *et al.*, 1988), do not form these structures upon addition of ethanol.

While the balloon-like structures are easily observed and thus easily assayed, the question is whether they are genuine, though oversized, mating structures, or are simply artifacts. Mating-structure balloons can be created from natural mating structures by squeezing activated cells under a cover slip. This led us to consider the possibility that ballooning occurs because of an elevated hydrostatic pressure (turgor) in the cell. Indeed, cells treated with agents that increase Ca^{2+} look swollen and lack the usual periplasmic space at the rear of each cell. They are apparently under pressure because the action of the contractile vacuole is impaired, probably due to high Ca^{2+} levels (A. Musgrave, unpublished results). It is attractive to assume that the $Ins(1,4,5)P_3/Ca^{2+}$ system in *Chlamydomonas* controls the water relations in the cell, much as it does in guard cells of plant stomata (Blatt *et al.*, 1990; Gilroy *et al.*, 1990) and in halophilic *Dunaliella* (Einspahr & Thompson, 1990). This opens up new vistas for research on signalling in *Chlamydomonas*.

References

Berridge, M. J. & Irvine, R. F. (1989). Inositol phosphates and cell signalling. *Nature* **341**, 197–205.

Blatt, M. R., Thiel, G. & Trentham, D. R. (1990). Reversible inactivation of K^+ channels in *Vicia* stomatal guard cells following the photolysis of caged inositol 1,4,5-trisphosphate. *Nature* **346**, 766–9.

Bloodgood, R. A. (1990). Glycoprotein dynamics in the *Chlamydomonas* flagellar membrane. In *Ciliary and Flagellar Membranes*, ed. R. A. Bloodgood, pp. 91–127. New York: Plenum Publ. Corp.

Bloodgood, R. A. & Levin, E. N. (1983). Transient increase in calcium efflux accompanies fertilization in *Chlamydomonas*. *Journal of Cell Biology* **97**, 397–404.

Bloodgood, R. A. & Salomonsky, N. L. (1990). Calcium influx regulates antibody-induced glycoprotein movements within the *Chlamydomonas* flagellar membrane. *Journal of Cell Science* **96**, 27–33.

Bloodgood, R. A. & Salomonsky, N. L. (1991). Regulation of flagellar glycoprotein movements by protein phosphorylation. *European Journal of Cell Biology* **54**, 85–9.

Brederoo, J., de Wildt, P., Popp-Snijders, C., Irvine, R. F., Musgrave, A. & van den Ende, H. (1991). Polyphosphoinositol lipids in *Chlamydomonas eugametos* gametes. *Planta* **184**, in press.

Brown, R. M., Johnson, C. & Bold, H. C. (1968). Electron and phase-contrast microscopy of sexual reproduction in *Chlamydomonas moewusii*. *Journal of Phycology* **4**, 100–20.

Bulté, L. & Bennoun, J. (1990). Translational accuracy and sexual differentiation in *C. reinhardtii*. *Current Genetics* **18**, 155–60.

Claes, H. (1980). Calcium ionophore-induced stimulation of secretory activity in *Chlamydomonas reinhardtii*. *Archives of Microbiology* **124**, 81–6.

Demets, R., Tomson, A. M., Stegwee, D. & van den Ende, H. (1988). Cell–cell adhesion in conjugating *Chlamydomonas* gametes: a self-enhancing process. *Protoplasma* **145**, 27–36.

Dentler, W. L. (1990). Linkages between microtubules and membranes in cilia and flagella. In *Ciliary and Flagellar Membranes*, ed. R. A. Bloodgood, pp. 91–127. New York: Plenum Publ. Corp.

Detmers, P. A., Goodenough, U. W. & Condeelis, J. (1983). Elongation of the fertilization tubule in *Chlamydomonas*: new observations on the core microfilaments and the effect of transient intracellular signals on their structural integrity. *Journal of Cell Biology* **97**, 522–32.

Einspahr, K. J. & Thompson, G. A. (1990). Transmembrane signalling via phosphatidylinositol 4,5-bisphosphate hydrolysis in plants. *Plant Physiology* **92**, 361–6.

Ellis, L., Morgan, D. O., Clauser, E., Edery, M., Jong, S. M., Wang, L. H., Roth, R. A. & Rutter, W. J. (1986). Mechanisms of receptor-mediated transmembrane communication. *Cold Spring Harbor Symposia on Quantitative Biology* **51**, 773–84.

Gilroy, S., Read, N. D. & Trewavas, A. J. (1990). Elevation of cytoplasmic calcium by caged calcium or caged inositol trisphosphate initiates stomatal closure. *Nature* **346**, 769–71.

Goodenough, U. W. (1989). Cyclic AMP enhances the sexual agglutinability of *Chlamydomonas* flagella. *Journal of Cell Biology* **109**, 247–52.

Hoek, J. B. & Rubin, E. (1990). Alcohol and membrane-associated signal transduction. *Alcohol and Alcoholism* **25**, 143–56.

Homan, W. L., Hodenpijl, P. G., Musgrave, A. & van den Ende, H. (1982). Reconstitution of biological activity in isoagglutinins from *Chlamydomonas eugametos*. *Planta* **155**, 529–35.

Homan, W. L., Musgrave, A., de Nobel, H., Wagter, R., de Wit, D., Kolk, A. & van den Ende, H. (1988). Monoclonal antibodies directed against the sexual binding site of *C. eugametos* gametes. *Journal of Cell Biology* **107**, 177–89.

Homan, W. L., Sigon, C., van den Briel, W., Wagter, R., de Nobel, H., Mesland, D. A. M. & van den Ende, H. (1987). Transport of membrane receptors and the mechanics of sexual cell fusion in *Chlamydomonas eugametos*. *FEBS Letters* **215**, 323–6.

Hunnicutt, G. R., Kosfiszer, M. G. & Snell, W. J. (1990). Cell body and flagellar agglutinins in *Chlamydomonas reinhardtii*: the cell body plasma membrane is a reservoir for agglutinins whose migration to the flagella is regulated by a functional barrier. *Journal of Cell Biology* **111**, 1605–16.

Hynes, R. O. (1987). Integrins: a family of cell surface receptors. *Cell* **48**, 549–54.

Irvine, R. F., Letcher, B. K., Drobak, A. P., Dawson, A. & Musgrave, A. (1989). Phosphatidylinositol (4,5)bisphosphate and phosphatidylinositol (4)phosphate in plant tissue. *Plant Physiology* **89**, 888–92.

Jentoft, P. (1990). Why are proteins *O*-glycosylated? *Trends in Biochemical Sciences* **15**, 291–4.

Johnson, J. D., Wong, M. L. & W. J. Rutter. (1988). Properties of the insulin receptor ectodomain. *Proceedings of the National Academy of Sciences, USA* **85**, 7615–20.

Kalshoven, H. W., Musgrave, A. & van den Ende, H. (1990). Mating receptor complex in the flagellar membrane of *Chlamydomonas eugametos* gametes. *Sexual Plant Reproduction* **3**, 77–87.

Kaska, D. D., Piscopo, I. C., & Gibor, A. (1985). Intracellular calcium redistribution during mating in *Chlamydomonas reinhardtii*. *Experimental Cell Research* **160**, 371–9.

Klar, A. J. S. (1987). Differentiated parental DNA strands confer developmental asymmetry on daughter cells in fission yeast. *Nature* **326**, 466–70.

Kooijman, R., de Wildt, P., Beumer, S., van de Vliet, G., Homan, W. L., Kalshoven, H., Musgrave, A. & van den Ende, H. (1989). Wheat germ agglutinin induces mating reactions in *Chlamydomonas eugametos* by cross-linking agglutinin-associated glycoproteins in the flagellar membrane. *Journal of Cell Biology* **109**, 1677–87.

Kooijman, R., de Wildt, P., Homan, W. L., Musgrave, A. & van den Ende, H. (1988). Light affects flagellar agglutinability in *Chlamydomonas eugametos* by modification of the agglutinin molecules. *Plant Physiology* **86**, 216–23.

Kooijman, R., de Wildt, P., van den Briel, W., Tan, S., Musgrave, A.

& van den Ende, H. (1990). Cyclic AMP is one of the intracellular signals during the mating of *Chlamydomonas eugametos*. *Planta* **181**, 529–37.

Marcantonio, E. E. & Hynes, R. O. (1988). Antibodies to the conserved cytoplasmic domain of the integrin beta-1 subunit react with proteins in vertebrates, invertebrates and fungi. *Journal of Cell Biology* **106**, 1765–72.

Mesland, D. A. M. (1976). Mating in *Chlamydomonas eugametos*. A scanning electron microscopic study. *Archives of Microbiology* **109**, 31–5.

Musgrave, A., de Wildt, P., Etten, I., Pijst, H., Scholma, C., Kooijman, R., Homan, W. L. & van den Ende, H. (1986). Evidence for a functional membrane barrier in the transition zone between the flagellum and the cell body of *Chlamydomonas eugametos* gametes. *Planta* **167**, 544–53.

Musgrave, A. & van den Ende, H. (1987). How *Chlamydomonas* court their partners. *Trends in Biochemical Sciences* **12**, 470–3.

Pasquale, S. M. & Goodenough, U. W. (1987). Cyclic AMP functions as a primary signal in gametes of *Chlamydomonas reinhardtii*. *Journal of Cell Biology* **105**, 2279–92.

Pijst, H. L. A., Ossendorp, F. A., van Egmond, P., Kamps, A., Musgrave, A. & van den Ende, H. (1984*a*). Sex-specific binding and inactivation of agglutination factor in *Chlamydomonas eugametos*. *Planta* **160**, 529–35.

Pijst, H. L. A., van Driel, R., Janssens, P. M. W., Musgrave, A. & van den Ende, H. (1984*b*). Cyclic AMP is involved in sexual reproduction of *Chlamydomonas eugametos*. *FEBS Letters* **174**, 132–6.

Samson, M. R., Klis, F. M., Homan, W. L., van Egmond, P., Musgrave, A. & van den Ende, H. (1987). Composition and properties of the sexual agglutinins of the flagellated green alga *Chlamydomonas eugametos*. *Planta* **170**, 314–21.

Samson, M. R., Klis, F. M. & van den Ende, H. (1988). Cysteine is a fusion inhibitor in sexual mating of the green alga *Chlamydomonas eugametos*. *Acta Botanica Neerlandica* **37**, 351–61.

Schuring, F., Brederoo, J., Musgrave, A. & van den Ende, H. (1990). Increase in calcium triggers mating structure activation. *FEMS Microbiology Letters* **71**, 237–40.

Schuring, F., Musgrave, A., Elders, R., Teunissen, Y., Homan, W. L. & van den Ende, H. (1991). Fusion-defective mutants of *Chlamydomonas eugametos*. *Protoplasma*, in press.

Snell, W. J. & Moore, W. S. (1980). Aggregation-dependent turnover of flagellar adhesion molecules in *Chlamydomonas* gametes. *Journal of Cell Biology* **84**, 203–10.

Tomson, A. M., Demets, R., Bakker, N. P. M., Stegwee, D. & van den Ende, H. (1985). Gametogenesis in liquid cultures of *Chlamydomonas eugametos*. *Journal of General Microbiology* **131**, 1553–60.

Tomson, A. M., Demets, R., Musgrave, A., Kooijman, R., Stegwee, D. & van den Ende, H. (1990*a*). Contact activation in *Chlamydomonas* gametes by increased binding capacity of sexual agglutination. *Journal of Cell Science* **95**, 293–301.

Tomson, A. M., Demets, R., Sigon, C. A. M., Stegwee, D. & van den Ende, H. (1986). Cellular interactions during the mating process in *Chlamydomonas eugametos*. *Plant Physiology* **81**, 521–6.

Tomson, A. M., Demets, R., van Spronsen, E. A., Brakenhoff, G. J., Stegwee, D. & van den Ende (1990*b*). Turnover and transport of agglutinins in conjugating *Chlamydomonas* gametes. *Protoplasma* **155**, 200–9.

Treier, U., Fuchs, S., Weber, M., Wakarchuk, W. W. & Beck, C. F. (1989). Gametic differentiation in *Chlamydomonas reinhardtii*: light dependence and gene expression patterns. *Archives of Microbiology* **152**, 572–7.

van den Ende, H., Musgrave, A. & Klis, F. M. (1990). The role of flagella in the sexual reproduction of *Chlamydomonas* gametes. In *Ciliary and Flagellar Membranes*, ed. R. A. Bloodgood, pp. 129–47. New York: Plenum Publ. Corp.

VanWinkle-Swift, K. P. & Thuerauf, D. J. (1991). The unusual sexual preferences of a *Chlamydomonas* mutant may provide insight into mating-type evolution. *Genetics* **127**, 103–15.

J. A. CALLOW, C. J. STAFFORD
AND J. R. GREEN

Gamete recognition and fertilisation in the fucoid algae

Introduction: *Fucus* as a model system

Our understanding of the molecular basis of recognition and signalling systems in plant cells is very poor, and information and concepts are often uncritically extrapolated from studies on animal cell systems. Most higher plant systems suffer from disadvantages such as limited accessibility to plasma membrane-based receptors because of intervening cell walls, thus necessitating protoplasting, with potentially undesirable side effects. Responses to stimuli are often slow, rendering experimentation more difficult. The analysis of the molecular organisation of plant plasma membranes has in itself been fraught with considerable difficulty because of problems in obtaining sufficient quantities of pure material and the lack of truly unequivocal criteria for purity.

In studies on higher plant reproductive systems, interactions between egg and sperm cells are particularly difficult to follow because the events are embedded within tissues, and although male and female gametes can now be isolated from angiosperms (for a recent review, see Theunis *et al.*, 1991), this is still fraught with considerable technical difficulty. In this context then, 'simpler' recognition systems presented by oogamous lower plants such as *Fucus* have much to offer. Naked gametes of these algae are released in large enough quantities to permit detailed biochemical studies. The interaction provides a rapid and simple bioassay permitting direct analysis of the activity of blocking agents such as antibodies, oligosaccharides, putative receptor fractions, etc. Reproductive mechanisms apart, studies on *Fucus* gametes provide information on the molecular architecture of natural protoplast surfaces (e.g. the existence of discrete subsets (glycoforms) of glycoproteins, types of oligosaccharide structure and presence of topographical domains) that will be of value in exploring the organisation and complexity of plant plasma membranes in general.

Society for Experimental Biology Seminar Series 48: *Perspectives in Plant Cell Recognition*, ed. J. A. Callow & J. R. Green. © Cambridge University Press 1992, pp. 19–31.

Fucus gametes

The general properties of the *Fucus* system have been reviewed several times (e.g. Callow, 1985; Callow *et al.*, 1985; Green *et al.*, 1990). In brief, fucoid eggs are brown, spherical, apolar, non-motile cells, 60–80 μm in diameter. Although the eggs are released from oogonia in polysaccharide mucilage, the latter may be washed away to provide a natural protoplast, devoid of any cell wall or other surrounding layers. The bright orange, carotenoid-containing sperm cells are more highly differentiated, being biflagellate cells, some 5 μm long. The shorter, anterior 'tinsel' flagellum bears mastigonemes, the longer, 'whiplash' flagellum, which provides the propulsive force, is smooth. The anterior region of the sperm body is extended into a peculiar flattened, microtubule-rich region (known as the proboscis) of unknown function. The presumption is that the anterior flagellum and associated proboscis provide direction to sperm movement.

Fertilisation

Sperm and egg cells are liberated at low tide and the process of fertilisation can be considered to involve a number of stages, each involving aspects of recognition.

1. *Chemoattraction.* This is mediated by hydrocarbon pheromones known as octatrienes (Muller & Seferiades, 1977), which are secreted by the egg cells. The polarised motility of sperm cells down the concentration gradient results in 'swarming' of large numbers of sperm around the egg. The pheromones are not species-specific and discrete receptors for these compounds have yet to be characterised.

2. *Sperm probing and binding to the egg cell.* Sperm cells that contact the egg surface appear to engage in active probing of the surface via the anterior flagellum without immediately causing a fertilisation potential (Brawley, 1990). Sperm of *Pelvetia canaliculata*, for example, appear to move over the surface of the egg for 10 s or more, attach to one spot then gyrate against the egg for another 10–20 s before evoking a fertilisation potential (Brawley, 1990). It has been speculated that this behaviour is evidence for sperm receptors on the egg plasma membrane being present in restricted domains.

3. *Plasmogamy.* The earliest perceived response of fucoid eggs to fertilising sperm is a membrane depolarisation from about -60 to -25 mV (Brawley, 1991). This activation, or fertilisa-

tion potential, serves as a 'fast block' to polyspermy and is well known in various animal egg systems where it is associated with the insertion of sperm-associated ion channels which depolarise the plasma membrane. This latter is associated with a large increase in intracellular calcium through a positive feedback loop involving phosphoinositide second messengers. In fucoid algae we know little of such mechanisms, although Na^+ dependence has recently been reported (Brawley, 1991).

The fertilisation potential lasts a few minutes and is followed by the secretion of β-linked polyuronides from pre-formed cytoplasmic vesicles to form a simple cell wall. This wall can be stained with the fluor Calcofluor White (Callow *et al.*, 1978) and is formed in *Fucus serratus* within 3–4 min of sperm being added, providing another, but slower, block to polyspermy. However, there appears to be some variation between different fucoid species in the speed of wall formation, and therefore in its significance in preventing polyspermy, and other possibilities for polyspermy blocks exist (Brawley, 1991).

Fertilisation can be prevented electrically by current-clamping of the egg cells but sperm still appear to bind to the egg surface (Brawley, 1990). If fusion (plasmogamy) of the egg and sperm protoplasts involves receptors different from those involved in initial binding, as suggested above, then this approach might be refined to allow these two aspects of recognition to be quantified. The widely used bioassay for successful fertilisation involving the detection of wall release using Calcofluor White fluorescence clearly integrates a number of stages at which discrete recognition events could be involved. We are attempting to develop an assay to quantify binding of egg membrane vesicles to sperm cells and membranes in order to examine this particular aspect of the recognition process.

Experiments to characterise receptors

The general working hypothesis is that species-specific binding and/or fusion of eggs and sperm is mediated by molecular associations involving surface receptors and complementary ligands. Experimental evidence to support this interpretation was obtained a number of years ago through testing for the inhibitory effects of certain surface probes, principally lectins. These were used to identify the presence and distribution of specific saccharides that may form components of surface glycoprotein receptors, or as functional probes to inhibit fertilisation in the standard bioassay. Preliminary identification of fractions with properties consistent with the presence of receptors was also obtained. These lines of evidence have been reviewed several times (see e.g. Callow, 1985) and will not be considered further here. More recently our approach to the elucidation of the molecular events involved in fertilisation has concentrated on a more comprehensive understanding of gamete plasma membranes, and the proteins and glycoproteins exposed on their surfaces through the use of highly specific monoclonal antibodies (MAbs) as molecular and functional probes.

Isolation and characterisation of egg plasma membranes

Bolwell *et al.* (1980) isolated a membrane vesicle fraction from *Fucus* eggs and showed that this competitively inhibited fertilisation in a species-specific manner. Ideally one would wish to use membrane vesicles such as these in binding assays that would enable a more critical dissection of recognition processes than that permitted by the use of the fertilisation bioassay. However, the membrane fractions were poorly characterised; more recently therefore, we have examined their properties more closely (Stafford *et al.*, 1992*a*). Treatment of *Fucus* eggs with borate buffer (pH 10.2) induces the budding of vesicles from the egg surface (Fig. 1). After removal of debris the vesicles are fractionated over a sucrose cushion to yield a preparation containing 4.6% of the original membrane protein. The preparation is only marginally contaminated with uronic acid (from cytoplasmic vesicles) and polyphenols (from polyphenol bodies). Absence of polyphenols is important, since they cause non-specific inhibition of fertilisation. The activity of the plasma membrane marker, Mg^{2+}-dependent, K^+-stimulated ATPase was enriched fivefold in the preparation (10% of the original activity). Chloroplast contamination is low (0.7% of total chlorophyll) and there is no detectable succinic dehydrogenase from mitochondria. Less than 1% of the activities of NADPH cytochrome *c* reductase and inosine diphosphatase were recovered in this fraction, indicating negligible contamination with endoplasmic reticulum or Golgi. Finally, we have used a MAb, FS2, to provide an independent assessment of the degree to which these egg vesicles are enriched with plasma membranes. This MAb binds specifically to egg plasma membranes *in situ*, as assessed by immunogold labelling (Stafford *et al.*, 1992*b*). Inhibition enzyme-linked immunosorbent assays (ELISAs) demonstrate a fivefold enrichment in FS2 binding compared with crude egg membranes.

The plasma membrane-enriched preparations are effective in inhibiting fertilisation in *F. serratus* by binding to sperm cells (C. J. Stafford, J. R. Green & J. A. Callow, unpublished results). This activity is largely reduced by pretreating vesicles with periodate (which oxidises carbohydrate residues). The inhibition is also species- and genus-specific, since there

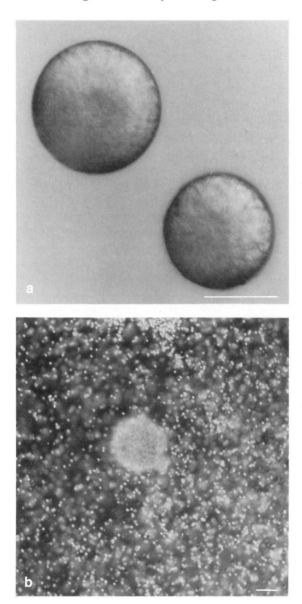

Fig. 1. Phase contrast images of *Fucus serratus* eggs, (a) before and (b) 15 min after treatment with borate buffer (pH 10.2). Most of the egg contents remain as a coherent mass and are removed by centrifugation. The bars represent 40 μm. (From Stafford *et al.*, 1992*b*.)

are no effects on fertilisation in either *Fucus vesiculosus* or *Ascophyllum nodosum*.

The membrane vesicles therefore have appropriate properties that encourage us to use them in biochemical studies as a source of egg surface antigens. They can also be used to develop a binding assay that can then be used as a routine biochemical screen for inhibitory MAbs that will provide a good supply of functional probes.

Use of MAb surface probes

MAbs are now frequently used to explore surface interactions between cells. They provide highly specific affinity probes for surface molecules, including those that interact in mutual recognition events. We have there-fore investigated the value of MAbs in exploring surface interactions between *Fucus* gametes. The specific questions we seek to address through MAb technology include the following. What kinds of protein and glycoprotein are to be found on gamete surfaces? What is the dif-ference between egg and sperm: are all surface antigens gamete-specific or are some common? Which antigens have functional properties as gamete 'receptors'? Are antigens homogeneously distributed over gamete surfaces or are they arranged in distinct surface domains? Are such domains significant in egg–sperm recognition: do they contribute to receptor functions? What interspecific differences are there in the presence/structure of surface antigens between different fucoids and how do these relate to species specificity in fertilisation?

Progress in our laboratory towards answering some of these questions is described below.

MAbs to sperm surface antigens

Jones *et al.* (1988, 1990) raised MAbs to whole sperm of *F. serratus*. From several hundred hybridomas, a panel of 12 MAbs (designated FS1-12) was selected by immunofluorescence on sperm and ELISA on egg vesicles to provide a range of specificities and characteristics (Table 1).

Domain specificity

Two broad groups of MAb binding can be recognised. Group I MAbs (FS1, FS3, FS4, FS6, FS8, FS9 and FS10) bind to antigens preferentially but are not located exclusively on the sperm body. Group II MAbs (FS2, FS5, FS7, FS11, FS12) bind to the anterior region of the sperm, notably the anterior flagellum. There is some overlap and variation between and within these groups. Thus, in group I, antigens to FS4 are fairly uniformly distributed over the whole sperm whilst FS9 and FS10 antigens are most

Table 1. *MAb binding to sperm and eggs of* Fucus serratus *as determined by indirect immunofluorescence*

MAb	Antibody type	Anterior flagellum	Sperm body	Posterior flagellum	Egg
FS4	IgM	++	+++	++	+
FS3	IgM	+	++	+	−
FS6	IgM	+	++	+	−
FS8	IgM	+	++	+	−
FS9	IgG2b	+	+++	+	+
FS10	IgM	+	+++	+	−
FS1	IgM	+/−(M)	++	−	−
FS7	IgG1	+++(M)	++	+	−
FS12	IgG1	++(M)	+	+	−
FS5	IgM	+++(M)	+/−	−	+
FS2	IgG1	++(M)	+/−	−	+
FS11	IgM	++	−	−	−

−, Absent; +/−, very weak; +, weak; ++, intermediate; +++, strong. Antibody binding to the mastigonemes (M) of the anterior flagellum was determined by immunogold labelling. Binding of MAbs to eggs was determined by ELISA. (From Green *et al.*, 1990, with data from Jones *et al.*, 1988.)

abundant on the sperm body, with weak but positive labelling of both flagella. FS1 labels the sperm body strongly, does not label the posterior flagellum or the surface membrane of the shaft of the anterior flagellum, but does label the mastigonemes of the anterior flagellum, albeit weakly (detected by immunogold labelling).

Group II MAbs all strongly label the anterior flagellum but whereas FS11 labelling is restricted to the shaft of the anterior flagellum, FS2 and FS5 label the mastigonemes on the anterior flagellum. FS7 and FS12 antigens are located primarily on the mastigonemes of the anterior flagellum but also label the anterior region of the sperm body and bind weakly to the posterior flagellum.

Thus, the sperm surface has a number of molecular domains.

Gamete specificity

Most of the MAbs raised against *F. serratus* sperm do not bind to egg vesicles, showing a clear difference between the surfaces of the two gametes. However, four antibodies (FS2, FS4, FS5 and FS9) bind to crude egg membrane vesicles (Table 1), indicating the presence of common epitopes. (The characteristics of these antigens are explored below.)

Species and genus specificity

When these MAbs are tested against sperm of other species or genera, a range of specificities is obtained. Most of the antibodies are genus-specific, recognising antigens of both *F. serratus* and *F. vesiculosus*, but one antibody, FS10, is species-specific. Two antibodies, FS4 and FS9 recognise antigens of another fucoid genus, *Ascophyllum nodosum*.

The molecular nature of the antigens/epitopes recognised by the MAbs has been determined by Western blotting, and periodate oxidation, protease and endo-F treatments of antigens (Jones *et al.*, 1990). Most of the MAbs recognise carbohydrate determinants on glycoproteins and some of the antibodies bind to common epitopes of several different antigens (FS2, FS5, FS7, FS11, FS12). However, the set of MAbs which bind preferentially to the sperm body (FS3, FS4, FS8, FS10) recognise carbohydrate epitopes of an immunodominant high molecular weight glycoprotein (average M_r 205,000 on one-dimensional polyacrylamide gels). However, the five MAbs which recognise this glycoprotein, though they compete, probably bind to adjacent epitopes, since there are differences in their binding specificities to eggs and sperm of other species and in their biochemical characteristics.

Probing of egg surfaces with MAbs

Three of the MAbs raised against *F. serratus* sperm which recognise common epitopes of egg antigens have been used to characterise the distribution of these antigens on the egg surface using immunofluorescence and confocal laser scanning microscopy (CLSM) and immunogold-electron microscopy (EM) (Stafford *et al.*, 1992). FS2, FS4 and FS5 label the egg surface in patches or domains but these patches are smaller, and more distinct for FS2 (Fig. 2) and FS5. Double-labelling experiments using fluorescein isothiocyanate and 30 nm gold-conjugated second antibodies (Stafford *et al.*, 1992) show clearly that FS5 binds to more discrete domains than does FS4 and that these domains are generally separate from each other (Fig. 2f). There are, however, some areas of double-labelling where both antibodies appear to bind. The lectins concanavalin A (Con A) and fucose-binding protein (FBP) also bind to small, even more discrete, domains on the egg surface but unequivocal double-labelling experiments to test congruence with antibody labelling have yet to be done (Stafford *et al.*, 1992). The domains are not induced by any form of patching or capping of antigens by the multivalent antibodies or lectins, since they are detectable on both fresh and fixed eggs. Nor is the cytoskeleton apparently involved in their organisation, since inhibitors of cytoskeleton function have no effect on their presence.

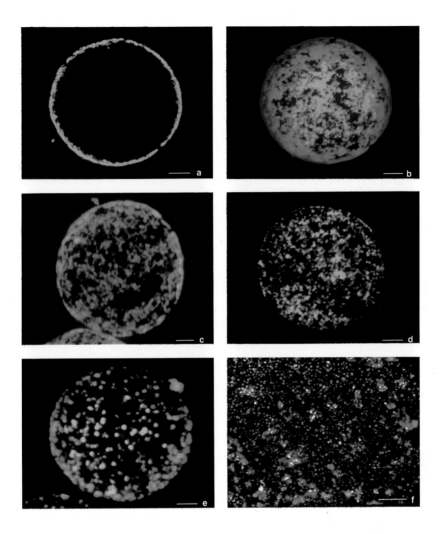

For caption, see overleaf

(Facing p.26)

Fig. 2. Fluorescence/reflectance confocal images of *Fucus serratus* eggs treated with various antibody or lectin probes. (a) to (c) Binding of the antibody is visualised using fluorescein isothiocyanate (FITC) rabbit-anti-mouse immuno-globulin (FITC-RAMIG). (d) to (e) FITC-labelled lectins were used. (a) An optical section through the equatorial region of an egg labelled with FS4. (b) to (f) Composite, integrated images of 10 optical sections, 3 μm apart, i.e. showing labelling patterns over egg hemispheres for FS4 (b), FS2 (c), Con A (d) and FBP (e). (f) A higher-powered image of the egg surface double-labelled with FS5 and FS4. FS5 binding was revealed with FITC-labelled second antibody (FITC-RAMIG); FS4 binding was revealed with gold-labelled second antibody (GAMIG 30). The colours shown in these confocal laser scanning microscopy images are false. In all cases binding of FITC-labelled probes is shown as green. In (f) binding of the gold-labelled probe is shown as purple. Areas where both probes bind appear white (i.e. green complements purple). The bars represent 10 μm (egg in f is flattened).

The existence of discrete antigen domains can also be demonstrated by EM-immunogold labelling of egg sections with FS2 (Fig. 3). In this case the domains seem to form small protuberances on the egg cell surface, although caution must be exercised in that these could be artefacts of specimen fixation and dehydration. There is negligible labelling of internal membranes.

The molecular characteristics of the egg surface antigens have been examined by Western blotting and antigen modification experiments with periodate and pronase (Stafford *et al.*, 1992). FS5 and FS2 have similar labelling patterns on Western blots recognising a wide range of proteins/ glycoproteins (FS5 data shown in Fig. 4). Other data (Stafford *et al.*, 1992) suggest that these two antibodies recognise the same or overlapping epitopes on a range of common protein/glycoprotein antigens. The antigens recognised by FS2/5 appear to be different from those recognised by FS4 (Fig. 4). The lectin Con A labels a subset of the glycoproteins labelled by FS2/5, and FBP labels primarily a glycoprotein of 62 kDa. FS2 and Con A also label a protein at the same position.

Overall these data show that different sets of proteins and glycoproteins on the egg plasma membrane are not distributed homogeneously but form more or less discrete domains. FBP, Con A and FS2/5 appear to recognise progressively larger families of glycoproteins held within small discrete domains on the egg surface and these are both spatially and in molecular characteristics largely distinct from a family of glycoproteins recognised by FS4. The functional significance of this is speculated upon below.

Finally, in comparing *Fucus* eggs and sperm it must also be remembered that MAbs are specific for individual epitopes and that, whilst clearly there are common epitopes present on egg and sperm, these are present in different molecules. Thus, FS4 binding to egg and sperm is periodate-sensitive but on sperm it recognises a 205 kDa glycoprotein whilst on eggs it recognises a set of glycoproteins none of which has an M_r greater than 160,000.

Functional significance of glycoprotein domains

There is, then, a complexity to antigen distribution on the egg surface which is perhaps not expected from such an undifferentiated cell. Surface antigen domains are quite common for highly differentiated cells, e.g. epithelial cells or animal sperm cells (Primakoff & Myles, 1983; Simons & Fuller, 1985) and these domains have functional significance. It is possible that the complex antigen- and lectin-binding domain structure of the *Facus* egg surface has some direct relevance to sperm binding and fusion and it was pointed out earlier that observations of the behaviour of sperm

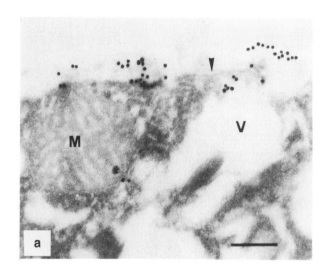

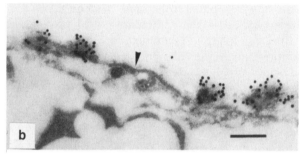

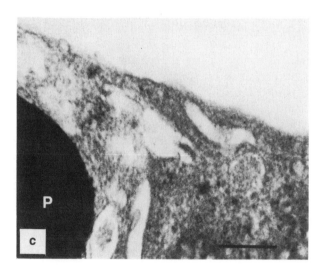

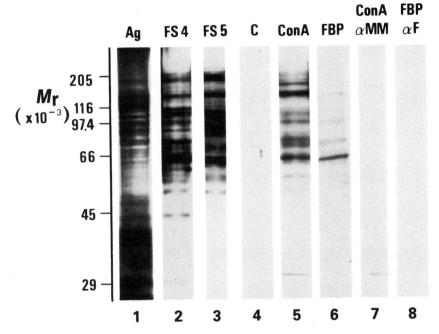

Fig. 4. Nitrocellulose blots of egg plasma membrane lysates run on sodium dodecyl sulphate/polyacrylamide gel electrophoresis. From left to right, Ag is silver-stained and shows a range of proteins and gly-coproteins. The next three tracks are probed with MAbs FS4, FS5 and a control antibody (C) raised to rat bone cells, each followed by alkaline-phosphatase labelled second antibody. Tracks 5 and 6 (from left) show the results of probing with lectins Con A and FBP labelled with horse-radish peroxidase, and tracks 7 and 8 are controls incubated in the presence of appropriate hapten sugars: methyl mannoside (α MM) and fucose (αF), respectively.

on reaching the egg surface are consistent with non-uniformly distributed sperm receptors. Evidence has recently been obtained from experiments with MAbs which is consistent with this interpretation.

It has been shown (C. J. Stafford *et al.*, unpublished results) that fertilisation in the standard bioassay can be inhibited by pretreating eggs

Fig. 3. (a) and (b) Localisation of FS2 binding by EM-immunogold labelling of egg sections. (c) A control labelled with a rat bone cell antibody. Egg plasma membrane is shown by arrowheads. M, mitochondria; V, peripheral vesicle; P, polyphenol body. The bars represent 200 nm. (From Stafford *et al.*, 1992*a*.)

with those antibodies that also recognise the discrete antigen domains, viz. FS2 and FS5, but there is no inhibition by FS4. Antibodies are large molecules and it is possible that they could bind to egg surface molecules not involved in recognition, thus causing steric hindrance to interacting receptors. However, recent experiments with smaller Fab fragments of FS5 suggest more direct effects, since these fragments still inhibit. It is also interesting that there is a correlation between the inhibitory effects of FS2/5, Con A and FBP and the presence of a common band at 62 kDa in blots of these MAbs and lectins against egg surface glycoproteins. These lines of evidence encourage future experiments on the FS2/5 antigens and their common ligands recognised by Con A and FBP, particularly in testing their functional significance as putative sperm receptors.

Acknowledgement

The authors thank the SERC for the award of a Research Studentship to C.J.S.

References

Bolwell, G. P., Callow, J. A. & Evans, L. V. (1980). Fertilization in brown algae. III. Preliminary characterisation of putative gamete receptors from eggs and sperm of *Fucus serratus*. *Journal of Cell Science* **43**, 209–24.

Brawley, S. H. (1990). Polyspermy blocks in fucoid algae and the occurrence of polyspermy in nature. In *Mechanisms of Fertilisation*, NATO ASI series, vol. H45, ed. B. Dale, pp. 419–31. Berlin, Heidelberg, New York: Springer-Verlag.

Brawley, S. H. (1991). The fast block against polyspermy in fucoid algae is an electrical block. *Developmental Biology* **144**, 94–106.

Callow, J. A. (1985). Sexual recognition and fertilization in brown algae. *Journal of Cell Science, Supplement* **1** (6th John Innes Symposium), 219–32.

Callow, J. A., Callow, M. E. & Evans, L. V. (1985). Fertilization in *Fucus*. In *Biology of Fertilization*, ed. C. B. Metz & A. Monroy, vol. 2, pp. 389–407. New York: Academic Press.

Callow, M. E., Evans, L. V., Bolwell, G. P. & Callow, J. A. (1978). Fertilisation in brown algae. I. SEM and other observations on *Fucus serratus*. *Journal of Cell Science* **32**, 45–54.

Green, J. R., Jones, J. L., Stafford, C. J. & Callow, J. A. (1990). Fertilisation in *Fucus*: exploring the gamete cell surfaces with monoclonal antibodies. In *Mechanisms of Fertilisation*, NATO ASI series, vol. H45, ed. B. Dale, pp. 189–202. Berlin, Heidelberg, New York: Springer-Verlag.

Jones, J. L., Callow, J. A. & Green, J. R. (1988). Monoclonal antibodies to sperm antigens of the brown alga *Fucus serratus* exhibit region, gamete-, and genus-preferential binding. *Planta* **176**, 298–306.

Jones, J. L., Callow, J. S. & Green, J. R. (1990). The molecular nature of *Fucus serratus* sperm surface antigens recognised by monoclonal antibodies FS1 to FS12. *Planta* **182**, 64–71.

Muller, D. G. & Seferiades, K. (1977). Specificity of sexual chemotaxis in *Fucus serratus* and *Fucus vesiculosus* (Phaeophyceae). *Zeitschrift für Pflanzen Physiology* **84**, 85–94.

Primakoff, P. & Myles, D. G. (1983). A map of the guinea pig sperm surface constructed with monoclonal antibodies. *Developmental Biology* **98**, 417–28.

Simons, K. & Fuller, S. D. (1985). Cell surface polarity in epithelia. *Annual Review of Cell Biology* **1**, 243–88.

Stafford, C. J., Green, J. R. & Callow, J. A. (1992). Organisation of glycoproteins into plasma membrane domains on *Fucus serratus* eggs. *Journal of Cell Science*, **101**, 439–51.

Theunis, C. H., Pierson, E. S. & Cresti, M. (1991). Isolation of male and female gametes in higher plants. *Sexual Plant Reproduction* **4**, 145–54.

G. W. GOODAY

The fungal surface and its role in sexual interactions

Introduction

The fungi exhibit a polyphyletic range of sexual differentiation, with a variety of mating systems promoting genetic exchange and gene flow (see Carlile, 1987; Prillinger, 1987). This review considers examples of the biochemical and cytological mechanisms of fusions of cells of different mating types (Table 1).

Membrane fusion of motile gametes

Allomyces macrogynus, a member of the Chytridiomycetes, produces five types of flagellate cell in its life cycle: male and female gametes, the resultant zygote, and haploid and diploid zoospores. Of these, fusion occurs only between a male and female gamete. Both gametes are motile by means of one posterior flagellum, but there are many differences between them that must be phenotypic sexual differences, since they are both produced by the same haploid plant and must be isogenic. Only the female produces the chemotactic attractant, sirenin, and only the male (the spermatozoid) is attracted by it (Carlile & Machlis, 1965; Pommerville, 1978, 1981). Likewise, but less obviously, the male produces an attractant, parisin, which attracts the female (Pommerville & Olson, 1987). Sirenin is a sesquiterpene (Fig. 1d) and parisin shows similar chemical characteristics. The male gamete swims much more actively than the female, is bright orange with an accumulation of γ-carotene, is much smaller than the female, and has fewer and smaller mitochondria (Fig. 1a,b). It also appears to lack the 'sidebody complex' or 'Stüben body', a complex of lipid granules and microbodies of unknown function that is present in the female, and has fewer γ-like particles (Pommerville & Fuller, 1976). An unexplained phenomenon is that, under some conditions, sirenin can induce the homotypic agglutination of male gametes,

Society for Experimental Biology Seminar Series 48: *Perspectives in Plant Cell Recognition*, ed. J. A. Callow & J. R. Green. © Cambridge University Press 1992, pp. 33–58.

Table 1. *Examples of sexual interactions in fungi*

Class	Genus	Type of interaction
Chytridiomycetes	*Allomyces*	Membrane fusion of motile gametes
Zygomycetes	*Mucor, Absidia*	Fusion of zygophore hyphae
Hemiascomycetes (Endomycetales)	*Saccharomyces, Hansenula, Pichia*	Fusion of conjugation tubes via agglutinins
	Schizosaccharomyces	Fusion of pairs of cells via fimbriae?
Blastobasidiomycetes	*Tremella*	Fusion of conjugation tubes via agglutinins
	Ustilago	Fusion of conjugation tubes via fimbriae and agglutinins?

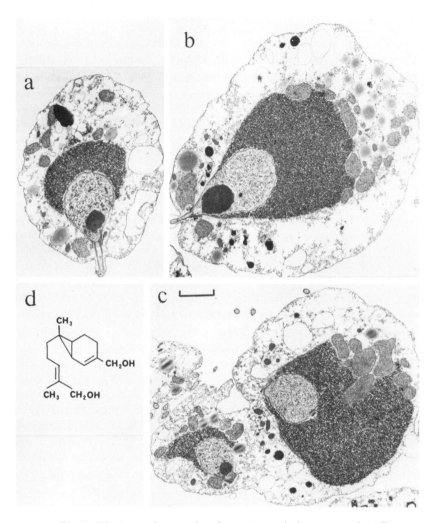

Fig. 1. Electron micrographs of gametes and plasmogamy in *Allomyces macrogynus*. (a) Male gamete. (b) The larger female gamete. (c) An early stage of fertilisation soon after fusion of gamete plasma membranes. The bars on (a) to (c) represent 1 μm. ((a), (b), (c) from Pommerville & Fuller, 1976.) (d) Chemical structure of the female's attractant, sirenin.

which then swim as a clump towards the female gamete (Pommerville, 1978).

Hatch (1938) gives us a light microscopic description of fertilisation: 'They came together smoothly, slid about over each other momentarily, and then fused with a rush. Upon this abrupt mingling of cytoplasms a mad swirling and streaming ensued which resulted in bizarre distortion for the young zygote. It seemed to boil and bubble over its whole surface. . . . [A]fter seven minutes of this ebullition, the zygote rounded up and swam away. . . .' He observed that fusion of young gametes (from 1 to 15 min after release from the gametangia) is 'quick, almost instantaneous', but it becomes increasingly slower as the gametes age, and eventually becomes impossible. It is unclear whether there is a block to polyspermy, but polynucleate zygotes are rare.

Pommerville & Fuller (1976) and Pommerville (1982) have described the fine structure of fertilisation in detail. The gametes do not fuse initially at their flagella, as do those, for example, of the alga *Chlamydomonas* (see van den Ende, this volume), but they join at a specific region of plasma membrane towards the end of the cell near the flagellum. Fusion is inhibited by: (1) agents, such as diphenhydramine and chloroquine, that stabilise membranes, suggesting that membrane fluidity is required; and (2) trypsin, implying that a protein or glycoprotein surface is involved. Treatment of female gametes, but not male gametes, with cytochalasin B, inhibited fusion, suggesting the involvement of microfilaments in the female. Increasing concentrations of Ca^{2+}, Mg^{2+} or Sr^{2+} decreased the time for the mating process from just over 1 min in standard conditions to half of that value. This time represents the events required for gamete contact, and alignment of plasma membranes and mating sites; time-lapse motion picture analysis suggested that the actual fusion process takes less than 50 ms (Pommerville, 1982). In a study of effect of ageing on gamete fusion ability, Pommerville showed that the female gametes lost about 80% of their ability in 2 h, while the male gametes lost only about 20%. In all treatments tested, the mating of the female gamete was more sensitive than that of the male gamete. At the 'fusion interface' many small cytoplasmic bridges are formed between the fusing cells, a row of vesicles appears (Fig. 1c) and, eventually, the fusion is complete and the resultant binucleate cell becomes spherical. The two nuclei line up, and nuclear fusion follows, first that of the two outer nuclear membranes and then separately those of the two inner nuclear membranes, to produce a series of bridges between the nuclei analogous to the earlier process of fusion of the cell membranes.

The two gametes are apparently naked, but presumably have some form of glycocalyx. These observations on their fusion strongly suggest that complementary sexual agglutinins are present on their surfaces.

Mating in zygomycetes

The occurrence of the zygospores in cultures of these fungi remained a sporadic and unpredictable phenomenon until Blakeslee in 1904 presented his results showing clearly that there are two patterns of mating, homothallic and heterothallic. In homothallic species, such as *Zygorhynchus moelleri*, zygospores result from the fusion of two sexual hyphae, zygophores, arising from adjacent hyphae of the same thallus. Most species of Mucorales, however, are heterothallic, with zygophores and then zygospores forming only when two compatible cultures are mated (Fig. 2). Blakeslee designated the two mating types (+) and (−), the plus sign signifying more luxuriant growth in his original cultures of *Mucor mucedo*. He further showed that 'a process of imperfect hybridization will occur between unlike strains of different heterothallic species in the same or even in different genera, or betwen a homothallic form and both strains of a heterothallic species', and that 'taking advantage of this character it has been possible to group together in two opposite series the strains of all the heterothallic forms under cultivation'. His final conclusions were:

(a) that the formation of zygospores is a sexual process;
(b) that the mycelium of a homothallic species is bisexual;
(c) while the mycelium of a heterothallic species is unisexual;
(d) and further, that in the (+) and (−) series of the heterothallic group are represented the two sexes.

Schipper & Stalpers (1980) suggested that homothallic and heterothallic strains of the mucors are related to each other as are those of *Saccharomyces cerevisiae*; i.e. all fertile strains have both (+) and (−) mating genes. In heterothallic strains one is permanently expressed throughout the thallus and the other is silent; in homothallic strains the two genes are on transposable elements and may be alternately switched on, so that there is a stochastically determined mixture of (+) nuclei and (−) nuclei in the thallus. Mating will occur between a branch containing

Fig. 2. Zygotropism and fusion of two zygophores of *Mucor mucedo*, taken at 4 min intervals. The bar represents 100 µm. For more details, see Gooday (1973, 1975).

(+) nuclei and one containing (−) nuclei. Thus homothallic and heterothallic strains need differ by only a gene regulating the mating-type switch. Such a system would explain the mutagen-induced mating-type switch observed in *Rhizomucor pusillus* (Nielsen, 1978) and the histochemical demonstration of (+)-type enzyme activity in only the terminal partner of two mating zygophores of the homothallic *Z. moelleri* (Werkman, 1976).

Zygophores are characteristically distinctive hyphae formed in response to stimulation of vegetative hyphae by the isoprenoid sex hormone, trisporic acid (Gooday 1978, 1983). That this hormone system is not species-specific, but is universal in the Mucorales, is shown by mutual zygophore formation (but of course not zygospore formation) when (+) and (−) strains of different species are grown together (Blakeslee & Cartledge, 1927; Schipper, 1987). In *M. mucedo* the zygophores are aerial hyphae that are readily distinguished from sporangiophores and vegetative hyphae. They are rarely branched, and have a smooth appearance. Plasmogamy occurs between mated pairs of zygophores. It is likely that there are zygophore-specific and mating-type-specific cell surface components involved in this process. Distinctive cell surface properties of the zygophores of *M. mucedo* have been demonstrated by Jones & Gooday (1977, 1978) in two studies using fluorescent lectin binding (Fig. 3) and immunofluorescence (Fig. 4). The zygophores of both (+) and (−) showed a much stronger binding of wheat-germ agglutinin (WGA) than did vegetative hyphae, particularly to their apical 120–15 µm, and also stronger binding of concanavalin A (Table 2). For the immunofluorescent study, four types of fluorescent antibody preparation were made: to whole zygophores from (+) and (−), and to whole vegetative mycelia from (+) and (−). Antisera to both (+) and (−) zygophores stained zygophores of both (+) and (−) *M. mucedo*, but not vegetative mycelium. Antisera to both (+) and (−) vegetative mycelia stained zygophores as well as vegetative mycelium. Thus, this study detected antigens that were specific to zygophore surfaces, i.e. were formed *de novo* during development of the zygophores, but did not detect any mating-type-specific antigens. The zygophores also had antigens common to vegetative mycelium. The zygophore antisera from *M. mucedo* were not species-specific as they also stained zygophores of *Mucor hiemalis*, but not those of *Phycomyces blakesleeanus*. These surface properties of zygophores remained apparent during sexual development after fusion. Progametangia and gametangia, therefore, stained much more strongly with WGA than did vegetative hyphae or suspensors (Fig. 3; Table 2). Concanavalin A also showed much stronger binding to gametangia than to other hyphal surfaces, even though much lesser in intensity than

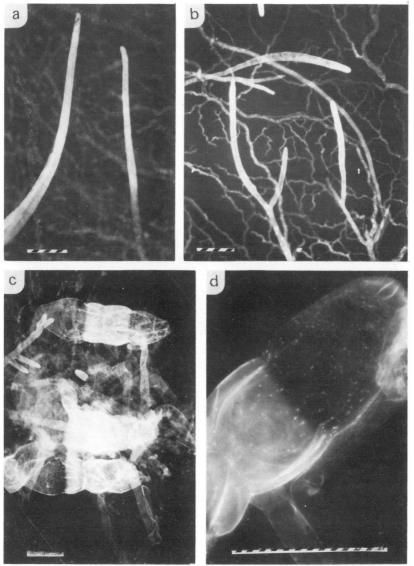

Fig. 3. Fluorescent lectin staining of *Mucor mucedo*. (a) Zygophores of (−) strain, treated with FITC-Con A. (b) Zygophores of (−) strain, treated with FITC-WGA. Bright spots in hyphae represent auto-fluorescence of carotene-containing droplets. (c) and (d) Developing zygospores, treated with FITC-WGA, showing staining of gametangia and characteristic speckled effect on suspensors and gametangia. The bars represent 50 μm. Preparation described by Jones & Gooday (1977).

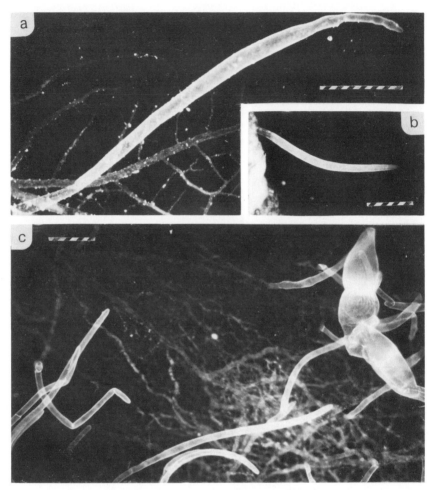

Fig. 4. Immunofluorescent labelling of *Mucor mucedo*. (a) (+)
Zygophore stained with antiserum to (+) zygophores; note the lack of
staining of vegetative hyphae. (b) (−) Zygophore staining with antiserum to (+) zygophores, together with staining of contents of cut
hyphae below the zygophore. (c) Zygophores and young zygospores
staining with antiserum to (+) zygophores; note the strong staining of
progametangia. The bars represent 50 μm. Preparations described by
Jones & Gooday (1978).

Table 2. *Binding of fluorescent lectins and antibodies to Mucor mucedo*

Cell type	Lectins			Antisera	
	WGA	Con A	SBA	(+) or (−) ZA	(+) or (−) CVA
Vegetative hyphal tips	+	−	−	−	+++
Mature vegetative hyphae	−	−	−	−	+++
Zygophores	++	+	−	++++	+++
Progametangia	++	+	−	++	+++
Gametangia	++++	++	−	−	+++
Suspensors	+	−	−	−	+++
Zygospores	−	−	−	−	+
Major specificity	GlcNAc	Man, Glc	GalNAc, Gal		

FITC-labelled lectins with appropriate controls for specificity; using (+) and (−) strains: wheat-germ agglutinin (WGA), concanavalin A (Con A), soya-bean agglutinin (SBA); see Jones & Gooday (1977). FITC-labelled antisera; to (+) or (−) zygophores (ZA) and to (+) or (−) complete vegetative mycelium (CVA); see Jones & Gooday (1978). Increasing intensity of staining, −, +, ++, +++, ++++.

WGA. The zygophore-specific fluorescent antibodies bound to young progametangia (Fig. 4), but this was quickly lost as they matured (Table 2).

Teepe *et al.* (1988) have characterised surface proteins from hyphae of *Absidia glaerca*, by extracting with buffer containing 1 M-lithium chloride. All major surface proteins with molecular masses above 22 kDa proved to be glycoproteins, as shown by their capability to bind concanavalin A in affinity blotting experiments. Only a few bands in the lower molecular mass range could be stained by this procedure. There were characteristic qualitative differences between surface proteins of aerial hyphae and those of submerged cultures. There were changes in abundance of particular proteins, some increasing and others decreasing, as the culture aged. There was a minor protein of molecular mass 29 kDa found only in preparations of mating cultures. This might be related to the zygophore-specific antigen described above for *M. mucedo*. Two proteins, of molecular mass 22 and 66 kDa, were more abundant in preparations from (−) mating type than from (+) mating type, while one protein, of molecular mass 15 kDa, was found only in preparations from (+) mating type. This proved to be a major protein of both aerial and submerged hyphae of (+) mycelium. It did not bind to concanavalin A or to antibodies against total surface proteins, and so is probably not a glycoprotein.

Wöstermeyer *et al.* (1990) have studied the progeny resulting from fusions of protoplasts of (+) and (−) strains with auxotrophic markers. Of 72 colonies tested, 11 were homothallic, 11 were (+) mating type, 43 were (−) mating type and 7 were non-mating. Of these, 7 homothallic, 5 (+) and 6 (−) strains were tested for the presence of the (+)-specific 15 kDa surface protein, by Western blotting with antiserum raised to purified protein. Only the five (+) strains were positive. This suggests that the expression of this surface protein is repressed in the mated (+/−) hybrid. In mixed cultures of (+) and (−) mycelia, however, the 15 kDa protein can be easily recognised, and so genetic information from the (−) mating type is required in the same cell for the repression of expression of this (+) mating-type protein.

These studies do not provide us yet with contenders for specific sexual agglutinins as seen in mating yeast cells. Perhaps such agglutinins are expressed only very locally at the sites of contact between compatible zygophores.

Plasmogamy of two aerial zygophores of *M. mucedo* is achieved by zygotropism. Two zygophores of opposite mating type will unerringly grow toward each other to meet and fuse from a distance of at least 2 mm (Fig. 2). This phenomenon was observed by Blakeslee (1904); 'a mutual attraction is exercised between the zygophoric hyphae of opposite mycelia, and they may be seen gradually to approach each other' and 'two minutes before contact occurred, and while the hyphae were separated by a distance equal to about a third of their width, very slight protrusions were observed on the sides mutually facing, seemingly as if the forces which were drawing the filaments laterally had effected a bulging of the delicate walls at the growing points'. The phenomenon was confirmed by Burgeff (1924), who introduced the term 'zygotropism'. There are clearly two effects involved here, both mediated by mating-type-specific volatile effectors: directed growth of each zygophore from a distance, and then, when very close a localised lateral budding or branching. Zygotropism is mating-type-specific, not species-specific, as it occurs between zygophores of different species. Thus, Schipper (1987) recorded the tall zygophores of *Mucor flavus* bending down to meet the short zygophores of *M. mucedo*. While they are growing, zygophores of *M. mucedo* have a constant growth rate, irrespective of their zygotropic behaviour (Gooday, 1975), but a zygophore under the influence of zygotropic attractant grows much longer than an unexposed one (Plempel, 1960, 1962; Banbury, 1955), and when it meets a zygophore of opposite mating type it stops elongating immediately on contact and starts to form a progametangium (Fig. 2; Gooday, 1975). Intriguingly, other 'unrequited' nearby zygophores that were growing towards this site also stop elongating when the successful fusion has been made (Gooday, 1975), a situation reminiscent of the blocking of egg surfaces, such as in animals and *Fucus*, after the successful entry of a sperm.

All evidence suggests that there are two complementary volatile chemicals regulating zygotropism, one produced by each mating type and active on the opposite mating type (Gooday, 1973, 1975; Mesland *et al.*, 1974). Mesland *et al.* (1974) suggest that these two are, or are closely related to, particular mating-type-specific precursors of the zygophore-inducing hormone trisporic acid. This is an appealing idea, as we do not have to invoke new biosynthetic pathways for the biosynthesis and reception of the zygotropic pheromones. It also explains the following observations: Fig. 2 of Mesland *et al.* (1974) shows (−) zygophores being attracted to a (+) sporangiophore, as in this experiment the (+) mycelium increasingly produces its trisporate precursors in response to the nearness of the (−) mycelium; Fig. 7 of Gooday (1975) shows zygophores of both mating types ceasing to be attractive on fusing with a compatible mate, as after fusion the precursors that are the attractants are now efficiently converted to trisporic acid by the partner, and so no longer can diffuse through the air.

Mating in ascomycetous yeasts

Of all the fungi, sexual interactions are best understood in brewer's/ baker's yeast *Saccharomyces cerevisiae*, as concerted efforts in studying its genetic system have given rise to much elegant experimental work. *Saccharomyces cerevisiae* exists in one of three cell types; haploids of opposite mating type (*MATa* and *MATα*), which mate to give diploid cells (*MATa/MATα*), which can sporulate to give haploid ascospores (Yanagishima, 1986; Cross *et al.*, 1988). Conjugation involves: the synchronisation of the cell cycles of adjacent *MATa* and *MATα* cells so that they become unbudded gametes under G1 arrest; growth of complementary conjugation tubes towards each other; fusion of the apices of these tubes; and lysis of the adjoining walls to allow plasmogamy and the karyogamy. Levi (1956) observed mating responses of yeast cells on an agar medium: *MATa* cells produced conjugation tubes when in the vicinity of *MATα* cells, or on medium on which *MATa* and *MATα* cells had previously mated. In liquid medium, sexual adhesion is dependent on random collisions (Campbell, 1973). The induction of morphogenesis from vegetative cell to gamete cell is regulated by a pair of complementary pheromones, α-factor, a peptide produced by *MATα* cells and affecting *MATa* cells, and a-factor, a lipopeptide, produced by *MATa* cells and affecting *MATα* cells (Cross *et al.*, 1988; Lipke, 1986). The cell responding to its complementary pheromone is termed a 'shmoo' (as it looks like Shmoo, a cartoon character of Al Capp). It has the following characteristics of relevance to its cell surface sexual interactions: it elongates towards its mating partner, the cell wall composition of the elongating tube is distinctly different from that of the original yeast cell, with more chitin and glucan and less mannoprotein, and the elongating tube becomes covered with a mating-type-specific agglutinin (Lipke *et al.*, 1976; Betz *et al.*, 1978; Fehrenbacher *et al.*, 1978; Schekman & Brawley, 1979; Moore, 1983). Several genes that function during cell fusion have been characterised (Suzuki & Yanagishima, 1985; McCaffrey *et al.*, 1987; Trueheart *et al.*, 1987; Cross *et al.*, 1988):

> *AGα1* codes for α-agglutinin on the *MATα* shmoos;
> *FUS1* and *FUS2* are involved in cell fusion;
> *CDC24* is involved in shmoo formation.

Mutations in these genes greatly reduce the efficiency of mating (Suzuki & Yanagishima, 1985; Trueheart *et al.*, 1987).

  The agglutinins are clearly of great importance in successful conjugation. The induction of **a**-agglutinin in *MATa* cells by α-factor is a particularly sensitive response, the K_{50} value (for 50% response) being

6×10^{-11} M, below values for cell division arrest and conjugation tube formation (Moore, 1983). The *MATa* and *MATα* agglutinins are complementary, as agglutination is seen only between haploid cells of opposite mating type (Yanagishima & Yoshida, 1981). When complementary competent cells are mixed, *MATa–MATα* cell pairs can be seen to form rapidly, but most cells form large aggregates, in about 1 h. When cells of one of the two mating types were marked by staining with a vital dye, and the composition of the aggregates was scored, the two cell types were seen in equal numbers. This remained so even when the initial ratio *MATa:MATα* was varied from 1:1 to 19:1 (Kawanabe *et al.*, 1979). Sakai & Yanagishima (1972) and Terrance & Lipke (1981) report considerable variation in agglutinability of different strains, with *MATa* strains often being strongly inducible and *MATα* strains often being constitutively agglutinable. Tokoyama & Yanagishima (1982) reported that a mutant of the *MATα2* gene, a negative regulator of a-specific genes, shows simultaneous expression of *MATa* and *MATα* agglutinins under certain conditions, but no agglutination ability was expressed, because of a mutual interaction of the two agglutinins. The *MATα* agglutinin has two classes of binding site on *MATa* cells, and so there may be another binding site for it as well as for the *MATa* agglutinin (Lipke *et al.*, 1987; Terrance & Lipke, 1987).

The complementary agglutinins were characterised by Yanagishima & Yoshida (1981) as glycoproteins with acidic pI values, which masked the action of agglutinability of opposite mating type cells and formed a molecular complex with each other. They are univalent, as they do not cause agglutination by themselves (Yanagishima & Yoshida, 1981; Yamaguchi *et al.*, 1984a; Terrance *et al.*, 1987; Orlean *et al.*, 1986). The purified *MATα* agglutinin is large (200–300 kDa) and approximately half of its mass is *N*-linked carbohydrate, which is not necessary for its activity (Yamaguchi *et al.*, 1982, 1984a; Terrance *et al.*, 1987; Hauser & Tanner, 1989). The protein moiety corresponds to 68.2 kDa. Hauser & Tanner (1989) and Lipke *et al.* (1989) have cloned and sequenced the gene. DNA sequence predicts a protein of 631 amino acid residues with 12 potential *N*-glycosylation sites. A proteolytic carboxy terminal deletion gave a polypeptide of 51 kDa that was still biologically active in preventing agglutination. The purified *a* agglutinin is much smaller, 2 kDa, and is solely *O*-glycosylated (Yamaguchi *et al.*, 1984a; Orlean *et al.*, 1986; Watzele *et al.*, 1988). It is a potential inhibitor of sexual agglutination, being active at 4×10^{-9} M. It is 29% carbohydrate, and mild treatment with 10 mM sodium iodate inactivates it, indicating the involvement of the *O*-linked sugars in the agglutination reaction (Watzele *et al.*, 1988). Using an endoplasmic reticulum accumulating secretory mutant (*sec18*),

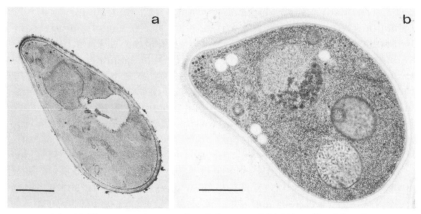

Fig. 5. Electron micrographs of shmoos of *MATa* cells of *Saccharomyces cerevisiae*, showing diffuse material coating the elongating cell walls which probably represents the agglutinin. (a) Prepared by chemical fixation ($KMnO_4$; OsO_4); kindly provided by C. E. Ballou (cf Lipke *et al.*, 1976). (b) Prepared by freeze-substitution; kindly provided by M. Baba (cf. Baba *et al.*, 1989). The bars represent 1 μm.

Watzele *et al.* showed that a mannosylated precursor of *a* agglutinin accumulated in cells induced with α-factor at the restrictive temperature, demonstrating the biogenesis of the glycoprotein.

Electron micrographs of *MATa* cells responding to α-factor, prepared by both chemical fixation and freeze-substitution, show fibrous material on the outer surface of conjugation tubes (Osumi *et al.*,1974; Lipke *et al.*, 1976; Baba *et al.*, 1989; Fig. 5). This material is presumably the sexual agglutinin. This region stains brightly with fluorescein-labelled concanavalin A, consistent with a localisation of mannose units (Tkacz & MacKay, 1979). Watzele *et al.* (1988) and Hauser *et al.* (1990) have used differently fluorescent antibodies to *MATa* and *MATα* agglutinins to demonstrate their localisation on the surfaces of mating cells. Immunofluorescence of *MATa* cells treated with α-factor showed that the *MATa* agglutinin was deposited almost exclusively on growing buds after only 15 min. As shmoos developed, the agglutinin was seen on the tip of the increasingly pear-shaped cell, just where the fibrous material is seen in electron micrographs (Fig. 5). In mating pairs of cells, the *MATa* agglutinin is seen on the *MATa* conjugation tube; and the developing zygotes show no fluorescent labelling. Ford & Pringle (1986) have investigated the polarity of shmoo formation in a yeast cell with respect to its previous budding history as assessed by Calcofluor White staining of its bud scars. The results were unequivocal; neither fusion sites nor shmoo tips were restricted to the budding pole, and seemed to be located randomly on the cell surface.

Other species of yeast show similar mating systems. The pheromones do have some cross-reactions between species and genera (Yanagishima, 1986; Yoshida *et al.*, 1989). Thus α-factor from *S. cerevisiae* will elicit responses to induction of agglutinability and/or 'shmooing' in *MATa* cells of *Saccharomyces kluyveri*, *Saccharomyces exiguus*, *Hansenula wingei* and *Hansenula anomala*. The agglutinins, however, are strictly species-specific (Yamaguchi *et al.*, 1984*b*).

Pierce & Ballou (1983) and Mendonca-Previato *et al.* (1982) have characterised sexual agglutinins from *S. kluyveri* and *Pichia amethionina*, respectively, and find them to be glycoproteins similar to those of *S. cerevisiae*, again having no cross-reactivity between species. The *S. kluyveri* 17-agglutinin has been characterised further by Weinstock & Ballou (1986) and Lasky & Ballou (1988). It is a protein heavily glycosylated with *N*-linked carbohydrate and probably also with some *O*-linked carbohydrate.

H. wingei can be isolated as a diploid or haploid yeast from coniferous bark beetles and their habitat. Some haploid strains, both as isolated from nature and resulting from sporulation of diploid strains in the laboratory, show a very strong sexual agglutination (Crandall & Brock, 1968*a*; Crandall *et al.*, 1974*b*). Most work has been done with mating strains *5* and *21*, obtained by heat treatment of ascospores from a diploid isolate (Wickerham, 1956). Each strain does not agglutinate separately, and neither will agglutinate with the *5×21* diploid. Each strain produces a specific surface component, 5-agglutinin and 21-factor, respectively, that is specifically adsorbed by cells of the opposite mating type.

The 5-agglutinin will cause strain *21* cells to agglutinate, and so is multivalent. It can be released from strain *5* cells by treatment with enzymes such as subtilisin or snail digestive juice. It can be purified by affinity chromatography on strain *21* cells. It is a mannan-protein that appears as aggregates of heterogeneous molecular weight. It is inactivated by treatment with pronase and exo-α-mannanase, and by disulphide-cleaving reagents such as mercaptoethanol (Taylor & Orton, 1967, 1968; Crandall & Brock, 1968*a,b*; Yen & Ballou, 1973). Yen & Ballou (1974) further characterised a preparation released by treatment with subtilisin and showed it to be about 85% carbohydrate, 10% protein, and 5% phosphate. The mannan chains are highly branched, containing (1-2)α and (1-6)α linkages, and exist as small oligosaccharides directly attached to the protein backbone through the serine and threonine units. This structure is distinct from that of the mannan-protein of the wall of *H. wingei*, which has long mannan chains of up to 200 mannose units attached to the protein via *N*-acylglycosylamine linkages (Ballou, 1976).

The 21-factor is released from strain *21* cells by treatment with trypsin (Crandall & Brock, 1968*a*). It does not agglutinate strain *5* cells and is apparently univalent, but it does have the properties required of a surface component responsible for agglutination of strain *21* cells with strain *5* cells. It is only released from strain *21* cells, which concomitantly lose their sexual agglutinability, it neutralises the activity of 5-agglutinin, and it is adsorbed only by strain *5* cells. It is also a mannan-protein (Crandall *et al.*, 1974*b*; Crandall, 1976). Burke *et al.* (1980) report its further characterisation, as having a mass of 27 kDa, and pI of 3.8, being rich in acidic amino acid residues, containing 5% mannose and a trace of glucosamine and being stable to reducing agents but inactivated by heat. They report that treatment of strain *21* cells with zymolyase, (1-3)β-glucanase, releases a larger molecule with an additional portion that may anchor the 21-factor to the wall.

The 21-factor and the 5-agglutinin form a complex when mixed *in vitro* (Crandall *et al.*, 1974*b*). The resultant complexes are large, corresponding to the combination of 63, 16 and 6 molecules of 21-factor, with one unit each of the three 5-agglutinin fractions used here, of apparent molecular weight of 13×10^5, 5.6×10^5 and 2.2×10^5, respectively. These complexes are soluble, probably as the 21-factor is univalent and so cannot form cross-links. They presumably represent the primary binding mechanism when a strain *5* cell touches a strain *21* cell. Taylor & Orton (1970, 1971), using ^{35}S-labelled 5-agglutinin, found a very high apparent free energy of association for its binding to strain *21* cells. At pH 4, the value is -14.5 kcal/mol (1 cal=4.184 J), which is higher than most antigen-hapten binding free energies. They suggest that this high value results from the additive effect of several binding sites on each molecule of 5-agglutinin.

The *5×21* diploid does not constitutively produce 5-agglutinin or 21-factor (Crandall & Brock, 1968*a*). However, certain conditions can initiate the phenotypic production of one or other factor so that the diploid cells will agglutinate with *21* or *5* cells. For example, 5-agglutinin production results from the addition of 0.4 mM sodium vanadate or sodium molybate; 21-agglutinin results from the addition of 1 mM disodium ethylenediaminetetraacetic acid (Na$_2$ EDTA) (Crandall & Caulton, 1973, 1975). These treatments have no appreciable effect on agglutinability of *5* cells or *21* cells, and so presumably are affecting just the relative expression of the two mating factor genes in the diploid, perhaps via the postulated complementary regulatory genes that are only active in the diploid in repressing each mating factor.

Mating has also been studied in homothallic strains of the fission yeast *Schizosaccharomyces pombe* (Calleja *et al.*, 1981). These show a mating-type switching system analogous to that of *Saccharomyces cerevisiae*, so that a homothallic clone of cells is made up of two cell types that are sexually complementary and interconvertible. When these cells are grown to stationary phase and the culture is aerated, cells form flocs within 1 h, and the flocs increase in size and number in the next 5 to 6 h. Within the flocs, adjacent complementary cells conjugate by fusion of their end walls and karyogamy follows. Before conjugation, the cells can be reversibly deflocculated by treatments such as heating, extremes of pH, mechanical shear and sonication, and irreversibly deflocculated by treatment with proteases. Calleja *et al.* (1977) implicate the involvement of 'sex hairs' in this sexual flocculation, these being analogous to the fungal fimbriae described by Poon & Day (1975; and see below). Individual pairs of cells fuse preferentially at their poles, with Calleja *et al.* (1981) finding a 2:1 preference for the old growing end of the cell rather than the young end. There seems to be no precise site, with most polar fusions being eccentrically positioned. The age of a fission yeast cell can be assessed by counting the number of division scar structures in thin sections of the cell walls, and Calleja *et al.* found no effect of age of cell on its participation in conjugation.

Mating in basidiomycetous yeasts

Conjugation in the heterobasidiomycete *Tremella mesenterica*, the 'yellow-brain fungus', is between two yeast cells having compatible mating types *ab* and *AB* (Bandoni, 1963). Just as for *S. cerevisiae*, the preparation for mating is induced by two complementary pheromones, one produced by each mating type and active on the opposite mating type. These are lipopeptides, called tremerogens *a*-13 and *A*-10 from *ab* and *AB* cells, respectively (Sakagami *et al.*, 1981). In response to the complementary pheromone, a yeast cell produces a conjugation tube. Mating is accomplished by fusion of the apex of the conjugation tube of the *ab* cell with one of an *AB* cell. Labelling of cell surface proteins by lactoperoxidase-catalysed iodination with ^{125}I showed dramatic changes in the pattern of surface proteins after formation of conjugation tubes in response to the pheromones (Miyakawa *et al.*, 1982, 1984). In *ab* cells the major ^{125}I-labelled surface protein was of mass 66 kDa. On treatment with tremerogen *A*-10, this was lost to the medium, and proteins with masses 73, 60 and 43 kDa appeared in the gamete cells. In contrast, in *AB* cells there was little change in intensity at 66 kDa, but appearance of proteins with masses of 110, 40 and 22 kDa on treatment with tremerogen *a*-13. Electron microscope autoradiography showed that the iodinated proteins were predominantly localised on the surface of the conjugation tubes, where counts showed that the density of the silver grains was 3.5-fold higher than that over the original walls of the yeast cells. The involvement of cell surface protein in the sexual agglutination has been studied by Miyakawa *et al.* (1987) Agglutination was scored microscopically. When pheromone-induced gametes of *ab* and *AB* were mixed, there was about 70% agglutination in 1 h, to give clumps of 80–100 cells. When vegetative cells were mixed agglutination was delayed for several hours until conjugation tubes had been formed. The sexual nature of this agglutination was confirmed by pre-staining the cell wall chitin of cells of one of the mating types with Calcofluor White, and mixing with unstained cells of opposite mating type, in ratios from 1:9 to 9:1. In all cases the agglutination complexes contained nearly equal numbers of cells of the two mating types. Some homotypic agglutination was seen with pheromone-induced gametes of *AB* cells, but the clumps were much smaller, consisting of only five cells or less, and this phenomenon was not shown by vegetative *AB* cells or by *ab* gametes or vegetative *ab* cells. The involvement of cell surface proteins in the sexual agglutination was shown by pre-treatment of the cells with proteases, which abolished it (pronase or chymotrypsin) or very much reduced it (trypsin or thermolysin). It was also inhibited by dithiothreitol or *N*-ethylmaleimide, suggesting the involvement of thiols and disulphide groups of proteins in the cell–cell recognition. Furthermore, agglutination was inhibited by polypeptides released to the medium by treatment of both *AB* and *ab* gamete cells with thermolysin. Miyakawa *et al.* (1987) suggested that these are proteolytic products of the specific sexual agglutinins, and that they retain their specific binding properties, so can mask the complementary cell surface recognition molecules. The inhibitory activities of these polypeptides from the two mating types were lost by mixing them and so they interact in a complementary fashion. They did not agglutinate cells of opposite mating type, suggesting that they had only one or a very few binding sites.

Conjugation in the anther smut fungus of the Caryophyllaceae, *Ustilago violacea*, is between two yeast-like cells, sporidia, of opposite mating type, a_1 and a_2. Cells may mate if they are in contact, or at distances up to 20 μm apart. If they are apart, the a_2 cell produces a conjugation tube which grows towards the a_1 cell to fuse with a peg that it produces. This gives a copulatory bridge which pushes apart the two mating cells (Poon *et al.*, 1974). In natural conditions, a dikaryotic mycelium then develops to infect the host plant, and karyogamy is delayed until sporulation in the anthers of the host's flowers. Light microscopy of early stages of this process shows that mating cells still at distances of more than 5 μm apart are connected by 'invisible fibres' (Day, 1976). These connections were shown by pairs of cells floating together, rotating in unison and maintaining a maximum distance apart. The connections were fragile, and could be broken by pressure on the cover-slip. Structures on the cell walls of these sporidia, described as fimbriae, have been observed by Poon & Day (1974, 1975), Day & Poon (1975), Svircev *et al.* (1986) and Day & Garber (1988). Cells of *Ustilago violacea* may carry more than 200 fimbriae, showing by shadow-cast electron microscopy as fibrils from 0.5 to over 10 μm long with a diameter of 6–7 nm. Cells can be defimbriated by sonication or other physical treatments, and the fimbriae regenerate at the rate of 1–2 μm h^{-1}. These fimbriae have the properties of the 'invisible fibres' connecting mating cells (Day, 1976). From his observations Day suggested that the fimbriae connect the cells, and guide the growth of the conjugation tube. Further, afimbriate strains cannot conjugate, and treatments such as growth at high temperature that inhibit fimbriation also inhibit conjugation (Day & Poon, 1975). Day & Poon, however, showed that fimbriae are not essential for the initial cell pairing of adjacent cells, which they demonstrated is mediated by non-fibrous globular material that is produced in large amounts by mating cells, and which is digestible by α-amylase. The fimbriae, in contrast, are digestible by proteases such as Pronase, and inhibitors of protein synthesis such as cycloheximide inhibit their regeneration (Poon & Day, 1975). They are composed of a 74 kDa protein that can assemble spontaneously into 7 nm fibrils which can attach to the cell surfaces (Day & Cummins, 1981; Gardiner & Day, 1985). Similar fimbriae have been described for several other fungi (Poon & Day, 1975; Day & Gardiner, 1987) and it may be that they have roles in other cell–cell communications as well as in mating of *U. violacea*.

Conclusions

Presented here have been just a handful of examples of the role of the cell surface in sexual interactions amongst the fungi. Undoubtedly every mating species of the fungi will prove to have its own system of cell surface communication between gametic cells. Given the increasing awareness of the importance of fungi in our environment, these promise to be fruitful areas of research.

Acknowledgements

I thank M. Baba, J. Pommerville and C. E. Ballou for photographs, and numerous authors for reprints and unpublished results.

References

Baba, M., Baba, N., Ohsumi, Y., Kanaya, K. & Osumi, M. (1989). Three dimensional analysis of morphogenesis induced by mating pheromone α factor in *Saccharomyces cerevisiae*. *Journal of Cell Science* **94**, 207–16.

Ballou, C. E. (1976). Structure and biosynthesis of the mannan component of the yeast cell envelope. *Advances in Microbiological Physiology* **14**, 93–158.

Banbury, G. H. (1955). Physiological studies in the Mucorales. III. The zygotropism of zygophores of *Mucor mucedo* Brefeld. *Journal of Experimental Botany* **6**, 235–44.

Bandoni, R. J. (1963). Conjugation in *Tremella mesenterica*. *Canadian Journal of Botany* **43**, 627–30.

Betz, R., Duntze, W. & Manney, T. R. (1978). Mating factor-mediated sexual agglutination in *Saccharomyces cerevisiae*. *FEMS Microbiology Letters* **4**, 107–10.

Blakeslee, A. F. (1904). Sexual reproduction in the Mucorineae. *Proceedings of the American Academy of Arts and Science* **40**, 205–319.

Blakeslee, A. F. & Cartledge, J. L. (1927). Sexual dimorphism in Mucorales. II. Inter-specific reactions. *Botanical Gazette (Chicago)* **84**, 51–7.

Burgeff, H. (1924). Untersuchungen über Sexualität und Parasitismus bei Mucorineen. I. *Botanische Abhandlungen* **4**, 1–135.

Burke, P., Mendonca-Previato, L. & Ballou, C. E. (1980). Cell–cell recognition in yeast: purification of *Hansenula wingei* 21-cell agglutination factor and comparison of the factors from three genera. *Proceedings of the National Academy of Sciences, USA* **77**, 318–22.

Calleja, G. B., Johnson, B. F. & Yoo, B. Y. (1981). The cell wall as sex organelle in fission yeast. In *Sexual Interactions in Eukaryotic Microbes*, ed. D. H. O'Day & P. A. Horgen, pp. 225–59. New York: Academic Press.

Calleja, G. B., Yoo, B. Y. & Johnson, B. F. (1977). Fusion and erosion of cell walls during conjugation in the fission yeast (*Schizosaccharomyces pombe*). *Journal of Cell Science* **25**, 139–55.

Campbell, D. A. (1973). Kinetics of the mating-specific aggregation in *Saccharomyces cerevisiae*. *Journal of Bacteriology* **166**, 323–30.

Carlile, M. J. (1987). Genetic exchange and gene flow: their promotion and prevention. In *Evolutionary Biology of the Fungi*, ed. A. D. M. Rayner, C. M. Brasier & D. Moore, pp. 203–14. Cambridge: Cambridge University Press.

Carlile, M. J. & Machlis, L. (1965). A comparative study of the chemotaxis of the motile phases of Allomyces. *American Journal of Botany* **52**, 484–6.

Crandall, M. A. (1976). Mechanisms of fusion in yeast cells. In

Microbial and Plant Protoplasts, ed. J. F. Peberdy, A. H. Rose, H. J. Rogers & E. C. Cocking, pp. 161–75. New York: Academic Press.

Crandall, M. A. & Brock, T. D. (1968*a*). Molecular basis of mating in the yeast *Hansenula wingei*. *Bacteriological Reviews* **32**, 139–63.

Crandall, M. A. & Brock, T. D. (1968*b*). Molecular aspects of cell contact. *Science* **161**, 473–5.

Crandall, M. A. & Caulton, J. H. (1973). Induction of glycoprotein mating factors in diploid yeast of *Hansenula wingei* by vanadium salts or chelating agents. *Experimental Cell Research* **82**, 159–67.

Crandall, M. A. & Caulton, J. H. (1975). Induction of haploid glycoprotein mating factors in diploid yeasts. *Methods in Cell Biology* **12**, 185–205.

Crandall, M. A., Egel, R. & MacKay, V. L. (1974*a*). Physiology of mating in three yeasts. *Advances in Microbial Physiology* **15**, 307–98.

Crandall, M. A., Lawrence, L. M. & Saunders, R. M. (1974*b*). Molecular complementarity of yeast glycoprotein mating factors. *Proceedings of the National Academy of Sciences, USA* **71**, 26–9.

Cross, F., Hartwell, L. H., Jackson, C. & Konopka, J. B. (1988). Conjugation in *Saccharomyces cerevisiae*. *Annual Review of Cell Biology* **4**, 429–57.

Day, A. W. (1976). Communication through fimbriae during conjugation in a fungus. *Nature* **262**, 583–4.

Day, A. W. & Cummins, J. E. (1981). The genetics and cellular biology of sexual development in *Ustilago violacea*. In *Sexual Interactions in Eukaryotic Microbes*, ed. D. H. O'Day & P. A. Horgen, pp. 379–402. New York: Academic Press.

Day, A. W. & Garber, E. D. (1988). *Ustilago violacea*, anther smut of the Caryophyllaceae. *Advances in Plant Pathology* **6**, 457–82.

Day, A. W. & Gardiner, R. B. (1987). Fungal fimbriae. In *The Expanding Realm of Yeast-like Fungi*, ed. G. S. de Hoog, M. T. Smith & A. C. M. Weijman, pp. 334–49. Amsterdam: Elsevier Scientific Publishers.

Day, A. W. & Poon, N. H. (1975). Fungal fimbriae. II. Their role in conjugation in *Ustilago violacea*. *Canadian Journal of Microbiology* **21**, 547–57.

Fehrenbacher, G., Perry, K. & Thorner, J. (1978). Cell–cell recognition in *Saccharomyces cerevisiae*: regulation of mating-specific adhesion. *Journal of Bacteriology* **134**, 893–901.

Ford, S. & Pringle, J. (1986). Development of spatial organization during the formation of zygotes and shmoos in *Saccharomyces cerevisiae*. *Yeast* **2**, S114.

Gardiner, R. B. & Day, A. W. (1985). Fungal fimbriae. IV. Composition and properties of fimbriae from *Ustilago violacea*. *Experimental Mycology* **9**, 334–50.

Gooday, G. W. (1973). Differentiation in the Mucorales. *Symposium of the Society for General Microbiology* **23**, 269–94.

Gooday, G. W. (1975). Chemotropism and chemotaxis in fungi and algae. In *Primitive Sensory and Communication Systems*, ed. M. J. Carlile, pp. 155–204. London: Academic Press.

Gooday, G. W. (1978). Functions of trisporic acid. *Philosophical Transactions of the Royal Society of London, series B* **284**, 509–20.

Gooday, G. W. (1983). Hormones and sexuality. In *Secondary Metabolism and Differentiation in Fungi*, ed. J. W. Bennett & A. Ciegler, pp. 239–66. New York: Academic Press.

Hatch, W. R. (1938). Conjugation and zygote germination in *Allomyces arbuscula*. *Annals of Botany, New Series* **2**, 583–614.

Hauser, K., Cappellaro, C., Mrsa, V. & Tanner, W. (1990). Glycoprotein in cell-wall recognition during mating of *S. cerevisiae*. *Proceedings of the 4th International Mycological Congress*, Regensburg, D-188/1.

Hauser, K. & Tanner, W. (1989). Purification of the inducible α-agglutinin of *S. cerevisiae* and molecular cloning of the gene. *FEBS Letters* **255**, 290–4.

Jones, B. E. & Gooday, G. W. (1977). Lectin binding to sexual cells in fungi. *Biochemical Society Transactions* **5**, 717–19.

Jones, B. E. & Gooday, G. W. (1978). An immunofluorescent investigation of the zygophore surface of Mucorales. *FEMS Microbiology Letters* **4**, 181–4.

Kawanabe, Y., Yoshida, K. & Yanagishima, N. (1979). Sexual cell agglutination in relation to the formation of zygotes in *Saccharomyces cerevisiae*. *Plant and Cell Physiology* **20**, 423–33.

Lasky, R. D. & Ballou, C. E. (1988). Cell–cell recognition in yeast: isolation of intact alpha-agglutinin from *Saccharomyces kluyveri*. *Proceedings of the National Academy of Sciences, USA* **85**, 349–53.

Levi, J. O. (1956). Mating reaction in yeast. *Nature* **177**, 753–4.

Lipke, P. N. (1986). Agglutination and mating activity of the MF-alpha 2-encoded alpha-factor analog in *Saccharomyces cerevisiae*. *Journal of Bacteriology* **168**, 1472–5.

Lipke, P. N., Taylor, A. & Ballou, C. E. (1976). Morphogenetic effects of α-factor on *Saccharomyces cerevisiae* a-cells. *Journal of Bacteriology* **127**, 610–18.

Lipke, P. N., Terrance, K. & Wu, Y. S. (1987). Interaction of alpha-agglutinin with *Saccharomyces cerevisiae* a-cells. *Journal of Bacteriology* **169**, 483–8.

Lipke, P. N., Wojciechowecz, D. & Kurjan, J. (1989). AGα1 is the structural gene for *Saccharomyces cerevisiae* α-agglutinin, a cell surface glycoprotein involved in cell–cell interactions during mating. *Molecular and Cell Biology* **9**, 3155–65.

McCaffrey, G., Clay, F. J., Kelsey, K. & Sprague, G. F. (1987). Identification of a gene required for cell fusion during mating in the yeast *Saccharomyces cerevisiae*. *Molecular and Cell Biology* **7**, 2680–90.

Mendonca-Previato, L., Burke, D. & Ballou, C. E. (1982). Sexual agglutination factor from the yeast *Pichia amethionina*. *Journal of Cellular Biochemistry* **19**, 171–8.

Mesland, D. A. M., Huisman J. G. & van den Ende, H. (1974). Volatile sexual hormones in *Mucor mucedo*. *Journal of General Microbiology* **80**, 111–17.

Miyakawa, T., Azuma, Y., Tsuchiya, E. & Fukui, S. (1987). Involvement of cell-surface proteins in sexual cell–cell interactions of *Tremella mesenterica*, a heterobasidiomycete fungus. *Journal of General Microbiology* **133**, 439–43.

Miyakawa, T., Kadota, T., Okubo, Y., Hatano, T., Tsuchiya, E. & Fukui, S. (1984). Mating pheromone-induced alteration of cell surface proteins in the heterobasidomycetous yeast *Tremella mesenterica*. *Journal of Bacteriology* **158**, 814–19.

Miyakawa, T., Okubo, Y., Matano, T., Tsuchiya, E., Yamashita, I. & Fukui, S. (1982). Appearance of new protein species on the cell surface during sexual differentiation in the heterobasidiomycetous yeast *Tremella mesenterica*. *Agricultural and Biological Chemistry* **46**, 2403–05.

Moore, S. A. (1983). Comparison of dose–response curves for α-factor-induced cell division arrest, agglutination, and projection formation in yeast cells. *Journal of Biological Chemistry* **258**, 13849–56.

Nielsen, R. J. (1978). Sexual mutants of a heterothallic *Mucor pusillus* species. *Experimental Mycology* **2**, 193–7.

Orlean, P., Ammer, H., Watzele, M. & Tanner, W. (1986). Synthesis of an *O*-glycosylated cell surface protein induced in yeast by α-factor. *Proceedings of the National Academy of Sciences, USA* **83**, 6263–6.

Osumi, M., Shimoda, C. & Yanagishima, N. (1974). Mating reaction in *Saccharomyces cerevisiae*. V. Changes in the fine structure during the mating process. *Archives of Microbiology* **97**, 27–38.

Pierce, M. & Ballou, C. E. (1983). Cell–cell recognition in yeast: characterisation of the sexual agglutination factors from *Saccharomyces kluyveri*. *Journal of Biological Chemistry* **258**, 3576–82.

Plempel, M. (1960). Die zygotropische Reaktion bei Mucorineen. I. *Planta, Berlin* **55**, 254–8.

Plempel, M. (1962). Die zygotropische Reaktion bei Mucorineen. III. *Planta, Berlin* **58**, 509–20.

Pommerville, J. (1978). Analysis of gamete and zygote motility in *Allomyces*. *Experimental Cell Research* **113**, 166–72.

Pommerville, J. (1981). The role of sexual pheromones in *Allomyces*. In *Sexual Interactions in Eukaryotic Microbes*, ed. D. H. O'Day & P. A. Horgen, pp. 53–77, New York: Academic Press.

Pommerville, J. (1982). Morphology and physiology in gamete mating and gamet fusion in the fungus *Allomyces*. *Journal of Cell Science* **53**, 193–209.

Pommerville, J. & Fuller, M. S. (1976). The cytology of the gametes and fertilization of *Allomyces macrogynus*. *Archives of Microbiology* **109**, 21–30.

Pommerville, J. & Olson, L. W. (1987). Evidence for a male-produced pheromone in *Allomyces macrogynus*. *Experimental Mycology* **11**, 245–8.

Poon, H. & Day, A. W. (1974). 'Fimbriae' in the fungus *Ustilago violacea*. *Nature* **250**, 648–9.

Poon, H. & Day, A. W. (1975). Fungal fimbriae. I. Structure, origin and synthesis. *Canadian Journal of Microbiology* **21**, 537–46.

Poon, H., Martin, J. & Day, A. W. (1974). Conjugation in *Ustilago violacea*. I. Morphology. *Canadian Journal of Microbiology* **20**, 187–91.

Prillinger, H. (1987). Yeasts and anastomoses: their occurrence and implications for the phylogeny of the Eumycota. In *Evolutionary Biology of the Fungi*, ed. A. D. M. Rayner, C. M. Brasier & D. Moore, pp. 355–77. Cambridge: Cambridge University Press.

Sakagami, Y., Yoshida, M., Isogai, I. & Suzuki, A. (1981). Structure of tremerogen *a*-13, a peptidal sex hormone of *Tremella mesenterica*. *Agricultural and Biological Chemistry* **45**, 1045–7.

Sakai, K. & Yanagishima, M. (1972). Mating reaction in *Saccharomyces cerevisiae*. II. Hormonal regulation of agglutinability of a-type cells. *Archives of Microbiology* **84**, 191–8.

Schekman, R. & Brawley, V. (1979). Localised deposition of chitin of the yeast cell surface in response to mating pheromone. *Proceedings of the National Academy of Sciences, USA* **76**, 645–9.

Schipper, M. A. A. (1987). Mating ability and the species concept in the zygomycetes. In *Evolutionary Biology of the Fungi*, ed. A. D. M. Rayner, C. Brasier & D. Moore, pp. 261–9. Cambridge: Cambridge University Press.

Schipper, M. A. A. & Stalpers, J. A. (1980). Various aspects of the mating system in Mucorales. *Personnia* **11**, 53–63.

Suzuki, K. & Yanagishima, N. (1985). An alpha-mating specific mutation causing specific defect in sexual agglutinability in the yeast *Saccharomyces cerevisiae*. *Current Genetics* **9**, 185–9.

Svircev, A., Smith, R., Gardiner, R. B., Racki, I. M. & Day, A. W. (1986). Fungal fimbriae. V. Protein-A-gold immunocytochemical labeling of the fimbriae of *Ustilago violacea*. *Experimental Mycology* **10**, 19–27.

Taylor, N. W. & Orton, W. L. (1967). Sexual agglutination in yeast. V. Small particles of 5-agglutinin. *Archives of Biochemistry and Biophysics* **120**, 602–8.

Taylor, N. W. & Orton, W. L. (1968). Sexual agglutination in yeast. VII. Significance of the 1.7 S component from reduced 5-agglutinin. *Archives of Biochemistry and Biophysics* **126**, 912–21.

Taylor, N. W. & Orton, W. L. (1970). Association constant of the sex-specific agglutinin in the yeast *Hansenula wingei*. *Biochemistry* **9**, 2931–4.

Taylor, N. W. & Orton, W. L. (1971). Cooperation among the active binding sites in the sex-specific agglutinin from the yeast *Hansenula wingei*. *Biochemistry* **10**, 2043–9.

Teepe, H., Böltge, J.-A. & Wöstermeyer, J. (1988). Isolation and electrophoretic analysis of surface proteins of the zygomycete *Absidia glauca*. *FEBS Letters* **234**, 100–6.

Terrance, K., Heller, P., Wu, Y. S. & Lipke, P. N. (1987). Identification of glycoprotein components of alpha-agglutinin, a cell adhesion protein from *Saccharomyces cerevisiae*. *Journal of Bacteriology* **169**, 475–82.

Terrance, K. & Lipke, P. N. (1981). Sexual agglutination in *Saccharomyces cerevisiae*. *Journal of Bacteriology* **148**, 889–96.

Terrance, K. & Lipke, P. N. (1987). Pheromone induction of agglutination in *Saccharomyces cerevisiae*. *Journal of Bacteriology* **169**, 4811–15.

Tkacz, J. S. & MacKay, V. L. (1979). Sexual conjugation in yeast. *Journal of Cell Biology* **80**, 326–33.

Tokoyama, H. & Yanagishima, N. (1982). Control of the production of the sexual agglutination substances by the mating type locus in *Saccharomyces cerevisiae*: simultaneous expression of specific genes for a and α-agglutination substances in MAT-alpha-2 mutant cells. *Molecular and General Genetics* **186**, 322–7.

Trueheart, J., Boeke, J. D. & Fink, G. R. (1987). Two genes required for cell fusion during yeast conjugation: evidence for a pheromone induced surface protein. *Molecular and Cellular Biology* **7**, 2316–28.

Watzele, M., Klis, F. & Tanner, W. (1988). Purification and characterisation of the inducible a agglutinin of *Saccharomyces cerevisiae*. *EMBO Journal* **7**, 1483–8.

Weinstock, K. & Ballou, C. E. (1986). Cell–cell recognition in yeast: molecular nature of the sexual agglutinin from *Saccharomyces kluyveri* 17-cells. *Journal of Biological Chemistry* **261**, 16174–9.

Werkman, T. B. A. (1976). Localization and partial characterization of a sex-specific enzyme in homothallic and heterothallic Mucorales. *Archives of Microbiology* **109**, 209–13.

Wickerham, L. J. (1956). Influence of agglutination on zygote formation in *Hansenula wingei*, a new species of yeast. *Comptes rendus des travaux du Laboratoire Carlsberg* **26**, 423–43.

Wöstermeyer, A., Teepe, H. & Wöstermeyer, J. (1990). Genetic interactions in somatic inter-mating type hybrids of the zygomycete *Absidia glauca*. *Current Genetics* **17**, 163–8.

Yamaguchi, M., Yoshida, K. & Yanagishima, N. (1982). Isolation and partial characterization of cytoplasmic alpha-agglutination substance in the yeast *Saccharomyces cerevisiae*. *FEBS Letters* **139**, 125–9.

Yamaguchi, M., Yoshida, K. & Yanagishima, N. (1984*a*). Isolation, and biochemical and biological characterization of an α-mating type specific glycoprotein responsible for sexual agglutination from the cytoplasm of a-cells of the yeast *Saccharomyces cerevisiae*. *Archives of Microbiology* **140**, 113–19.

Yamaguchi, M., Yoshida, K. & Yanagishima, N. (1984*b*). Mating-type specific differentiation in ascosporogenous yeasts on the basis of mating-type specific substances responsible for sexual cell-recognition. *Molecular and General Genetics* **194**, 24–30.

Yanagishima, N. (1986). Sexual differentiation and interactions in yeasts. *Microbiological Sciences* **3**, 45–9.

Yanagishima, N. & Yoshida, K. (1981). Sexual interactions in *Saccharomyces cerevisiae* with special reference to the regulation of sexual agglutinability. In *Sexual Interactions in Eukaryotic Microbes*, ed. D. H. O'Day & P. A. Horgen, pp. 261–95. New York: Academic Press.

Yen, P. H. & Ballou, C. E. (1973). Composition of a specific intercellular agglutination factor. *Journal of Biological Chemistry* **248**, 8316–18.

Yen, P. H. & Ballou, C. E. (1974). Partial characterization of the sexual agglutination factor from *Hansenula wingei* Y-2340 type 5 cells. *Biochemistry* **13**, 2428–37.

Yoshida, K., Hisatomi, T. & Yanagishima, N. (1989). Sexual behaviour and its pheromonal regulation in ascosporogenous yeasts. *Journal of Basic Microbiology* **29**, 99–128.

A. CHABOUD, R. PEREZ, C. DIGONNET
AND C. DUMAS

Gamete recognition in angiosperms: model and strategy for analysis

Summary

Double fertilisation in angiosperms remains an enigma. The two male gametes from the pollen grain or tube fuse with the two female gametes in the embryo sac included in female tissues. Structural data in *Plumbago* showed that the plastid-rich male gamete preferentially fuses with the egg cell leading to the embryo (Russell, 1985). In maize BMS line, the male gamete bearing the non-disjointed B-chromosome preferentially fuses with the egg cell (Roman, 1948). Is this preferential fusion the result of recognition events at the gamete level? In order to answer this question, one needs to work with male and female gametes isolated from their gametophytes. For this purpose, we chose *Zea mays* (maize) as a plant model, because it offers large amounts of pollen and female flowers. We succeeded in isolating intact, viable and functional gametes (Dupuis *et al.*, 1987; Wagner *et al.*, 1989*b*; Roeckel, 1990). What, then, is the best strategy for intergametic recognition studies? We decided to construct a library of monoclonal antibodies directed against cell surface determinants of isolated male gametes. These immunological probes will be used to assess membrane dimorphism of the two male gametes of one pollen grain. We also attempted to develop an *in vitro* model of inter-gametic fusion. Experiments of fusion inhibition by monoclonal anti-bodies in this *in vitro* model would allow us to sort specific cell surface determinants involved in gametic recognition.

Introduction

Fertilisation, which has been defined as a multistep phenomenon, starting with the interaction between pollen and stigma and ending with the fusion of the male and female gametes (Clarke *et al.*, 1985), is a more complex system in higher plants than in lower plants. So, while gametic recog-

Society for Experimental Biology Seminar Series 48: *Perspectives in Plant Cell Recognition*, ed. J. A. Callow & J. R. Green. © Cambridge University Press 1992, pp. 59–77.

nition is relatively well documented in some lower plants (for reviews, see Callow *et al.*, and van den Ende, this volume), double fertilisation and potential gametic recognition in higher plants remain to be explored in detail (Guilluy *et al.*, 1990). Until recently, these cell recognition studies have been hampered because, in angiosperms, both types of gamete are housed in complex paternal and maternal sporophytic tissues and double fertilisation itself is also basically internal (see Knox *et al.*, 1986).

The ultimate aim of this chapter is, after a brief review of fertilisation and gametic recognition in angiosperms, to point out how recent developments in isolating male and female gametes from their gametophytes provide new strategies for analysing gametic recognition.

General considerations on fertilisation

Male and female gametogenesis

In angiosperms, male and female developmental programmes are devoted to the preparation of male genetic information carried by male gametes for discharge near the two female gametes within the embryo sac, as a prelude to double fertilisation.

Male gamete formation and maturation occur either in bicellular pollen grains during tube growth through the style several hours after pollen germination (about 70% of angiosperms), or in tricellular pollen grains during pollen development within the anther (about 30% of angiosperms) (Knox *et al.*, 1986). An illustration of tricellular pollen at anthesis and a schematic representation of its formation are given in Figs. 1 and 2.

In bicellular, as in tricellular pollen species, pollen development is a dramatic differentiation process, with many discriminant steps leading to mature pollen (Willemse, 1988). Steps in the male developmental programme can be characterised by several changes in cytomorphology (for a review, see Giles & Prakash, 1987), protein pattern and protein synthesis (for a review, see Roeckel *et al.*, 1990a) and gene activity (for a review, see Mascarenhas, 1990). In particular, after generative cell division, major transitions occur in stage-specific protein patterns in maize (Frova *et al.*, 1987; Delvallée & Dumas, 1988) or wheat (Vergne & Dumas, 1988) and stage-specific protein neo-synthesis in *Brassica oleracea* (Detchepare *et al.*, 1989). Thus, we could expect that some of these changes in gene expression are related to the formation and differentiation of male gametes which occurs in this late period of development.

In angiosperms, the female gametophytic generation is represented by the embryo sac deeply embedded in the protective sporophytic envelope of the ovule as illustrated in Fig. 3. It is therefore more difficult to observe and handle than is its male counterpart, the pollen grain. Consequently,

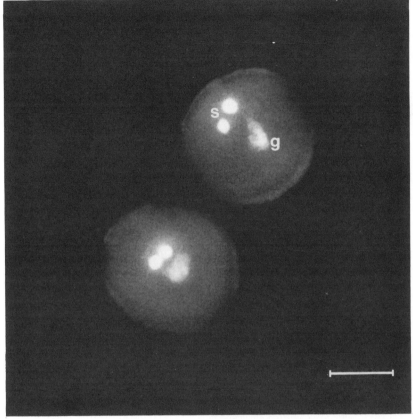

Fig. 1. DNA-specific 4,6′-diamidino-2-phenylindole dihydrochloride (DAPI) fluorescence showing nuclei of pollen in *Brassica oleracea*. g, Generative nucleus; s, sperm nuclei. The bar represents 50 μm.

the biology of the embryo sac is less well studied, with a consequent lack of physiological and biochemical data.

In most angiosperms (70–80%), embryo sacs are of the *Polygonum* type (Maheshwari, 1950). They typically contain the female gametes (egg cell at the origin of the zygotic embryo and a large central cell with two polar nuclei) and several accessory cells (generally two synergids and three antipodial cells) that appear to be held together by cellular connections (for reviews, see Jensen, 1974; Willemse & Van Went, 1984). In most species, cell walls surrounding the egg and central cells are not continuous, allowing direct membrane–membrane contacts between gametes (Van Lammeren, 1986).

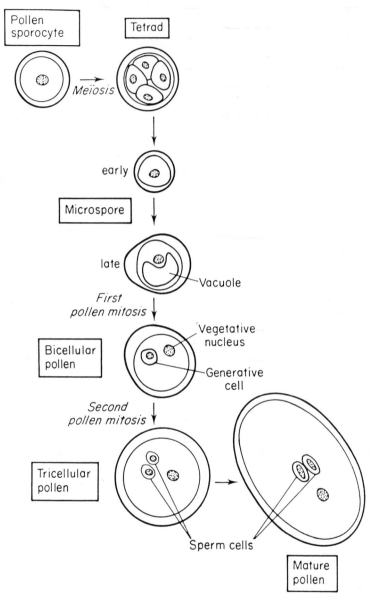

Fig. 2. Diagrammatic representation of cytological stages of tricellular pollen development. (From Roeckel *et al.*, 1990*a*).

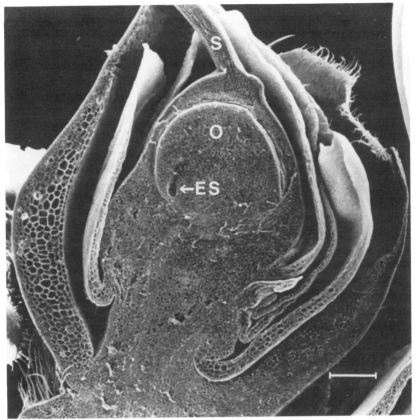

Fig. 3. Scanning electron micrograph of the median-cut *Zea mays* pistil-late spikelet. S, Silk; O, ovule; ES, embryo sac. The bar represents 2 mm. (From Wagner *et al.*, 1990.)

Pollination and double fertilisation events

Double fertilisation was first described by Nawashin (1898) and Guignard (1899) and involves the fusion of the two male gametes with the two female gametes within the embryo sac: one fusion initiates embryo forma-tion and a second fusion triggers formation of the secondary endosperm. But, from the arrival of the pollen grain on the pistil to the time when double fertilisation actually takes place, several stages can be recognised viz. pollen adhesion, pollen hydration, pollen germination, pollen tube growth toward the ovule and pollen tube discharge into one synergid of the embryo sac (for a review, see Knox *et al.*, 1986).

The two fusions characteristic of double fertilisation occur at adjacent cells close together in time and without competition from other male gametes, since in most cases, only one pollen tube discharges into a synergid (for a review, see Knox *et al.*, 1986). Transmission electron microscopy studies show that the male gametes are discharged free from the outer vegetative membrane and that fusion occurs through interrupted cell walls in the female gametophyte (Jensen, 1974; Mogensen, 1978, 1982; Russell, 1983). So, it is reasonable to suppose that higher plant gametes might have cell surface determinants that function in fertilisation as lower-plant gametes.

Gametic recognition: fact or hypothesis?

The process of double fertilisation previously described may be 'random' or 'directed'. In the random hypothesis, either male gamete can fuse with either female target cell, suggesting that the two gametes of one pollen grain are identical. In the directed hypothesis, each male gamete of a pollen grain is 'programmed' for specific fusion with either the egg cell or the central cell, implying that gametic recognition occurs. In the literature this last hypothesis is supported by only two examples, in *Zea* and *Plumbago*, where male gamete dimorphism is associated with preferential fusion.

Genetic indication of gametic recognition in *Zea mays*

Male gamete nuclear dimorphism was genetically established by Roman (1948) in certain lines of maize in which B-chromosomes were frequently (frequency = 50–98%) transmitted unequally into the sperm descendants of the generative cell (Fig. 4). Some male gametes of maize would contain two B-chromosomes (or more) and their 'sister' cells would contain less than their expected complement. This form, since termed 'nuclear heterospermy' (Russell, 1985), was also linked to functional differences, since the male gamete with the excess complement of B-chromosomes was genetically more likely (frequency = 66%) to fuse with the egg cell (Fig. 4). Roman (1948) attributed 'directed fertilisation', as he termed it, to the order of sperm cell arrival in the female gametophyte. Since B-chromosomes from two different chromosomal origins seemed to influence fertilisation, he doubted that they carried the genes required to assure their transmission. In line TB-4a, where B-chromosomes were

Fig. 4. Preferential fusion in *Zea mays*. Schematic presentation of the behaviour of the B^A translocated chromosome during pollen development and double fertilisation in maize lines harbouring a supernumerary (B) chromosome. CC, central cell; EC, egg cell. (From Roman, 1948.)

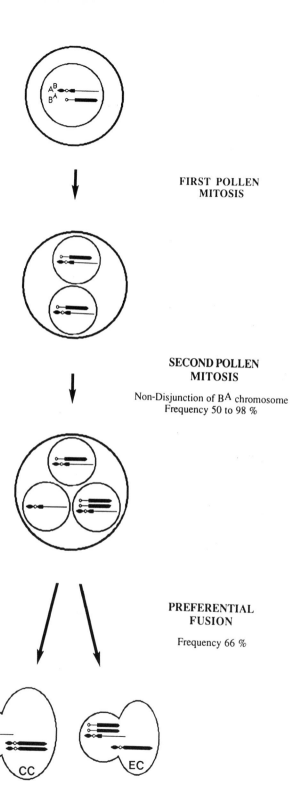

FIRST POLLEN
MITOSIS

SECOND POLLEN
MITOSIS

Non-Disjunction of B^A chromosome
Frequency 50 to 98 %

PREFERENTIAL
FUSION

Frequency 66 %

derived from chromosome 4, 76.8% of the resultant embryos expressed the B-chromosome, and in line TB-7b, derived from chromosome 7, 67.1% of the embryos expressed it (Roman, 1948).

Cytological indication of gametic recognition in *Plumbago zeylanica*

The two male gametes may be dimorphic in terms of the number of mitochondria or plastids, and thereby differ in their cytoplasmic genetics. This trend reaches its logical extreme in *Plumbago zeylanica*, in which the male gametes may differ to the extent of one of them totally lacking plastids (Russell, 1985; Fig. 5). Consequently, the fate of maternal and paternal organelles can be traced by organellar differences and, therefore, not only is the fate of the dimorphic male gametes known, but the participation of their cytoplasms is well documented (Russell, 1983). Preferential fertilisation results in the fusion of the plastid-rich gamete into the embryo in over 94% of the cases, strongly supporting the presence of factors that may recognise and discriminate between the male gametes of this plant. However, according to currently available data (Corriveau & Coleman, 1988), this strong pattern of plastid dimorphism seems to be restricted to the immediate subfamily to which *Plumbago* belongs.

Gametic recognition: a question to be analysed

Differences in the nuclear genetics of male gametes are extremely rare, since they must occur during generative cell mitosis when the gametes form. More common is the situation of 'cytoplasmic heterospermy'. The majority of the flowering plants studied to date express a weak form of cytoplasmic dimorphism (for a review, see Knox *et al.*, 1986) in which the two male gametes may differ in size, shape and mitochondrial content, but not plastid content. Whether these express preferentiality in fertilisation remains to be demonstrated. Moreover, evidence in barley suggests that fertilisation may exclude the male cytoplasm entirely (Mogensen, 1988). In these plants and others in which the male gametes do not appear to be dimorphic, demonstration of any preference of fertilisation between the two male gametes will be essentially impossible.

Therefore, the only alternative approach to address the question of gametic recognition during double fertilisation is to get direct access to

Fig. 5. Preferential fusion in *Plumbago zeylanica*. Schematic presentation of the behaviour of cytoplasmic organelles (mitochondria and plastids) during pollen development and double fertilisation. CC, central cell; EC, egg cell. (From Russell, 1985.)

FIRST POLLEN
MITOSIS

SECOND POLLEN
MITOSIS

Sperm dimorphism in organelle content

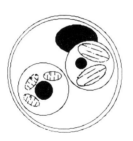

PREFERENTIAL
FUSION

Frequency 94 %

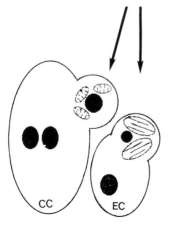

CC   EC

gametic interactions via isolation of male and female gametes from their gametophytes.

What is the best model for gametic recognition analysis?

Gamete isolation

Is double fertilisation the result of recognition events at the gamete level? In order to answer this question, one needs to work with the largest amounts of intact, viable and functional male and female gametes isolated from their gametophytes. Several attempts to do this have been carried out for different species in the last decade (see Tables 1 and 2).

Female gamete isolation requires essentially an enzymic digestion followed by some mechanical disruption to ensure proper liberation of the embryo sac (Table 1). After this treatment, the method currently used to test viability is based on the evaluation of both membrane integrity and enzyme activity. The fluorochromatic reaction test (Heslop Harrison & Heslop Harrison, 1970) is based on penetration of fluorescein diacetate (non-fluorescent and apolar) into the cells, and its hydrolysis by an esterase to yield fluorescein (fluorescent and polar). Viable cells show bright fluorescence, and are noted FCR$^+$; this method has been used by the majority of the authors (Table 1). It is possible, then, to microdissect the viable isolated embryo sac to obtain the female target cells (egg cell and central cell) for the male gametes. The yield is mostly rather low (about 5–10%) but is nevertheless satisfactory for maize (for example, 1–200 living isolated embryo sacs per day and per experimenter).

For male gametes, as they are embedded in the vegetative cells of pollen grains or pollen tubes, their release can be obtained by osmotic shock, grinding the pollen grain, or by wall degrading enzymes (for an analysis of different protocols, see Roeckel *et al.*, 1990a). The methods used must be adapted to the species studied to allow pollen grains to break without damaging male gametes. The use of an osmotic shock utilises the sensitivity of certain types of pollen to rapid changes in water flux. In *Zea mays*, an osmotic shock must be coupled with a pH shock in order to release the male gametes through the germinal aperture. For other genera such as *Brassica* or *Gladiolus* it is necessary to use such methods as grinding or enzymic treatment. Grinding methods, employing tissue homogenisers or glass beads take advantage of the smaller size of the male gametes, which renders them more difficult to crush. Another step of the isolation procedure is to separate released male gametes from organelles and cytoplasmic debris. Larger contaminants can be removed by filtration. Smaller contaminants can be separated from male gametes by gradient centrifugation. Nevertheless, for further cellular recognition

Table 1. *Isolation of viable female gametophyte i.e. embryo sac in different species*

Species	Isolation method	Viability	References
Dicotyledons			
Nicotiana tabacum	Enzymic digestion, squashing	+ (FCR)	Hu *et al.*, 1985
Antirrhinum majus	Enzymic digestion	+ (FCR)	Zhou & Yang, 1985
Torenia fournieri	Enzymic digestion	+ (FCR)	Mol, 1986
Helianthus annuus	Enzymic digestion	+ (FCR)	Zhou, 1987
Plumbago zeylanica	Enzymic digestion	+ (FCR)	Huang *et al.*, 1989
Monocotyledons			
Lilium longiflorum	Enzymic digestion, suction	+ (FCR)	Wagner *et al.*, 1989b
Zea mays	Enzymic digestion, dissection	+ (FCR, TEM)	Wagner *et al.*, 1988
			Wagner *et al.*, 1989c

FCR, fluorochromatic reaction test (according to Heslop Harrison & Heslop Harrison, 1970); TEM, transmission electron microscopy.

Table 2. *Isolation of male gametes in different species. Comparison of yield obtained and viability of isolated cells*

Species	Yield[a]	Viability[b] (%)	References
Dicotyledons			
Plumbago zeylanica	1.7×10^5 cells, 60–75%	95	Russell, 1986
Rhododendron spp.	90–270 pairs per style[c]	nd	Shivanna et al., 1988
Brassica oleracea	2%	nd	Roeckel et al., 1988
Beta vulgaris	7×10^4 cells	30	Nielsen & Olesen, 1988
Gerbera jamesonii	6×10^4 cells	nd	Southworth & Knox, 1989
Spinacia oleracea	10^5 cells, 5–10%	90	Theunis & Van Went, 1989
Monocotyledons			
Zea mays	3×10^6 cells, 20–30%	80–90	Dupuis et al., 1987
			Roeckel et al., 1988
Zea mays	1.5×10^6 cells	50	Cass & Fabi, 1988
Gladiolus gandavensis	65–84 pairs style[c]	nd	Shivanna et al., 1988
Lolium perenne	2%	nd	Van der Maas & Zaal, 1990

[a]Yield was recalculated from quantitative data of papers, as total number of isolated cells obtained at the end of a single isolation procedure and/or percentage of isolated cells recovered from the initial number of male gametes within pollen grains.
[b]Viability percentage of isolated cells at the end of isolation procedure.
[c]In this study, male gametes have been isolated from pollen tubes growing in style segments by the *in vivo/in vitro* method, and yield has been estimated by the number of isolated pairs of male gametes per style.
nd, not determined.

studies, the key point is to isolate good quality male gametes in large numbers. Comparison of the results obtained with several species (Table 2) indicate clearly that maize is the best-suited species from this point of view.

Maize as a model

Maize is well suited for this type of research, because of its economic importance as well as the considerable amount of genetic and cytogenetic information currently available (Sheridan, 1982). In addition, the recent development of molecular biology has provided a lot of valuable information on restriction fragment length polymorphism, and pollen-specific genes have been isolated (for a review, see Mascarenhas, 1990). Maize bears its male and female flowers on separate structures, which facilitates the collection of both pollen and pistil. Moreover, several grams of pollen grains can be produced by a single tassel and the large ear with 500 or more individual spikelets is a good starting material to obtain large numbers of isolated embryo sacs.

Consequently, our group has chosen maize as a plant model, and we have succeeded in isolating intact, viable and functional maize gametes (Dupuis *et al.*, 1987; Wagner *et al.*, 1989c). We have also used numerous methods for assessing the quality and the functional state of isolated male gametes. Transmission electron microscopy (Dupuis *et al.*, 1987) allowed direct visualisation of both plasma membrane configuration and intactness (Fig. 6), and also the condition of cellular organelles (Wagner *et al.*, 1989a). The functional state of the plasma membrane was demonstrated by measurement of transmembrane currents recorded by preliminary patch-clamp experiments in cell-attached configuration (Roeckel *et al.*, 1990b). Further studies have shown the presence of ATP (a nucleotide known to be an indicator of life and used to test the fertility of human semen; Comhaire *et al.*, 1983) within isolated male gametes from *Z. mays*, indicating the metabolic potential of these cells (Roeckel *et al.*, 1990c). Morever, [^{35}S]methionine labelling experiments clearly indicate that these isolated cells are also able to synthesise new proteins (P. Roeckel, unpublished results). However, the ability of isolated male gametes to fertilise the egg or central cell will be the ultimate biotechnological assay.

Strategy for gametic recognition analysis

Our success in isolating intact, viable and functional maize gametes offers the possibility of direct access to the interacting gamete membranes which are supposed to bear the specific determinants of interest in double

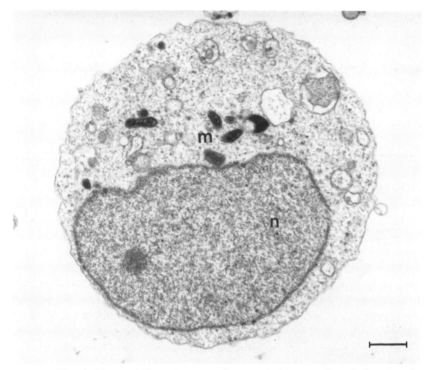

Fig. 6. Transmission electron micrograph feature of an isolated male gamete in *Zea mays*. n, Male gamete nucleus; m, mitochondrial complex. The bar represents 1 μm.

fertilisation. Thus, we are currently designing an experimental strategy based on the construction of a monoclonal antibody library directed against the cell surface of isolated viable male gametes, and the development of a model of intergametic fusion *in vitro*.

Monoclonal antibodies were first raised to isolated sperm of *Brassica campestris* pollen (Hough *et al.*, 1986) but more detailed investigations on the potential use of monoclonal antibodies for male gametes characterisation were reported by Pennell *et al.* (1987). In this study, male gametes of *P. zeylanica* appeared to be strongly immunogenic, as a good proportion of reactive lines were sperm-specific. However, the strongly reactive lines were directed against nuclear or cytoplasmic components of the gamete, and many more reacted with distinctly particulate or soluble components of pollen cytoplasm. This study clearly indicates that more purified male gamete preparations free from pollen cytoplasmic contaminants will be necessary before more discriminating screening procedures to select gamete-specific cell surface determinants are possible. Similarly, electrophoretic comparison of protein patterns of homogenates of isolated male gametes and respective pollen preparations (Knox *et al.*, 1988; Geltz & Russell, 1988) suggests that a high degree of overlap exists between gamete-rich and pollen fractions. Our own first trials of mice immunisations pointed out the same limiting factor, so that the large-scale purification of isolated male gametes is the current objective in the laboratory. However, as in maize, gamete volume represents only 0.02–0.2% of the pollen volume; considerable effort may be needed to obtain purified fractions suitable for analysis. Nevertheless, as soon as this problem is solved, a monoclonal antibody library directed against the cell surface of isolated male gametes will be constructed and used for cell-sorting the population of isolated gametes. This will allow assessment of membrane dimorphism of the two male gametes of one pollen grain. The development of an *in vitro* model of intergametic fusion without any artificial fusogen, which is currently in progress, will allow selection of fusion-inhibiting monoclonal antibodies, and thus permit the eventual isolation of specific cell surface determinants involved in gametic recognition. This approach should lead to the study of the ontogenesis and the role in the double fertilisation process of the gamete-specific determinants identified.

Acknowledgements

We thank P. Audenis for photographic work. This work was carried out in the framework of contract no. 0203-F of the Biotechnology Action Program of the EEC. The financial support of INRA to the laboratory is gratefully acknowledged.

References

Cass, D. D. & Fabi, G. C. (1988). Structure and properties of sperm cells isolated from the pollen of *Zea mays*. *Canadian Journal of Botany* **66**, 819–25.

Clarke, A. E., Anderson, M. A., Bacic, T., Harris, P. J. & Mau, S. L. (1985). Molecular basis of cell recognition during fertilization in higher plants. *Journal of Cell Science suppl.* **2**, 261–85.

Comhaire, F., Vermeulen, L., Ghedirak, K., Mas, J., Irvine, S. & Callipolitis, G. (1983). ATP in human semen: a quantitative estimate of fertilizing potential. *Fertility and Sterility* **40**, 500–4.

Corriveau, J. L. & Coleman, A. W. (1988). A rapid screening method to detect potential biparental inheritance of plastid DNA and results for over 200 selected angiosperms. *American Journal of Botany* **75**, 1443–58.

Delvallée, I. & Dumas, C. (1988). Anther development in *Zea mays*. Changes in proteins, peroxidases and esterases patterns. *Journal of Plant Physiology* **132**, 210–17.

Detchepare, S., Heizmann, P. & Dumas, C. (1989). Changes in protein patterns and protein synthesis during anther development in *Brassica oleracea*. *Journal of Plant Physiology* **135**, 129–37.

Dupuis, I., Roeckel, P., Matthys-Rochon, E. & Dumas, C. (1987). Procedure to isolate viable sperm cells from corn (*Zea mays* L.) pollen grains. *Plant Physiology* **85**, 876–8.

Frova, C., Binelli, G. & Ottaviano, E. (1987). Isozyme and HSP gene expression during male gametophyte development in maize. In *Isozymes: Current Topics in Biological and Medical Research*, vol. 15 *Genetics, Development and Evolution*, ed. M. C. Rattazi, J. G. Scandalios & G. S. Whitt, pp. 97–120. New York: Alan R. Liss Inc.

Geltz, N. R. & Russell, S. D. (1988). Two dimensional electrophoretic studies of the proteins and polypeptides in mature pollen grains and the male germ unit of *Plumbago zeylanica*. *Plant Physiology* **88**, 764–9.

Giles, K. H. & Prakash, J. (1987). Pollen: cytology and development. *International Review of Cytology* **107**.

Guignard, L. (1899). Sur les anthérozoides et la double copulation sexuelle chez les végétaux angiospermes. *Revue de Génétique Botanique* **11**, 129–35.

Guilluy, C. M., Gaude, T., Digonnet-Kerhoas, C., Chaboud, A., Heizmann, P. & Dumas, C. (1990). New data and concepts in angiosperm fertilization. In *Mechanism of Fertilization*, ed. B. Dale, NATO ASI series, vol. H 45, pp. 254–70. Berlin, Heidelberg, New York: Springer-Verlag.

Heslop Harrison, J. & Heslop Harrison, Y. (1970). Evaluation of pollen viability by enzymatically induced fluorescence, intracellular hydrolysis of fluorescein diacetate. *Stain Technology* **45**, 115–20.

Hough, T., Sing, M. B., Smart, I. J. & Knox, R. B. (1986). Immunofluorescent screening of monoclonal antibodies to surface antigens of animal and plant cells bound to polycarbonate membranes. *Journal of Immunological Methods* **92**, 103–7.

Hu, S. Y., Li, L. G. & Zhu, C. (1985). Isolation of viable embryo sacs and their protoplasts of *Nicotiana tabacum*. *Acta Botanica Sinica* **27**, 337–44.

Huang, B. Q., Strout, G. W. & Russell, S. D. (1989). Isolation of fixed and viable eggs, central cells, and embryo sacs from ovules of *Plumbago zeylanica*. *Plant Physiology* **90**, 9–12.

Jensen, W. A. (1974). Reproduction in flowering plants. In *Dynamic Aspects of Plant Ultrastructure* ed. A. W. Robards, pp. 481–503. New York: McGraw-Hill.

Knox, R. B., Southworth, D. & Singh, M. B. (1988). Sperm cells determinants and control of fertilization in plants. In *Eucaryote Cell Recognition: Concepts and Model Systems*, ed. G. P. Chapman, C. C. Ainsworth & C. J. Chatham, pp. 175–93. Cambridge: Cambridge University Press.

Knox, R. B., Williams, E. G. & Dumas, C. (1986). Pollen, pistil and reproductive function in crop plants. *Plant Breeding Reviews* 4, 8–79.

Maheshwari, P. (1950). *An Introduction to the Embryology of Angiosperms*. New York: McGraw-Hill.

Mascarenhas, J. P. (1990). Anther and pollen expressed genes. *Annual Review of Plant Physiology and Plant Molecular Biology* 41, 317–38.

Mogensen, H. L. (1978). Pollen tube-synergid interactions in *Proboscidea louisianica* (Martineaceae). *American Journal of Botany* 65, 953–64.

Mogensen, H. L. (1982). Double fertilization in barley and the cytological explanation for haploid embryo formation, embryoless caryopses, and ovule abortion. *Carlsberg Research Communications* 47, 313–54.

Mogensen, H. L. (1988). Exclusion of male mitochondria and plastids during syngamy as a basis for maternal inheritance. *Proceedings of National Academy of Sciences, USA* 85, 2594–7.

Mol, R. (1986). Isolation of protoplasts from female gametophytes of *Torenia fournieri*. *Plant Cell Reports* 3, 202–6.

Nawashin, S. G. (1898). Resultate einer Revision der Befruchtungs von gänge bei *Lilium martagon* and *Fritillia tenella*. *Izo. Imp. Akad. Nank.* 9, 377–82.

Nielsen, J. E. & Olesen, P. (1988). Isolation of sperm cells from the trinucleate pollen of sugar beet (*Beta vulgaris*). In *Plant Sperm Cells as Tools for Biotechnology*, ed. H. J. Wilms & C. J. Keijzer, pp. 111–12. Wageningen: Pudoc.

Pennell, R. I., Geltz, N. R., Koren, E. & Russell, S. D. (1987). Production and partial characterization of hybridoma antibodies elicited to the sperm of *Plumbago zeylanica*. *Botanical Gazette* 148, 401–6.

Roeckel, P. (1990). Transformation du maís par la voie sexuée mâle: utilisation du pollen ou des gamètes isolés comme vecteurs d'ADN transformant. Thèse de Doctorat, Université Lyon 1, France.

Roeckel, P., Chaboud, A., Matthys-Rochon, E., Russell, S. & Dumas, C. (1990*a*). Sperm cell structure, development and organisation. In *Microspores. Evolution and Ontogeny*, ed. S. Blackmore & R. B. Knox, pp. 281–307. London: Academic Press.

Roeckel, P., Dupuis, I., Detchepare, S., Matthys-Rochon, E. & Dumas, C. (1988). Isolation and viability of sperm cells from corn (*Zea mays*) and kale (*Brassica oleracea*) pollen grains. In *Plant Sperm cells as Tools for Biotechnology*, ed. H. J. Wilms & C. J. Keijzer, pp. 105–10. Wageningen: Pudoc.

Roeckel, P., Dupuis, I., Matthys-Rochon, E., Pilsduski, R., Chaboud, A., Rougier, O. & Dumas, C. (1990*b*). Isolement et caractérisation des gamètes mâles de *Zea mays*. Colloque Société Française de Physiologie Végétale, Lyon (France), 23 March 1990. [Poster]

Roeckel, P., Matthys-Rochon, E. & Dumas, C. (1990*c*). Pollen and isolated sperm cell quality in *Zea mays*. In *Characterization of Male Transmission Units in Higher Plants*, ed. B. Barnabas & K. Liszt, pp. 41–8. Budapest: MTA Copy.

Roman, H. (1948). Directed fertilization in maize. *Proceedings of National Academy of Sciences, USA* **34**, 36–42.

Russell, S. D. (1983). Fertilization in *Plumbago zeylanica*: gametic fusion and fate of the male cytoplasm. *American Journal of Botany* **70**, 416–34.

Russell, S. D. (1985). Preferential fertilization in *Plumbago*: ultrastructural evidence for gamete-level recognition in an angiosperm. *Proceedings of National Academy of Sciences, USA* **82**, 6129–32.

Russell, S. D. (1986). Isolation of sperm cells from the pollen of *Plumbago zeylanica*. *Plant Physiology* **81**, 317–19.

Sheridan, W. F. (1982). *Maize for Biological Research*. Charlottesville, VA: Publications of the Plant Molecular Biology Association University Press of North Dakota.

Shivanna, K. R., Xu, H., Taylor, P. & Knox, R. B. (1988). Isolation of sperms from the pollen tubes of flowering plants during fertilization. *Plant Physiology* **87**, 647–50.

Southworth, D. & Knox, R. B. (1989). Flowering plant sperm cells: isolation form pollen of *Gerbera jamesonii* (Asteraceae). *Plant Science* **60**, 273–7.

Theunis, C. H. & Van Went, J. L. (1989). Isolation of sperm cells from mature pollen grain of *Spinacia oleracea* L. *Sexual Plant Reproduction* **2**, 97–102.

Van der Maas, H. M. & Zaal, M. A. C. M. (1990). Sperm cell isolation from pollen of perennial ryegrass (*Lolium perenne* L.). In *Characterization of Male Transmission Units in Higher Plants*, ed. B. Barnabas & K. Liszt, pp. 31–6. Budapest: MTA Copy.

Van Lammeren, A. A. M. (1986). A comparative ultrastructural study of the megagametophytes in two strains of *Zea mays* L. before and after fertilization. *Agricultural University Wageningen Papers*. 86-1.

Vergne, P. & Dumas, C. (1988). Isolation of viable wheat gametophytes of different stages of development and variations in their protein patterns. *Plant Physiology* **88**, 969–72.

Wagner, V. T., Dumas, C. & Mogensen, H. L. (1989*a*). Morphometric analysis of isolated *Zea mays* sperm. *Journal of Cell Science* **93**, 179–84.

Wagner, V. T., Dumas, C. & Mogensen, H. L. (1990). Quantitative three-dimensional study on the position of the female gametophyte and its constituent cells as a prerequisite for corn (*Zea mays*) transformation. *Theoretical and Applied Genetics* **79**, 72–6.

Wagner, V. T., Kardolus, J. P. & Van Went, J. L. (1989*a*). Isolation of the lily embryo sac. *Sexual Plant Reproduction* **2**, 219–24.

Wagner, V. T., Song, Y. C., Matthys-Rochon, E. & Dumas, C. (1988). The isolated embryo sac of *Zea mays*: structural and ultrastructural observations. In *Sexual Reproduction in Higher Plants* ed. M. Cresti, P. Gori & E. Pacini, pp. 233–8. Berlin, Heidelberg, New York: Springer-Verlag.

Wagner, V. T., Song, Y. C., Matthys-Rochon, E. & Dumas, C. (1989*b*). Observations on the isolated embryo sac of *Zea mays*. *Plant Science* **59**, 127–32.

Wagner, V. T., Song, Y. C., Matthys-Rochon, E. & Dumas, C. (1989*c*). Observations on the isolated embryo sac of *Zea mays*. *Plant Science* **59**, 127–32.

Willemse, M. T. M. (1988). The plant sperm cell as a result of discriminating steps during gametogenesis. In *Plant Sperm Cells as Tools for Biotechnology* ed. H. J. Wilms & C. J. Keijzer, pp. 11–16. Wageningen: Pudoc.

Willemse, M. T. M. & Van Went, J. L. (1984). The female gametophyte. In *Embryology of Angiosperms*, ed. B. M. John, pp. 159–96. Berlin, Heidelberg, New York: Springer-Verlag.

Zhou, C. (1987). A study of fertilization events in living embryo sacs isolated from sunflower ovules. *Plant Science* **52**, 147–51.

Zhou, C. & Yang, H. Y. (1985). Observations on enzymatically isolated, living and fixed embryo sacs in several angiosperm species. *Planta* **165**, 225–31.

F. C. H. FRANKLIN, R. M. HACKETT
AND V. E. FRANKLIN-TONG

The molecular biology of self-incompatible responses

Introduction

The process of pollination and fertilisation in flowering plants involves a series of interactive events between male and female cells. One of the earliest stages in the process of fertilisation is the recognition, and acceptance or rejection, of pollen grains alighting on the stigma of the recipient plant. Self-incompatibility (SI) involves these processes. Prevention of self-fertilisation is accomplished by the inhibition of pollen that has the same incompatibility phenotype as that of the stigma on which it lands. These highly specific recognition events are both developmentally expressed and tissue-specific. Investigation of the molecular basis of the expression and regulation of the S-genes, and the mode of action of their products, therefore, provides a model system for the study of gene expression and cellular recognition in flowering plants.

There is currently considerable interest in the elucidation of the molecular basis of SI and much work has been carried out in an attempt to identify the molecules involved in this interaction, especially those on the female side. S-linked glycoproteins from styles and stigmas, and the genes that encode them, have been identified and cloned. Less progress has been made with the pollen component. We aim to look at what is currently known about SI, with a view to examining what is known about the mechanism of this response.

What is known about the pistil and pollen components?

Identification and characterisation of stigmatic S-linked glycoproteins

There have been a number of studies carried out on proteins which have been isolated from stigmatic/stylar tissues. The general features of these molecules are outlined below.

Society for Experimental Biology Seminar Series 48: *Perspectives in Plant Cell Recognition*, ed. J. A. Callow & J. R. Green. © Cambridge University Press 1992, pp. 79–103.

The expression of the S-associated proteins should be subject to both tissue-specific and developmental control. All of the S-specific glycoproteins have been detected only in female reproductive tissue and specifically only in those regions of tissue where the growth of incompatible pollen tube is arrested; that is, they are tissue-specific. In addition, they are detected only in tissue which has reached maturity so far as expression of self-incompatibility is concerned. Thus, the S-associated components fulfil the criterion of their association with breeding behaviour, since the tissue specificity and developmental expression strongly correspond with the expression of SI.

Where it has been investigated, for instance in self-compatible *Nicotiana tabacum* and self-compatible *Petunia* cultivars, the glycoprotein has usually not been detected. However, recent investigations have revealed exceptions: one is a pseudo-self-compatible *Petunia hybrida* (Clark *et al.*, 1990) and another is a self-compatible potato line in *Solanum tuberosum* and in the tetraploid which was expected to be self-compatible (Kirch *et al.*, 1989). These patterns suggest that self-compatibility in these cases is not due to the loss of these major style-specific glycoproteins.

Some of the S-glycoproteins have been shown to have the biological activity expected, in that they are capable of inhibiting pollen tube growth, in some cases S-specifically.

All work on the stigmatic S-gene product in sporophytic systems to date has been in *Brassica oleracea* (Nasrallah & Wallace, 1967; Nasrallah *et al.*, 1970, 1972) and *Brassica campestris* (Nishio & Hinata, 1978, 1982; Isogai *et al.*, 1987; Takayama *et al.*, 1987).

Most work on the gametophytic female (stylar or stigmatic) S-gene product has concentrated on Solanaceae. Stylar glycoproteins which segregate with S-genotypes have been isolated from *Nicotiana alata* (Clarke *et al.*, 1985; Anderson *et al.*, 1986; Jahnen *et al.*, 1989); a wild tomato, *Lycopersicon peruvianum* (Mau *et al.*, 1986); potato, *Solanum tuberosum* (Kirch *et al.*, 1989); a wild relative of potato, *Solanum chacoense* (Xu *et al.*, 1990a); *Petunia hybrida* (Broothaerts *et al.*, 1989). Two rather different species, which also have gametophytic control of SI are *Prunus avium* and *Papaver rhoeas*. In both species, biologically active stigmatic glycoproteins associated with S-genotype have been identified and characterised (Mau *et al.*, 1982; Franklin-Tong *et al.*, 1988, 1989).

A summary of the sporophytic and gametophytic S-associated proteins which have been identified or isolated is presented in Table 1. All of these putative S-gene products are glycoproteins. Allelic variation is ascribed to variation in isoelectric point (pI), which almost always has a relatively high value. These differences in charge are thought to reflect functional

Table 1. *Examples of S-linked glycoproteins*

Species	M_r $(\times 10^{-3})$	pI	Glycosylation	Basis of identification
Nicotiana alata	30–34	8.5–9.5	3–4 chains	Linkage studies; activity in bioassay[a]
Lycopersicon peruvianum	27–28	7.5–9.5	nd	Sequence identity with *N. alata* amino acid N terminus
Petunia hybrida	27–33	8.6–9.3	+	Sequence identity with *N. alata* amino acid N terminus
P. inflata	24–25	~9.3	+	Sequence identity with *N. alata* amino acid N terminus
Solanum tuberosum	23–27	8.3–9.1	+	Sequence identity with *N. alata* amino acid N terminus
S. chacoense	29–31	~8.6	nd	Sequence identity with *N. alata* amino acid N terminus
Prunus avium	37–39	8.8	nd	Sequence identity with *N. alata* amino acid N terminus
Papaver rhoeas	22	7.4–8.6	+	Activity in bioassay[b]
Brassica oleracea	54–62	5.7–11.1	5–7 chains	Linkage studies[b]
B. campestris	48–53	5.7–8.4	5–7 chains	

nd, not determined.

[a]Incomplete specificity of activity; some non-specific inhibition between *S*-alleles.
[b]Multigene family, apparently not all *S*-linked.

specificity-determining bases. Although amino acid sequence identity has been reported among S-glycoproteins within both gametophytic and sporophytic systems, there is no sequence identity observed between the two systems. These differences may reflect differences in the fundamental functions or mechanisms of these SI systems.

These glycoproteins have been analysed in some detail with respect to the carbohydrate chains. In *Brassica*, the carbohydrate chains are not unique to S-glycoproteins, but are found widely in other plant glycoproteins (Takayama *et al.*, 1986). It was suggested that, although the carbohydrate chains may play an important role in recognition, the S-allele specificity is probably not located in the carbohydrate portion of the S-glycoprotein. Data from *N. alata* suggest that there is more than sufficient variability in the glycan chains to encode allelic specificity (Woodward *et al.*, 1989). It is not known whether the carbohydrate or the polypeptide domains of the S-glycoproteins determine S-specificity, but it is clear that there is ample potential in both parts of the molecule.

Cloning the pistil part of the S-gene

The first report of an S-locus-specific glycoprotein (SLG) being cloned was in *Brassica oleracea* (Nasrallah *et al.*, 1985). Soon after, a cDNA clone from *N. alata* was isolated (Anderson *et al.*, 1986). Since these two reports, several more related SLG genes have been cloned and sequenced in *Brassica* and all segregate with the S-locus (Nasrallah *et al.*, 1987; Trick & Flavell, 1989; Scutt *et al.*, 1990). Recently, several gametophytic S-genes have been cloned, using part of the *N. alata* sequence (*P. hybrida*, Clark *et al.*, 1990; *S. chacoense*, Xu *et al.*, 1990*b*; *Lycopersicon peruvianum*, E. Walker & F. C. H. Franklin, unpublished results).

Several properties of these clones suggested that they encode the S-gene. Tissue specificity and developmental expression of the genes correlated with the site and timing of the S-gene in both systems. Southern blot analysis provided evidence that the cDNA clones show linkage to the S-locus. In *B. oleracea* extensive restriction fragment length polymorphisms (RFLPs) were found, and they segregated with the S-locus (Nasrallah *et al.*, 1985).

In *Brassica*, a second stigma-specific gene was found to be coordinately expressed at levels higher than the S-linked transcript. This gene was denoted SLR (S-locus related), while the other is SLG (S-locus glycoprotein). These two genes apparently share 70% DNA sequence identity (Lalonde *et al.*, 1989; Trick & Flavell, 1989). The patterns of expression of the two genes are identical with respect to their tissue specificity and developmental timing (Lalonde *et al.*, 1989). However, while the SLG segregates with the S-gene, the SLR is not linked to S and the SLR gene is highly expressed in both self-incompatible and self-compatible strains of *Brassica* (Lalonde *et al.*, 1989; M. Trick, personal communication), and consequently cannot contribute to S-allele specificity.

In the gametophytic S-genes, homology appears to be variable. In *N. alata* only a small degree of sequence identity was found between S-alleles (Anderson *et al.*, 1986), suggesting that the S-alleles are quite divergent. Similarly, the *N. alata* S_2-allele has a greater degree of identity with *L. peruvianum* S-allele than with another S-allele from *N. alata* (E. Walker & F. C. H. Franklin, unpublished results). Two of the *Petunia* S-alleles share extensive sequence similarity, while the other is quite different (Clark *et al.*, 1990). There appears to be no sequence identity between the genes from *Brassica* and *Nicotiana* (Nasrallah *et al.*, 1987; Bernatzky *et al.*, 1988). This indicates that the two systems are different, perhaps having evolved independently. It may also mean that SI may operate rather differently in these two systems.

Identification and characterisation of the pollen component

Despite the progress achieved by the characterisation and cloning of stigmatic components involved in the SI reaction in these species, comparable progress towards the identification of pollen components which participate in the reaction and the elucidation of the mechanism(s) whereby pollen tube inhibition occurs has not yet been achieved.

It appears that, in sporophytically controlled systems, the exine layer of the pollen is involved in the SI reaction. Callose deposition may be induced at the stigma surface by application of a protein fraction derived from the exine of ungerminated incompatible pollen, which suggests that the exine contains information relating to the incompatibility genotype (Heslop Harrison *et al.*, 1974, 1975). In gametophytically controlled systems it is thought that the proteins of the intine are involved in the SI reaction (Knox & Heslop Harrison, 1971; Knox, 1973).

Very early on in the study of the biochemical nature of SI, studies were made to try to detect the pollen component. At that time, since it was thought that because the S-gene controlled SI in both pistil and pollen then the products would be identical, pollen tissue was investigated to see if it contained specifically serologically cross-reacting proteins (Lewis, 1964). Lewis (1952) reported that experiments with antisera to pollen extracts from *Oenothera organensis* gave results which suggested that pollen produces S-allele-specific proteins. Studies which examined pollen proteins in more detail were carried out (Lewis *et al.*, 1967), but did not identify a protein specifically associated with SI. More recently, Brown & Crouch (1985) analysed pollen proteins from *O. organensis* and found that polyclonal antibodies raised against pollen proteins detected S-linked differences in peptide patterns on sodium dodecyl sulphate/ polyacrylamide gel blots. Study of the tissue and developmental expression of these proteins revealed that one of the proteins was also present in stigma and stylar tissue in much lower amounts, but not in other tissues. Linskens (1960) reported antigens which were S-allele specific in *Petunia*. Gaude & Dumas (1986, 1987) looked for a correlation of electrophoretic banding patterns with S-genotype in extracts from *Brassica* pollen. Two pollen genotypes exhibited glycoprotein bands which appeared to be genotype-specific, while the other two genotypes studied did not. The genotype-specific glycoproteins appeared to be developmentally regulated, in that the concentration of these glycoproteins increased with pollen maturation. However, the S-specificity remains to be demonstrated more clearly. These remain the only reports of proteins in pollen which are apparently S-specific. Thus, although early studies looked promising, further investigations in a variety of species apparently did not reveal a pollen component which was either identical with or equivalent to the stigmatic glycoprotein. No S-associated proteins were found in pollen tissue of *Petunia* (Broothaerts *et al.*, 1989); Kamboj & Jackson, 1986, *N. alata* (Bredemeijer & Blaas, 1981; Anderson *et al.*, 1986;) or *Brassica* (Nishio & Hinata, 1978; Nasrallah, 1979; Gaude & Dumas, 1986, 1987).

There is an indication that pollen may have components with the same structural components, probably saccharide chains, as S-glycoproteins in the pistil, which interfere with the immunological detection of the recognition molecules in the pollen. K. Okazaki (1986, cited by Isogai *et al.*, 1987) found that, when polyclonal antibody was absorbed with an extract from pollen of an identical phenotype as the stigma, it became more specific to S-glycoproteins and unreactive to the acidic glycoproteins commonly found in stigmas of different homozygotes. This may explain some of the difficulties found in the attempt to identify pollen S-specific glycoproteins.

These studies looking for a pollen-specific equivalent of the stigmatic S-specific gene product suggested that it is likely that the pollen component of SI is different and distinct from the stigma/stylar component. There are several reasons why it is now expected that the S-gene products in the pistil and pollen are different. First, it is difficult to envisage how two identical molecules could interact to cause inhibition of pollen tube growth. Although Lewis (1964) postulated that the dimerisation of two identical molecules might result in an inhibitory complex which could possibly cause the incompatibility reaction, it is now thought that this is not a viable mechanism for eukaryotes. A second objection lies in the structure of the molecules. It would be expected that, since the stigmatic-stylar S-glycoprotein is secreted, the pollen S-gene product would be membrane- or wall-bound. An interaction of this sort would enable signal transduction, which would set off a set of events within the pollen, leading to inhibition. Although the stigma and pollen components are unlikely to be identical, this does not rule out the concept of the S-gene being identical. There may be differential splicing of the same transcriptional unit, which would result in different products being generated from the same gene. Alternatively, the products of the gene may be processed differently, for instance, there may be different post-translational modifications, such as glycosylation.

In *Brassica*, components with sequence identity to stigma SLG, present in very low concentrations, have been reported to be detected on native and Western blots of anther and pollen extracts probed with anti-SLG antibodies (Nasrallah & Nasrallah, 1986). There are also indications that there may be very low expression of S-sequences (several hundred-fold lower than in stigmas) in postmeiotic anthers, but not in premeiotic, later postmeiotic anthers or mature pollen (Nasrallah & Nasrallah, 1989). However, evidence has also been presented which indicates that the SLG sequence (or any cross-hybridising sequence) from *Brassica* is not detectable in pollen or anther tissue at any stage of development (Scutt *et al.*, 1990).

Despite this observation, we have recently transformed *N. tabacum* plants with a deletion series of the *Brassica* SLR promoter. Using ß-glucuronidase (GUS) as a reporter, we have shown that there is expression in the stigma and upper style and also in mature (but not immature) pollen (R. M. Hackett *et al.*, unpublished results). This suggests that the SLR product can potentially be expressed in both the male and female tissues. It has also recently been shown that the promoter of the *Brassica* SLG gene in transformed *N. tabacum* can direct expression in both the pollen and the stigma (Thorsness *et al.*, 1991) and also when incorporated into *Arabidopsis thaliana* (Toriyama *et al.*, 1991). It also demonstrates that expression of the SLR gene in the male tissue is gametophytic when it is transformed into a gametophytic species. The SLG gene, when incorporated into *N. tabacum*, is also expressed in a manner analogous to the SI system in *N. alata* rather than that in *B. oleracea* (Kandaswamy *et al.*, 1990; Moore & Nasrallah, 1990). In contrast, the promoter of the SLG gene fused to GUS has also been incorporated into *A. thaliana*, where expression was as expected in *Brassica* (Toriyama *et al.*, 1991). It has been suggested that there may be certain factors present in the tobacco pistil that activate the S-gene to be expressed like a *Nicotiana* S-associated gene. It therefore appears that all members of the S-gene family are likely to play a role in pollen–stigma interaction and that the stigmatic component or a modified form of it is expressed in both of these tissues. This also suggests that there is not such a clear distinction between the two forms of the genes (SLR and SLG), thus complicating the explanation of how SI functions in *Brassica*.

What is known about the mechanism of the SI reaction?

Models for a mechanism for SI

Since it was first proposed by Prell (1921), the oppositional mechanism (inhibition of like genotypes) has been generally accepted as an explanation of the incompatibility response (de Nettancourt, 1977; Lawrence *et al.*, 1985), rather than the alternative complementary system. We assume the oppositional hypothesis as a paradigm for the operation of SI, since it provides a more economical explanation for the number of S-gene products required to be carried by the pistil.

A common feature of all the models postulated is the implication that the pollen and pistil components are encoded by a single gene or supergene (the S-gene). This view is prevalent, since it appears to be impossible to generate new functional S-alleles by mutation (de Nettancourt, 1977). This has always been explained by arguing that new functional S-alleles would require mutations in the gene at the sites controlling the specificities of both the pollen and the pistil at the same time, thus making it very unlikely. Nevertheless, S-allele function can be lost independently in pollen and stigma (Lewis, 1949, 1951; Hinata & Okazaki, 1986). What is still not clear is whether the pollen and stigma components are encoded by the same gene or two different genes that are tightly linked.

Most models for SI assume that S-allele products in both the pollen and the pistil are identical (Sampson, 1960; Lewis, 1964; Linskens, 1976). A possible mechanism (as mentioned earlier) to explain SI on the basis of Lewis' (1964) dimer hypothesis was suggested by de Nettancourt (1977), whereby each S-allele would encode an S-specific polypeptide in both the pollen and in the pistil, and dimerization of identical polypeptides would cause the cessation of pollen tube growth.

It is generally expected, for reasons outlined in pp. 83–5, that the S-gene products in the pistil and pollen are different.

In contrast to this like–like model, East (1929) very early in the history of investigation into SI, likened the response to an antibody–antigen interaction. van der Donk (1975) put forward a model postulating that the polypeptides produced by the pollen and stigma/style were different. He suggested that the pistil and pollen recognition molecules acted as both effector and receptor molecules: the style-specific molecules activated a set of genes necessary for pollen tube growth, while pollen-specific polypeptides, if recognised by complementary stylar polypeptides, would inactivate these and thus stop pollen tube growth.

Ferrari & Wallace (1977) suggested a model for SI in *Brassica*, which assigned separate functions to the male and female components: the pollen recognition molecule acted as a receptor, while the stigma recognition molecule acted as an effector. They postulated that these S-specific molecules acted only to differentiate self from non-self, while pollen germination and growth was controlled by regulation of the activity of other pre-synthesised enzymes.

The biochemistry of pollen and pistil components needs to be considered in any model for the mechanism of the SI reaction. Unfortunately, not very much was known about the interacting surfaces until quite recently. Although the exine and intine layers in pollen cell walls have been implicated in the SI reactions in sporophytic and gametophytic systems, respectively, no pollen-specific components linked to the S-locus have been unequivocally isolated to date. In the pistil, more progress has been made and several S-associated molecules have been isolated from a number of species. With some knowledge of the molecules involved in one-half of the reaction it is a little easier to postulate a model for the SI reaction.

Since the stigmatic component appears to be a glycoprotein and the pollen component appears to be different, the most likely mechanism is that of a 'lock and key' type of system involving a lectin–glycoprotein interaction as the basic mechanism for the SI reaction (Anderson *et al.*, 1983). The pistil glycoproteins could easily provide the specificity required, since enormous diversity is possible in the saccharide or the protein part of the molecule. However, evidence of the involvement of sugar residues in the recognition process is equivocal.

Comparisons have often been made between host–pathogen and pollen–pistil interactions, since they both involve very specific recognition processes (Heslop Harrison *et al.*, 1974; Teasdale *et al.*, 1974; Hogenboom, 1975, 1983; Linskens, 1976; Lewis 1980; Hodgkin *et al.*, 1988). The pistil–pollen SI system appears to be very like the 'gene-for-gene' host–parasite system in that matched dominant genes confer incompatibility in both systems. What is thought to be a major difference between SI and host–pathogen systems is the fact that in SI a single locus is thought to determine the specificity of both pollen and pistil, while in the latter systems different genes operate in the host and the pathogen. There is still no conclusive evidence about the S-locus to indicate whether this is the case or not. The host–pathogen response is an active process, involving production of new mRNAs and protein synthesis. There is evidence that the SI response is also like this. Franklin-Tong *et al.* (1990) have shown that gene transcription specific to an incompatible response takes place and that new products are formed as a result of this, although the nature of these is not known yet.

Ribonuclease activity

It has recently been demonstrated by McClure *et al.* (1989) that not only do the S-glycoproteins of *N. alata* have significant structural identity with the active site domain of the extracellular fungal ribonucleases (T_2 from *Aspergillus oryzae* and Rh from *Rhizopus niveus*), but they are also ribonucleases themselves. The specific activity of these ribonucleases was 100 to 1000-fold higher than that in the self-compatible *N. tabacum*. This, taken together with the observation that a purified ribonuclease (bovine pancreatic RNase A) inhibited pollen tube growth of *N. alata* non-specifically, suggested that the S-glycoproteins have both inherent ribonuclease activity and that this activity is involved in the inhibition of pollen tube growth in the SI reaction. Similar activity was found in *Lycopersicon peruvianum* (McClure *et al.*, 1989) and S-associated glycoproteins from styles of *Petunia hybrida* have also been shown to have ribonuclease activity, although the RNase specific activities for the three S-genotypes examined varied (Broothaerts & Vendrig, 1990; Clark *et al.*, 1990).

A new model, therefore, has been suggested for the functioning of the stigmatic S-gene product (McClure *et al.*, 1989). Instead of the traditional receptor–ligand model, it was proposed that as a ribonuclease it may function cytotoxically, assuming that there was a molecule carried by compatible pollen which could recognise and inactivate it. Thus, the S-specific stylar ribonucleases would, on contact with pollen tubes in the style, be transported into the pollen tube cytoplasm where it would be free to degrade RNA. This would, in turn, interfere with protein synthesis and cause inhibition of pollen tube growth.

Given that *P. rhoeas* has, genetically at least, the same SI system as *N. alata*, it might be expected that similar ribonuclease activity would be present in the stigmas of *P. rhoeas*. However, several differences with respect to ribonuclease activity have been found between *N. alata* and *P. rhoeas* which suggest that the two species have different SI mechanisms (Franklin-Tong *et al.*, 1991). These are discussed below.

First, a comparison of the RNase activities in fractions from pistil tissue revealed that *P. rhoeas* stigmas have very low levels of ribonuclease activity. Since the levels of this activity in stigmas from *P. rhoeas* are even lower than those found in styles of *N. tabacum* (self-compatible) by McClure *et al.* (1989), it suggests that high levels of ribonuclease activity in styles may not confer SI universally.

Second, immature stigmas from *P. rhoeas* exhibit ribonuclease activity at a level indistinguishable from that of the mature self-incompatible stigmas. Since it is well established that the timing of S-gene expression corresponds to the onset of SI during flower maturation; immature pistils, which are phenotypically self-compatible contain only low amounts of the S-associated proteins; ribonuclease activity does not appear to be correlated with acquisition of SI in this species. There also appears to be little tissue specificity in the distribution of ribonuclease activity in poppies.

Third, fractions from *P. rhoeas* which had S-specific pollen inhibitory activity had RNase activities which were identical in some cases with other fractions which had no S-specific activity (see Table 2). Thus, the stigmatic S-gene product, with its specific pollen inhibitory activity, was clearly not correlated with high ribonuclease activity in *P. rhoeas*. Since it is commonly accepted that the specific biological activity is one of the most important criteria for identifying the S-gene product, we must conclude that in *P. rhoeas* the stigmatic S-protein is not a ribonuclease.

Finally, the effect of bovine pancreatic RNase A on pollen tube growth of *P. rhoeas* was investigated since the ribonuclease activities in poppy tissue were so low. No pollen tube inhibition was detected at ribonuclease levels equivalent to those contained in styles of *N. alata*. Thus, in contrast to *N. alata*, there is no evidence that ribonucleases are capable of inhibiting pollen tube growth in *P. rhoeas*.

Table 2. *RNase is not associated with pollen inhibitory activity in* Papaver rhoeas

Fraction[a]	RNase activity units min^{-1} mg^{-1}	Pollen tube length[b]	
		Compatible	Incompatible
ConA-UB	0.34	24.9 (76%)	24.8 (77%)
ConA-B	0.35	30.3 (92%)	12.1 (38%)

[a]ConA-UB, concanavalin A unbound fraction; ConA-B, concanavalin A bound fraction.
[b]Pollen tubes were measured after 3 h of growth ($n = 12$). Pollen tube length is expressed in graticule units, where 32 units = 1 mm; pollen tube length expressed as a percentage of control pollen tube length follows in brackets.

Thus, ribonuclease assays have revealed that, in contrast to the S-linked glycoprotein of *N. alata*, there is no detectable ribonuclease activity that correlates with the presence of the functional S-gene product in *P. rhoeas*. These differences between these two gametophytically controlled self-incompatible species suggest that the ribonuclease model almost certainly is not a universal mechanism for the SI reaction. Indeed, the ribonuclease model may be unique to solanaceous species, since the S-associated glycoproteins in *B. oleracea*, at least from inspection of their nucleotide sequences for conserved motifs, do not appear to be ribonucleases either.

Perhaps one important difference between the two species that may be of relevance and should not be overlooked is the site of the inhibition reaction. This is stylar in *N. alata* and the other solanaceous species and stigmatic in *P. rhoeas* and *B. oleracea*. Thus, although it is rather early to make any generalisations, it appears that there may be significant differences between the mechanisms of the incompatibility reaction, depending on whether the site of inhibition is stigmatic or stylar, not just differences in the genetic control of the systems.

Approaches using metabolic studies to elucidate a mechanism for SI

Since the ribonuclease model is not a universal system for the SI reaction, what other possible mechanisms could be involved? By studying the effect of metabolic inhibitors on SI, it may be possible to elucidate some of the mechanisms which are required for pollen germination and tube growth, and also the incompatibility reaction.

The question of whether inhibition of pollen tube growth occurs at a particular stage in pollen tube development can be explored using an *in vitro* system. We have carried out experiments to investigate the stage at which pollen of *P. rhoeas* is inhibited by the addition of stigmatic extracts at several times during pollen growth. Results (Fig. 1) indicate that pollen, of this species at least, does not need to be at a specific stage of growth, since each sample is inhibited (V. E. Franklin-Tong *et al.*, unpublished results).

There have been several studies which have attempted to elucidate some of the molecular events involved in the SI response. Many have involved studies of pollen–stigma interactions *in vivo*. Early experiments by Ascher (1971) on *Lilium longiflorum* and by Kovaleva *et al.* (1978) on *Petunia hybrida* involved the injection of RNA synthesis inhibitors into the styles and/or pollen of these flowers. However, experiments of this sort have the disadvantage that their interpretation is complicated, since

Pollen tube growth following delayed addition of stigma extract

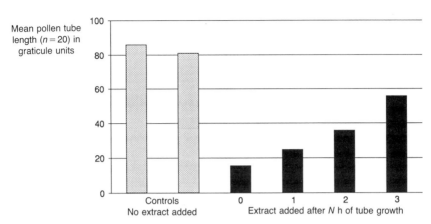

Fig. 1. The SI response may be elicited at any stage of pollen tube growth within 3 h of germination. All pollen was grown for 5 h. Incompatible stigma extracts were added at time 0, 1 h, 2 h and 3 h after initiation of growth. Controls consisted of extraction buffer alone and were tested side-by-side with the extracts using the same pollen sample.

the effects of the inhibitors on the pollen and pistil is confounded. Other studies, using pollen grown *in vitro* have revealed much information about pollen metabolism, although it is clear that generalisations are difficult.

Protein synthesis

Germination of the pollen grain is accompanied by the initiation of protein synthesis; germination and early pollen tube growth are often dependent on translation of preformed mRNA (Mascarenhas, 1975). In some species, the protein synthesis inhibitor, cycloheximide, has been found to inhibit pollen germination completely; in others, it inhibits pollen tube growth, and in some species it appears to inhibit neither phase of growth (Knox, 1984). In *P. rhoeas*, although cycloheximide did not prevent germination, it was a potent inhibitor of pollen tube growth (Table 3). Although this reveals the independence of pollen germination and protein synthesis in this species, it also means that the inhibitor would be of little value in elucidating the mechanism of SI. However, this is not the case in some species; in *B. oleracea*, for example, cycloheximide

Table 3. *Pollen challenged with stigmatic extracts in the presence of metabolic inhibitors in* Papaver rhoeas

	Pollen tube length[a]	
	Self (S_1S_2) pollen	Compatible pollen
S_1S_2 stigmatic extract alone	22.5	59.58
S_1S_2 stigmatic extract + CH	5.58	5.75
S_1S_2 stigmatic extract + TM	57.08	61.67
S_1S_2 stigmatic extract + ActD	48.33	62.50
Control: H_2O only	82.08	68.33

CH, cycloheximide; TM, tunicamycin (glycosylation inhibitor); ActD, actinomycin D (transcription inhibitor).
[a]Mean pollen tube length at 3 h is expressed in units ($n = 12$; 1 unit = 32 μm) in the body of the table.

affected pollen germination *in vitro* in a manner similar to that of high concentrations of incompatible-stigma extract (Ferrari & Wallace, 1976). They postulated that, while pollen germination does not require protein synthesis, regulation of the SI response does. Experiments by Sarker *et al.* (1988) suggested that the control of pollen hydration in *Brassica* requires continued protein synthesis. It is, perhaps, possible that the mechanism of the sporophytic SI system differs fundamentally from that of gametophytic systems which involve the inhibition of pollen tube growth.

Glycosylation

All of the molecules associated with SI on the pistil side of the reaction are glycosylated. It has not been established, however, whether the glycosylation of these stigmatic-stylar molecules actually plays a role in the SI reaction. The relevance to specificity of the sugar moieties involved in the pollen–stigma interaction has been investigated. Sharma & Shivanna (1983) and Sharma *et al.* (1984) examined the effectiveness of lectins and sugars in overcoming the SI reaction *in vitro* in *Petunia* and *Eruca*. The addition of some of these molecules resulted in the loss of inhibition of self-pollen, but the response of compatible pollen was not affected. Although the result is relatively crude, it demonstrates a clear involvement of sugar moieties in the recognition/inhibition response.

Recently, results from *semi vivo* experiments on *Brassica* which resulted in the overcoming of the SI reaction by tunicamycin, which is a glycosylation inhibitor, have been interpreted as suggesting that glycosylation of the stigmatic glycoproteins is required for the operation of SI in this species (Sarker *et al.*, 1988).

In *P. rhoeas* tunicamycin has very little detectable effect on pollen tube growth, which suggests that glycosylation of a product is not essential for pollen tube growth. However, tunicamycin treatment of pollen partially alleviates an incompatible response (Table 3). This demonstrates the importance of the glycosylation of pollen components involved in the SI reaction (Franklin-Tong *et al.*, 1990) and suggests that a pollen component which is required for an SI inhibitory response is not present in its active form in ungerminated pollen. A possible speculative interpretation is that the pollen 'receptor' with which the stigmatic S-glycoprotein interacts needs to be glycosylated. Inhibition of glycosylation during germination (assuming that this component is newly synthesized during germination) may prevent the recognition process. Alternatively, if the pollen products that are induced as a consequence of an incompatible interaction require the generation and attachment of glycosyl groups, they would be rendered inactive by the tunicamycin. Our observation, that pollen components need to be glycosylated *de novo* in order for the SI reaction to take place, thus provides an important contribution to the understanding of the mechanism of SI.

RNA synthesis

It has been established for some time that no new RNA is synthesised during the first stage of pollen tube growth in certain species (Mascarenhas, 1975). In *P. rhoeas*, translation studies *in vitro* confirm this, since the complement of mRNA species in poppy pollen prior to, and following, germination are indistinguishable (Franklin-Tong *et al.*, 1990). Additionally, actinomycin D, a transcription inhibitor, has no effect on pollen tube growth, indicating that at anthesis mature pollen contains the full complement of mRNA species that are required to specify the proteins that will be needed for pollen tube extension during the first few hours of growth.

Experiments *in vivo* have shown that SI can be overcome by the injection of RNA synthesis inhibitors into buds of the *Petunia* (Kovaleva *et al.*, 1978). However, although this demonstrates that RNA synthesis during development is necessary for the generation of SI, this could be due to any number of events, which could be non-specific to the SI system. van der Donk (1974) has demonstrated that differential synthesis of RNA occurs in self- and cross-pollinated styles of *Petunia hybrida*, and concludes that differential gene activity accompanies compatible and incompatible pollen tube growth. However, since these were *in vivo* pollinations, the effect could not be attributed to a specific tissue. Thus, although the observation that transcription inhibitors can overcome inhibition of incompatible pollen is not new, this inhibition has not been shown, until quite recently, to be specific to the SI response. We have shown (Franklin-Tong *et al.*, 1990) that when pollen from *P. rhoeas* is challenged with the stigmatic component in the presence of actinomycin D the effect of SI appears to be alleviated, though not completely (Table 3). This suggests that transcription in pollen is required for the SI response to operate. This requirement for transcription appears to be specific to an incompatible response, since there is no effect on fully compatible pollen.

Identification of pollen proteins associated with an incompatible response

The ability of actinomycin D partially to alleviate an incompatible reaction strongly suggests that pollen gene expression is induced during the response. This *de novo* gene expression presumably produces a product (or products) that causes inhibition of pollen tube growth, which should be detectable and specific to an incompatible response. *In vitro* translations from RNA extracted from ungerminated and germinated pollen, and from pollen in which a compatible and an incompatible response had been induced have been analysed (Franklin-Tong *et al.*, 1990). No detectable differences were observed between the samples from ungerminated (Fig. 2d) and germinated pollen (Fig. 2c) or from compatible pollen (Fig. 2b). Thus, mature, ungerminated *P. rhoeas* pollen contains all the mRNA species that are required to fulfil its role during at least early pollen tube growth. This explanation is consistent with the lack of effect of actinomycin D on normal pollen tube growth. In contrast, analysis of the translation products from the RNA isolated from pollen participating in an incompatible reaction (see Fig. 2a) clearly reveals the presence of novel proteins which are absent in the other samples. These are identified as a cluster of about 16 new proteins of molecular weight approximately 21,000–23,000, with variable, but essentially neutral pI values. Since the new proteins appear in a cluster, it is possible that they are isoforms of one or more protein species. Thus, new RNA transcripts specific to an incompatible reaction have been identified (Franklin-Tong *et al.*, 1990). This finding, taken together with the ability of actinomycin D to alleviate the incompatibility response, is a clear indication that the *de novo* transcription of pollen genes which are specific to this response play an important role in the recognition and/or inhibition of incompatible pollen in this species.

Because these pollen 'response' genes appear to be switched on as a result of the SI response in *P. rhoeas*, we are cloning these components.

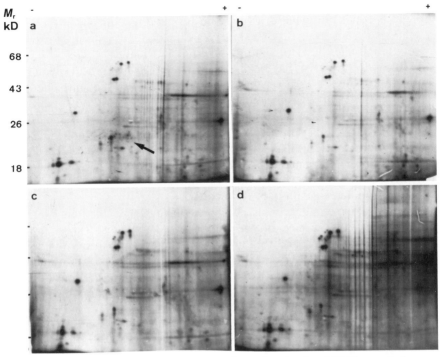

Fig. 2. Autoradiographs of two-dimensional gels of RNA translations from pollen growth *in vitro* (a) Pollen: incompatible reaction. (b) Pollen: fully compatible reaction. (c) Pollen control. (d) Dry pollen. The arrow in (a) indicates the region of new proteins produced as a result of the incompatible reaction. (Copyright: Franklin-Tong, Lawrence & Franklin, 1990; *New Phytologist* **116**, 319–24.)

A cDNA library has been generated from pollen challenged with an incompatible stigma extract, and this is currently being screened using differential and subtractive methods. Clones which are up- and down-regulated as a response to the SI reaction have been identified and are currently being characterised. Information as to the nature of these genes will substantially improve our knowledge as to how the SI mechanism works in this species.

Cell–cell signalling in SI

It is commonly assumed that SI must involve cell–cell recognition and signalling, since there must be some way in which the interaction between stigma and pollen is mediated. The induction of events within the pollen which occur as the result of an external interaction between the stigmatic S-glycoprotein and what is assumed to be some type of pollen receptor suggests the involvement of a signal transduction mechanism. However, very few studies have looked into how this is achieved.

Recently, a cDNA has been cloned from *Zea mays* that encodes a putative protein kinase receptor. This molecule has sequence identity with the SLGs of *Brassica* (Walker & Zhang, 1990). Although maize is not self-incompatible, this finding supports the hypothesis that the SLG, or a related protein, is involved in fertilisation, and possibly pollen recognition. Since there is a high level of this protein in other tissues, it is likely that this protein has another function. It suggests that there is a ligand-binding function on this molecule, which could be involved in cell–cell recognition and signalling.

Because experiments with *P. rhoeas* have shown that an external stimulus at the surface of the pollen results in gene expression in incompatible pollen, and that growth is inhibited as a result of this, it appears that the SI response may involve a classic signal transduction system. This is, to our knowledge, one of the first demonstrations of cell–cell signalling as a result of a specific stimulus involving a rapid response (within minutes) in plants. We are therefore investigating a role for inositol lipids and protein phosphorylation in the SI reaction in *P. rhoeas*.

Because pollen from *P. rhoeas* may be challenged with a stigmatic extract even after several hours growth (see Fig. 1) it has been possible to label pollen with [^{32}P]orthophosphate to equilibrium and then to challenge with stigmatic extract and observe any changes in phosphorylation. Preliminary results are encouraging, in that there appear to be rapid, transient changes in the state of protein phosphorylation as a result of an incompatible pollen–stigma interaction, with clear differences in the phosphorylation patterns in pollen which has been challenged with an incompatible and a compatible stigma extract within 1 min of the interaction (V. E. Franklin-Tong, unpublished results). Results to date suggest that activation of both protein kinases and phosphatases is involved in the SI response, which suggests a role for signal transduction in this system.

Conclusions

It is apparent that the control of SI is not as simple as it appears from the genetic models of the SI systems. From recent studies it is now clear that there are at least two different mechanisms for the operation of SI, and there may well be more. Moreover, there is strong evidence that not all gametophytic systems have the same mechanism for pollen tube inhibition.

Studies in *N. alata* and other gametophytically controlled solanaceous species provide evidence that the stylar component is a ribonuclease, and that this functions to inhibit incompatible pollen tube growth. It appears that the stigmatic component in *Brassica*, which has sporophytic control of SI, is not a ribonuclease. Similarly, *P. rhoeas*, which has gametophytic control, also has no ribonuclease activity. Thus, as far as the stigmatic side of SI is concerned, there is no consistent picture of how these molecules function. In *P. rhoeas* there are pollen 'response' genes which are activated as a consequence of an incompatible reaction. These are currently being cloned and characterised. From work in other species it appears that both transcription and glycosylation may play important roles in the SI reaction, but there is not much detailed information as yet.

Acknowledgements

We thank the AFRC for supporting this research on the Plant Molecular Biology Programme and also the Gatsby Charitable Foundation. Thanks are also due to Dave Marshall for producing Fig. 1.

References

Anderson, M. A., Cornish, E. C., Mau, S.-L., Williams, E. G., Hoggart, R., Atkinson, A., Bonig, I., Grego, B., Simpson, R., Roche, P. J., Haley, J. D., Penschow, J. D., Niall, H. D., Tregear, G. W., Cochlan, J. P., Crawford, R. J. & Clarke, A. E. (1986). Cloning of a cDNA for a stylar glycoprotein associated with expression of self-incompatibility in *Nicotiana alata*. *Nature* **321**, 38–44.

Anderson, M. A., Hoggart, R. M. & Clarke, A. E. (1983). The possible role of lectins in mediating plant cell–cell interactions. In *Chemical Taxonomy, Molecular Biology and Function of Plant Lectins*, ed. M. Etzler & I. Goldstein, pp. 143–61. New York: Alan R. Liss.

Ascher, P. D. (1971). The influence of RNA synthesis inhibitors *in vivo*, pollen tube growth and the self-incompatibility reaction in *Lilium longiflorum* Thunb. *Theoretical and Applied Genetics* **41**, 75–80.

Bernatzky, R., Anderson, M. A. & Clarke, A. E. (1988). Molecular genetics of self-incompatibility in flowering plants. *Developmental Genetics* **9**, 1–12.

Bredemeijer, G. M. & Blaas, J. (1981). S–specific proteins in styles of self-incompatible *Nicotiana alata*. *Theoretical and Applied Genetics* **59**, 185–90.

Broothaerts, W. J., van Laere, A., Witters, R., Preaux, G., Decock, B., van Damme, J. & Vendrig, J. C. (1989). Purification and N-terminal sequencing of style glycoproteins associated with self-incompatibility in *Petunia hybrida*. *Plant Molecular Biology* **14**, 93–102.

Broothaerts, W. J. & Vendrig, J. C. (1990). Self-incompatibility proteins from styles of *Petunia hybrida* have ribonuclease activity. *Physiologia Plantarum* **79**, Fasc 2 (2) A43.

Brown, S. M. & Crouch, M. (1985). Molecular genetic analysis of pollen antigens linked to the self-incompatibility locus in *Oenothera organensis*. *Genetics* **110**, s4, *supplement* **3**, pt 2.

Clark, K. R., Okuley, J. J., Collins, P. D. & Sims, T. L. (1990). Sequence variability and developmental expression of S-alleles in self-incompatible and pseudo-self-incompatible *Petunia*. *Plant Cell* **2**, 815–26.

Clarke, A. E., Anderson, M. A., Bacic, T., Harris, P. J. & Mau, S.-L., (1985). Molecular basis of cell recognition during fertilization in higher plants. *Journal of Cell Science, supplement* **2**, 261–85.

de Nettancourt, D. (1977). *Incompatibility in Angiosperms*. Berlin, Heidelberg, New York: Springer-Verlag.

East, E. M. (1929). Self-sterility. *Bibliographica Genetica* **5**, 331–70.

Ferrari, T. E. & Wallace, D. H. (1976). Pollen protein synthesis and control of incompatibility in *Brassica*. *Theoretical and Applied Genetics* **48**, 243–9.

Ferrari, T. E. & Wallace, D. H. (1977). A model of self-recognition and regulation of the incompatibility response of pollen. *Theoretical and Applied Genetics* **50**, 211–25.

Franklin, F. C. H., Franklin-Tong, V. E., Thorlby, G. J., Howell, E. C., Atwal, K. K. & Lawrence, M. J. (1991). Molecular basis of the incompatibility mechanism in *Papaver rhoeas L. Plant Growth Regulation*, in press.

Franklin-Tong, V E., Atwal, K. K., Howell, E. C., Lawrence, M. J. & Franklin, F. C. H. (1991). Self-incompatibility in *Papaver rhoeas*: there is no evidence for the involvement of stigmatic ribonuclease activity. *Plant, Cell and Environment*, **14**, 423–9.

Franklin-Tong, V. E., Lawrence, M. J. & Franklin, F. C. H. (1988). An *in vitro* bioassay for the stigmatic product of the self-incompatibility gene in *Papaver rhoeas L. New Phytologist* **110**, 109–18.

Franklin-Tong, V. E., Lawrence, M. J. & Franklin, F. C. H. (1990). Self-incompatibility in *Papaver rhoeas* L.: inhibition of incompatible pollen tube growth is dependent on pollen gene expression. *New Phytologist* **116**, 319–24.

Franklin-Tong, V. E., Ruuth, E., Marmey, P., Lawrence, M. J. & Franklin, F. C. H. (1989). Characterization of a stigmatic component from *Papaver rhoeas* L. which exhibits the specific activity of a self-incompatibility (S-) gene product. *New Phytologist* **112**, 307–15.

Gaude, T. & Dumas, C. (1986). Pollen–stigma interactions and S-products in *Brassica*. In *Biotechnology and Ecology of Pollen*, ed. D. L. Mulcahy, J. G. Mulcahy & E. Ottaviano, pp. 211–13. Berlin, Heidelberg, New York: Springer-Verlag.

Gaude, T. & Dumas, C. (1987). Molecular and cellular events of self-incompatibility. *International Review of Cytology* **107**, 333–66.

Heslop Harrison, J., Knox, R. B. & Heslop Harrison, Y. (1974). Pollen wall proteins: exine-held fractions associated with the incompatibility response in Cruciferae. *Theoretical and Applied Genetics* **44**, 133–7.

Heslop Harrison, J., Knox, R. B., Heslop Harrison, Y. & Mattsson, O. (1975). Pollen wall proteins: emission and role in incompatibility responses. In *The Biology of the Male Gamete*, ed. J. G. Duckett & P. A. Racey. *Biological Journal of the Linnean Society, supplement* **1**, pt 7, 189–202.

Hinata, K. & Okazaki, K. (1986). Role of stigma in the expression of self-incompatibility in crucifers in view of genetic analysis. In *Biotechnology and Ecology of Pollen*, ed. D. L. Mulcahy, G. Mulcahy & E. Ottaviano, pp. 185–90. Berlin, Heidelberg, New York: Springer-Verlag.

Hodgkin, T., Lyon, G. D. & Dickinson, H. G. (1988). Recognition in flowering plants: a comparison of the *Brassica* self-incompatibility system and plant pathogen interactions. *New Phytologist* **110**, 557–69.

Hogenboom, N. G. (1975). Incompatibility and incongruity: two different mechanisms for the non-functioning of intimate partner relationships. *Proceedings of the Royal Society of London, Series B.* **188**, 361–75.

Hogenboom, N. G. (1983). Bridging a gap between related fields of research: pistil–pollen relationships and the distinction between incompatibility and incongruity in non-functioning host–parasite relationships. *Phytopathology* **73**, 381–3.

Isogai, A., Takayama, S., Tsukomoto, C., Ueda, Y., Shiozawa, H., Hinata, K., Okazaki, K. & Suzuki, A. (1987). S-locus-specific glycoproteins associated with self-incompatibility in *Brassica campestris*. *Plant Cell Physiology* **28**, 1279–91.

Jahnen, W., Batterham, M. P., Clarke, A. E., Moritz, R. L. & Simpson, R. J. (1989). Identification, isolation and N-terminal sequencing of style glycoproteins associated with self-incompatibility in *Nicotiana alata*. *Plant Cell* **1**, 493–9.

Kamboj, R. K. & Jackson, J. F. (1986). Self-incompatibility alleles control a lower molecular weight basic protein in pistils of *Petunia hybrida*. *Theoretical and Applied Genetics* **71**, 815–19.

Kandaswamy, M. K., Dwyer, K. G., Paolillo, D. J., Doney, R C., Nasrallah, J. B. & Nasrallah, M. E. (1990). Brassica S-proteins accumulate in the intercellular matrix along the path of pollen tubes in transgenic tobacco pistils. *Plant Cell* **2**, 39–49.

Kirch, H. H., Uhrig, H., Lottspeich, F., Salamini, F. & Thompson, R. D. (1989). Characterization of proteins associated with self-incompatibility in *Solanum tuberosum*. *Theoretical and Applied Genetics* **78**, 581–8.

Knox, R. B. (1973). Pollen wall proteins: pollen stigma interactions in ragweed and *Cosmos* (Compositae). *Journal of Cell Science* **12**, 421–44.

Knox, R. B. (1984). Pollen–pistil interactions. *Encyclopaedia of Plant Physiology: Cellular Interactions. N.S.* **17**, 508–608.

Knox, R. B. & Heslop Harrison, J. (1971). Pollen wall proteins: the fate of intine-held antigens on the stigma in compatible and incompatible pollinations of *Phalaris tuberosa. Journal of Cell Science* **9**, 239–51.

Kovaleva, L. V., Milyaeva, E. L. & Chailakhyan, M. K. H. (1978). Overcoming self-incompatibility by inhibitors of nucleic acid and protein metabolism. *Phytomorphology* **28**, 445–9.

Lalonde, B. A., Nasrallah, M. E., Dwyer, K. G., Chen, C.-H., Barlow, B. & Nasrallah, J. B. (1989). A highly conserved *Brassica* gene with homology of the S-locus-specific glycoprotein structural gene. *Plant Cell* **1**, 249–58.

Lawrence, M. J., Marshall, D. F., Curtis, V. E. & Fearon, C. H. (1985). Gametophytic self-incompatibility reexamined: a reply. *Heredity* **54**, 131–8.

Lewis, D. (1949). Incompatibility in flowering plants. *Biological Reviews* **24**, 472–96.

Lewis, D. (1951). Structure of the incompatibility gene: III. Types of spontaneous and induced mutation. *Heredity* **5**, 399–414.

Lewis, D. (1952). Serological reactions of pollen incompatibility substances. *Proceedings of the Royal Society of London, Series B* **140**, 127–35.

Lewis, D. (1964). A protein dimer hypothesis on incompatibility. In *Genetics Today, Proceedings of the 11th Congress on Genetics.* The Hague, 1963, ed. S. J. Geerts, Vol. 3, pp. 653–6.

Lewis, D. (1980). Are there inter-relations between the metabolic role of boron, synthesis of phenolic phytoalexins and the germination of pollen? *New Phytologist* **84**, 261–70.

Lewis, D., Burrage, S. & Walls, D. (1967). Immunological reactions of single pollen grains, electrophoresis and enzymology of pollen protein exudates. *Journal of Experimental Botany* **18**, 371–8.

Linskens, H. F. (1960). Zur frage der entstehung der abwehr-korper bei der inkompatibilitatsreaktion von Petunia III. Mitteilung: serologische teste mit leitgewebs- und pollen-extracten. *Zeitschrift für Botanie* **48**, 126–35.

Linskens, H. F. (1976). Specific interactions in higher plants. In *Specificity in Plant Diseases*, ed. R. K. S. Wood & A. Graniti, pp. 311–25. New York: Plenum Press.

Mascarenhas, J. P. (1975). The biochemistry of Angiosperm pollen development. *Botanical Review* **41**, 259–314.

Mau, S.-L., Raff, J. W. & Clarke, A. E. (1982). Isolation and partial characterization of components of *Prunus avium* L. styles, including an antigenic glycoprotein associated with a self-incompatibility genotype. *Planta* **156**, 505–16.

Mau, S.-L., Williams, E. G., Atkinson, A., Cornish, E. C., Grego, B., Simpson, R., Kheyr-pour, A. & Clarke, A. E. (1986). Style proteins of a wild tomato (*Lycopersicon peruvianum*) associated with expression of self-incompatibility. *Planta* **169**, 184–91.

McClure, B. A., Haring, V., Ebert, P. R., Anderson, M. A., Simpson, R. J., Sakiyama, F. & Clarke, A. E. (1989). Style self-incompatibility gene products of *Nicotiana alata* are ribonucleases. *Nature* **342**, 955–7.

Moore, H. M. & Nasrallah, J. B. (1990). A *Brassica* self-incompatibility gene is expressed in the stylar transmitting tissue of transgenic tobacco. *Plant Cell* **2**, 29–38.

Nasrallah, M. E. (1979). Self-incompatibility antigens and S-gene expression in *Brassica*. *Heredity* **43**, 259–63.

Nasrallah, J. B., Kao, T.-H., Goldberg, M. L. & Nasrallah, M. E. (1985). A cDNA clone encoding an S-locus-specific glycoprotein from *Brassica oleracea*. *Nature* **318**, 263–7.

Nasrallah, J. B., Kao, T.-H., Goldberg, M. L. & Nasrallah, M. E. (1987). Amino acid sequence of glycoproteins encoded by three alleles of the S-locus of *Brassica oleracea*. *Nature* **326**, 617–19.

Nasrallah, J. B. & Nasrallah, M. E. (1989). The molecular genetics of self-incompatibility in *Brassica*. *Annual Review of Genetics* **23**, 121–39.

Nasrallah, M. E., Barber J. & Wallace, D. H. (1970). Self-incompatibility proteins in plants: detection, genetics and possible mode of action. *Heredity* **25**, 23–7.

Nasrallah, M. E. & Nasrallah, J. B. (1986). Immunodetection of S-gene products on nitrocellulose electroblots. In *Biotechnology and Ecology of Pollen*, ed. D. L. Mulcahy, G. Mulcahy & E. Ottaviano, pp. 197–201. Berlin, Heidelberg, New York: Springer-Verlag.

Nasrallah, M. E. & Wallace, D. H. (1967). Immunochemical detection of antigens in self-incompatibility genotypes of cabbage. *Nature* **213**, 700–1.

Nasrallah, M. E., Wallace, D. H. & Savo, R. M. (1972). Genotype, protein and phenotype relationships in self-incompatibility of *Brassica*. *Genetical Research* **20**, 151–60.

Nishio, T. & Hinata, K. (1978). Stigma proteins in self-incompatible *Brassica campestris* L. and self-incompatible relatives, with special reference to S.-allele specificity. *Japanese Journal of Genetics* **53**, 27–33.

Nishio, T. & Hinata, K. (1982). Comparative studies on S-glycoproteins purified from different S-genotypes in self-incompatible *Brassica* species. I. Purification and chemical properties. *Genetics* **100**, 641–7.

Prell, H. (1921). Das problem der unfruchbarkeit. *Naturwissenshaft Wochschrift N.F.* **20**, 440–6.

Sampson, D. R. (1960). An hypothesis of gene interaction at the S-locus in self-incompatibility systems of angiosperms. *American Naturalist* **94**, 283–92.

Sarker, R. H., Elleman, C. J. & Dickinson, H. G. (1988). The control of pollen hydration in *Brassica* requires continued protein synthesis whilst glycosylation is necessary for intra-specific incompatibility. *Proceedings of the National Academy of Sciences, USA* **85**, 4340–4.

Scutt, C. P., Gates, P. J. Gatehouse, J. A., Boulter, D. & Croy, R. R. D. (1990). A cDNA encoding an S-locus specific glycoprotein from *Brassica oleracea* plants containing the S5 self-incompatibility allele. *Molecular and General Genetics* **220**, 409–13.

Sharma, N., Bajaj, M. & Shivanna, K. R. (1984). Overcoming self-incompatibility through the use of lectins and sugars in *Petunia* and *Eruca*. *Annals of Botany* **55**, 139–41.

Sharma, N. & Shivanna, K. R. (1983). Lectin-like components of pollen and complementary saccharide moiety of the pistil are involved in self-incompatibility reaction. *Current Science* **52**, 913–16.

Takayama, S., Isogai, A., Tsukomoto, C., Ueda, Y., Hinata, K., Okazaki, K., Koseki, K. & Suzuki, A. (1986). Structure of carbohydrate chains of S-glycoproteins in *Brassica campestris* associated with self-incompatibility. *Agricultural and Biological Chemistry* **50**, 1673–6.

Takayama, S., Isogai, A., Tsukomoto, C., Ueda, Y., Hinata, K., Okazaki, K. & Suzuki, A. (1987). Sequences of S-glycoproteins, products of the *Brassica campestris* self-incompatibility locus. *Nature* **326**, 102–5.

Teasdale, J., Daniels, D., Davis, W. C., Eddy, R. & Hadwiger, L. A. (1974). Physiological and cytological similarities between disease resistance and cellular incompatibility responses. *Plant Physiology* **54**, 690–5.

Thorsness, M. K., Kandaswamy, M. K., Nasrallah, M. E. & Nasrallah, J. B. (1991). A *Brassica* S-locus gene promoter targets toxic gene expression and cell death to the pistil and pollen of transgenic *Nicotiana*. *Developmental Biology* **143**, 173–84.

Toriyama, K., Thorsness, M. K., Nasrallah, J. B. & Nasrallah, M. E. (1991). A *Brassica* S-locus gene promoter directs sporophytic expression in the anther tapetum of transgenic *Arabidopsis*. *Developmental Biology* **143**, 427–31.

Trick M. & Flavell R. B. (1989). A homozygous S-genotype of *Brassica oleracea* expresses two S-like genes. *Molecular and General Genetics* **218**, 112–17.

van der Donk, J. A. W. M. (1974). Differential synthesis of RNA in self- and cross-pollinated styles of *Petunia hybrida* L. *Molecular and General Genetics*. **131**, 1–8.

van der Donk, J. A. W. M. (1975). Recognition and gene expression during the incompatibility reaction in *Petunia hybrida*. *Molecular and General Genetics* **141**, 305–16.

Walker, J. C. & Zhang, R. (1990). Relationship of a putative receptor protein kinase from maize to the S-locus glycoproteins of *Brassica*. *Nature* **345**, 743–6.

Woodward, J. R., Bacic, A., Jahnen, W. & Clarke, A. E. (1989). N-linked glycan chains in S-allele-associated glycoproteins from *Nicotiana alata*. *Plant Cell* **1**, 511–14.

Xu, B., Grun, P., Kheyr-pour, A. & Kao, T.-H. (1990*a*). Identification of pistil-specific proteins associated with three self-incompatibility alleles in *Solanum chacoense*. *Sexual Plant Reproduction* **3**, 54–60.

Xu, B., Mu, J., Nevins, D., Grun, P. & Kao, T.-H. (1990*b*). Cloning and sequencing of cDNAs encoding two self-incompatibility associated proteins in *Solanum chacoense*. *Molecular and General Genetics* **224**, 341–6.

R. I. PENNELL

Cell surface arabinogalactan proteins, arabinogalactans and plant development

The plant cell surface and plant development

In normal development in multicellular plants, the directions of cell division and of cell enlargement play essential roles in the form assumed by a plant and its organs. Tissue development requires that cellular controls such as division and enlargement act locally. Some tissues (such as the vascular tissues of roots and shoots) display radial symmetry, while others (such as the component tissues of angiosperm ovules) are symmetrical only about a longitudinal axis. However, the structural basis of pattern control and of cell determination cannot be explained purely in terms of the direction of division and of the composition of cell walls (affecting enlargement). Patterning requires that certain cells or cell aggregates close to apical meristems become determined to particular developmental fates, and then differentiate accordingly. The position of these cells in the plant body must be controlled carefully, since all of organogenesis depends upon their differentiation. Somehow, cells acquire positional information early on.

Analysis of the control of plant cell fate is an intractible experimental problem. Genes have been identified that are involved with the control of complex developmental processes such as flowering (Coen *et al.*, 1991), but these appear to affect floral organ identity. Other genes encode structural proteins and glycoproteins in plant cell walls that are developmentally regulated and hence candidates for intercellular interactions (Hong *et al.*, 1989; Keller *et al.*, 1989; Ye & Varner, 1991), but these seem to affect the mechanical properties of the cells rather than control their fate. Monoclonal antibodies (MAbs) and specific antisera have also been raised against plant cell surface components. Unlike the cDNA probes that have been studied, some MAbs and antisera identify complex regulations of specific glycoproteins that relate not to organ identity but

Society for Experimental Biology Seminar Series 48: *Perspectives in Plant Cell Recognition*, ed. J. A. Callow & J. R. Green. © Cambridge University Press 1992, pp. 105–21.

to developmental events such as positional sensing and cell determination. In particular, arabinogalactan proteins (AGPs) at the outer surfaces of plasma membranes display expression patterns which appear to be quite novel (Knox *et al.*, 1989; Pennell & Roberts, 1990). In this sense, plasma membrane AGPs bear comparison with components of animal cell glycocalyces and suggest functional similarity between plant and animal cell surfaces (Roberts, 1989), the chemical differences between them arising from the less prominent part played by nitrogen in plant metabolism. This article describes the emergence of AGPs as molecules with developmental significance.

AGP composition

α-L-Arabinofuranose is a component of many kinds of glycoprotein and polysaccharide in plants, and most or all of these compounds reside at the cell surface. The best characterised are from flowering plants, from certain conifers, and from *Chlamydomonas*.

Arabinosylated glycoproteins are characteristic of plant cell surfaces and plant exudates. Most contain hydroxyproline-rich polypeptides and are therefore collectively termed hydroxyproline-rich glycoproteins (HRGPs; Showalter & Varner, 1989). There are three classes of HRGPs in vascular plants: the AGPs (Clarke *et al.*, 1979; Fincher *et al.*, 1983), the solanaceous ('potato') lectins in members of the Solanaceae (Allen & Neuberger, 1973; Allen *et al.*, 1978), and the extensins (Cassab & Varner, 1988). They differ in the extent and composition of glycosylations, in the presence of certain sugar and amino acid motifs, in the linkages between the sugars, and in the linkages between the sugars and the proteins (Table 1). Although the *Chlamydomonas* cell wall glycoproteins are also arabinosylated, they form crystalline arrays that are quite dissimilar from any cell surface glycoprotein structure in any vascular plant (Roberts, 1974). Arabinosylated polysaccharides are also components of plant cell walls and exudates, and like the arabinosylated glycoproteins fall into chemically distinct groups such as the rhamnogalacturonans (Ishii *et al.*, 1989) and the glucuronomannan gums (Aspinall, 1969). However, only in the arabinogalactans (AGs) do arabinose and galactose form the predominant sugars, and hence only the AGs resemble the AGPs chemically.

In AGPs and AGs, carbohydrate analysis indicates that the polysaccharides belong to one of two structural groups in which backbone β-D-Gal*p* is either 4-linked ('type 1') or is 3- and 6-linked ('type 2'; Clarke *et al.*, 1979). Indeed, the similarities between the carbohydrates of the AGPs and AGs are so great that the apparent difference between them –

Table 1. *Chemical characteristics of plant hydroxyproline-rich glycoproteins*

HRGP	Approximate M_r	Approximate carbohydrate (%)	Component sugars	Sugar motifs	Arabinosyl linkages	Amino acid motifs	Amino acid glycosides
AGPs							
Apoplast	30000–1000000[7]	90[7]	Araf, Galp, [Rhap, Fucp, Xylp, Manp, Glcp, GalpA, GlcpA][8]	ND	4-, 5-[7]	(Ala-Hyp)$_{1-3}$[9]	Hyp-Araf,[21] Hyp-Galp,[19] Ser/Thr-Galp[10]
Plasma membrane	10000–180000[15, 16]	70[15]	Araf, Galp, [Rhap, Fucp, Manp, Glcp][15]	ND	5-[16]	ND	ND
Solanaceous lectins	50000[3]	50[3]	Araf, [Galp][2]	(β-Araf)$_{1-3}$,[4] α-Araf-(β-Araf)$_3$[4]	2-, 3-[4]	ND	Hyp-Araf, Ser-Galp[2]
Extensins	86000[20]	65[19]	Araf, [Galp][5]	(α-Araf)$_{1-4}$[13]	2-, 3-[1]	Ser-(Hyp)$_4$[18]	Hyp-Araf,[11] Ser-Galp[12]
Chlamydomonas 2BII	s	23[17]	Araf, Galp, Manp, [Xylp, Glcp][6, 17]	ND	2, 4-[6]	ND	Hyp-Glcp, Hyp-Galp, Hyp-Araf[14]

HRGP, hydroxyproline-rich glycoproteins; [], low abundance; ND, no data; s, subunit composition (Roberts, 1979). superscript numbers indicate references as: 1, Akiyama *et al.*, 1980; 2, Allen *et al.*, 1978; 3, Allen & Neuberger, 1973; 4, Ashford *et al.*, 1982; 5, Cassab *et al.*, 1985, 6, Catt *et al.*, 1985, 7, Clarke *et al.*, 1979; 8, Fincher *et al.*, 1983, 9, Gleeson *et al.*, 1989; 10, Hillestead *et al.*, 1977; 11, Lamport, 1967; 12, Lamport *et al.*, 1973, 13, Lamport & Miller, 1971; 14, Miller *et al.*, 1972, 15, Norman *et al.*, 1990; 16, Pennell *et al.*, 1989; 17, Roberts, 1979; 18, Smith *et al.*, 1986; 19, Strahm *et al.*, 1981; 20, Stuart & Varner, 1980; 21, Yamagishi *et al.*, 1976.

the presence of polypeptides – may be no more than a difference in analytical sensitivity. This seems increasingly likely, since AGPs and AGs commonly have identical solubility properties, are localised in cells in the same domains and are secreted in the same kinds of exudate.

Chemical characteristics of AGPs and AGs

AGPs and AGs both contain sugar residues that they do not share with other arabinosylated HRGPs and hence are diagnostic when identified by sugar analysis or methylation analysis. Differences are evident in the proportions of arabinose and galactose to those of other sugars, in diversity of the sugar composition, and in certain linkages in the terminal arabinans (Table 2). Of the characteristic residues discovered so far, the elaborate and diverse terminal and subterminal glycosylations and preponderance of arabinose and galactose (Clarke et al., 1979) are most readily recognised, since they require only sugar analysis. Although neither amino nor imino acid motifs are thought to be AGP-specific, an Ala-Hyp motif has been described in the Gladiolus style AGP (Gleeson et al., 1989). Chemical analysis of extracellular and apoplastic AGPs and AGs is relatively easy, since they are secreted by plant cells growing in suspension culture or they can be solubilised from them and from plant tissues with aqueous buffers or with mild alkali (Clarke et al., 1979) and can therefore be easily purified. Analysis of plasma membrane AGPs is more difficult, however, and requires use of detergents and immunoaffinity chromatography (Norman et al., 1990). Different detergents are differentially effective in plasma membrane AGP extractions, sodium dodecyl sulphate and CHAPS being most effective for Nicotiana suspension cultures (Norman et al., 1990).

Table 2. *Diagnostic features of AGP and AG chemistry*

	Reference
Abundance of Ara*f*, Gal*p*; diversity of terminal and sub-terminal sugars including Rha*p*, Xyl*p*, Man*p*, Glc*p*, Glc*p*A, sugar linkages and sugar methylations	Clarke *et al.*, 1979
α-1-Ara*f*(1→5)-1-Ara*f*	Pennell *et al.*, 1989
β-D-Gal*p*(1→3)-1-Ara*f*	Clarke *et al.*, 1979
β-D-Gal*p*-4-0-Hyp	Strahm *et al.*, 1981
Ala-Hyp repeats	Gleeson *et al.*, 1989

Biochemical characterisation of AGPs and AGs

Artificial carbohydrate antigens and AGP-specific MAbs permit rapid and generally unambiguous identification of AGPs in plant tissues and in plant extracts (Table 3). The phenylglycosides termed Yariv reagents (Yariv *et al.*, 1967), and in particular the β-glucosyl Yariv reagents have been used almost exclusively for 20 years (Fincher *et al.*, 1983). Only β-linked glycosides are bound by AGPs (Yariv *et al.*, 1967); hence the erstwhile term for them, 'β-lectins'. Yariv reagents are red in colour and therefore make adequate AGP stains in tissue sections, in agarose gels and in Western blots. However, the nature of the interaction between the Yariv reagents and the AGPs is unknown, and it is not clear to what extent different Yariv reagents identify different AGPs. Since it has recently been demonstrated that several developmentally regulated AGPs differ from one another in ways that are invisible with methods other than immunochemical methods (Knox *et al.*, 1989), Yariv reagents would seem to have limited value in future AGP research. This is also the case for various lectins, galactose-binding antisera and the myeloma protein J539 (Fincher *et al.*, 1983).

Several AGPs (including all those associated with plasma membranes which are regulated developmentally) have been discovered not by chemical analysis but by antibody characterisation. Antibody characterisation is a valuable substitute for chemical analysis in the identification of unknown cell surface antigens present in small amounts or only transiently during development. Antibody characterisation involves analysis of antibody specificity or analysis of the binding site of an antibody in cross-reactivity assays or in epitope studies (Table 3). In cross-reaction studies, reaction of an antibody with a commercially available AGP such as gum arabic, but lack of reaction with purified potato lectin or purified extensin is generally sufficient to claim specificity. Gum arabic is particularly useful in this way, since it has very large and complex glycan chains (Churms *et al.*, 1978; Clarke *et al.*, 1979) which appear to

Table 3. *Biochemical identification of AGPs and AGs*

β-Glucosyl Yariv reagent binding
MAb characterisation
Binding by characterised AGPs (e.g. gum arabic); no binding by characterised solanaceous lectins or extensins
Binding inhibitions by characterised AGPs, arabinose, galactose and uronic acids
Binding inhibitions by characterised AGPs; plasma membrane localisation

contain many of the AGP epitopes present at plasma membranes and hence will bind most or all AGP-directed antibodies regardless of specificity. In epitope analysis, MAb specificity can be deduced from inhibition characteristics, usually in an enzyme-linked immunosorbent assay (ELISA). However, while gum arabic is a suitable antigen to use in cross-reaction studies, gum arabic inhibition in an ELISA is insufficient evidence for antibody characterisation since gum arabic contains many terminal sugars (including arabinose) which are common not only to other AGPs but also to other cell surface glycoproteins and polysaccharides. Additional inhibition by hapten monosaccharides and disaccharides such as arabinose and any uronic acid implies that the antibody is directed specifically against AGPs, since only in AGPs do these two sugar moieties co-exist in significant amounts (Clarke et al., 1979). Biochemical characterisation of AGP-directed antibodies may be inconclusive. However, if a partially characterised antibody can be shown to bind a plasma membrane antigen, the site of AGPs (Pennell et al., 1989), but not of extensins (Stafstrom & Staehelin, 1988), it is reasonable to conclude that it is AGP-specific.

Derivation of AGP- and AG-directed MAbs and antisera

Immunochemistry has made great impact on AGP research, and has revitalised interest in the functional significance of AGP and AG regulation during growth and development. Together with the (less-specific) Yariv reagents, antibodies now form the principal tools with which AGPs are studied (Table 4).

Five MAbs have provided specific information on the biochemistry, cell biology and developmental biology of cell surface AGPs and AGs. The first of these was PCBC3 (Anderson et al., 1984). The immunisation from which PCBC3 was derived marked a turning point in the study of AGPs, since it was shown to have primary affinity for arabinose and secondary affinity for galactose was derived with two similar MAbs (PCBC1 and PCBC2, recognising galactose and arabinose) from an immunisation with a crude salt extract of Nicotiana styles. These data provided the first evidence for the 'immunodominant' properties of certain AGP arabinogalactosides. (Although arabinose is known to occur in animals (Varma et al., 1977) it is not abundant and its presence is developmentally restricted. Plant arabinans therefore are thought to be recognised as very 'foreign' molecules when injected into animals, and hence AGPs are very immunogenic). Since protoplast agglutination with Yariv reagents suggested that certain AGPs were plasma membrane components (Larkin, 1977) several groups then performed immunisations

Table 4. *AGP- and AG-directed probes*

Probe	Derivation	Epitope components	Localisation	Reference
Lectins, artificial carbohydrates				
Galactose-binding proteins	*Arachis, Ricinus, Abrus, Ulex, Tridacna* lectins; myeloma J539 proteins	β-D-Gal*p*	Extracellular, plasma membrane	Fincher *et al.*, 1983
Yariv reagents	Chemical conjugation; β-glycosides, diazotized phloroglucinal	NA	Extracellular, cell wall (?)	Yariv *et al.*, 1967
Antisera				
Anti-AGP antiserum	Rabbit immunisation; *Gladiolus* style AGP	NA	Extracellular	Gleeson & Clarke, 1980
Anti-α-1-Ara*f* antiserum	Rabbit immunisation; BSA-conjugated α-1-Ara*p*	NA	Plasma membrane	Northcote *et al.*, 1989
Anti-75 kDa antiserum (?)	Rabbit immunisation; gel-excised *Lycopersicon* plasma membrane 75 kDa band	NA	Plasma membrane 75 kDa	Grimes & Breidenbach, 1987
MAbs				
PCBC3	Mouse immunisation; buffer extract *Nicotiana* styles	α-1-Ara*f*	Extracellular, cell wall	Anderson *et al.*, 1984
PN 16.4B4	Mouse immunisation; *Nicotiana* suspension cell membranes	u	Plasma membrane 135–180 kDa	Norman *et al.*, 1986
MAC 207	Rat immunisation; *Pisum* peribacteroid membranes	α-1-Ara*f*, β-D-Glc*p*A	Extracellular 70–100 kDa, plasma membrane 10–70 kDa	Pennell *et al.*, 1989
JIM4	Rat immunisation; *Daucus* suspension cell protoplasts	u	Extracellular 70–100 kDa, plasma membrane	Knox *et al.*, 1989
JIM8	Rat immunisation; *Beta* suspension cell protoplasts	β-D-Gal*p*	Plasma membrane 68–160 kDa	Pennell *et al.*, 1991
MAC 64, 202, 209 (?)	Rat immunisations; *Pisum* peribacteroid membranes	u	Plasma membrane 50–95 kDa	Bradley *et al.*, 1988
MVS-1(?)	Mouse immunisation; *Glycine* suspension cell protoplasts	u	Plasma membrane 400 kDa	Villanueva *et al.*, 1986

BSA, bovine serum albumen; NA, not applicable; u, unknown; ?, speculation.

with protoplasts in the belief that plasma membrane AGPs would preferentially elicit anti-AGP immunoglobulins (protoplasts being free from all other arabinosylated HRGPs, which are wall components; Stafstrom & Staehelin, 1988). Protoplast immunisations have now given rise to three plasma membrane AGP-directed MAbs. These are PN 16.4B4 (Norman *et al.*, 1986), JIM4 (Knox *et al.*, 1989) and JIM8 (Pennell *et al.*, 1991). It is likely that all of these MAbs were preferentially elicited by plasma membrane arabinogalactosides. However, not all anti-AGP MAbs have been derived from such crude techniques. The fifth AGP-specific MAb, which led to the identification of AGPs as plasma membrane components, was derived from a more refined immunisation with peribacteroid membranes isolated from root nodules of *Pisum* (Bradley *et al.*, 1988). This is MAC 207 (Pennell *et al.*, 1989). Other MAbs from the same series of peribacteroid membrane immunisations may also be directed against AGP epitopes, as judged from their characteristics in Western blots (Bradley *et al.*, 1988). Although it is possible that peribacteroid membranes in legume root nodules have altered and specific AGP composition, it is not known how this may have prompted elicitation of the MAC 207 immunoglobulin.

Immunisations with crude preparations of plant extracts have in general not been deployed as successfully to make specific antisera as they have to make MAbs; immunisation with plant extracts generally seems to require fractionation of the immune response by fusion and cloning to generate AGP-specific probes. However, one AGP-specific antiserum – inhibitable by galactose and by arabinose – has been developed from an immunisation with the AGP-rich *Gladiolus* style extract (Gleeson & Clarke, 1980). Another antiserum, from an immunisation with bovine serum albumin-conjugated α-L-Ara$p$, is inhibited by arabinose and recognises plant cell membranes (Northcote *et al.*, 1989); these characteristics are enough to regard it as AGP-specific. Immunisations with membrane fractions prepared from *Lycopersicon* suspension cultures and *Avena* roots have elicited antisera which blot with the characteristics of several AGP-specific MAbs (Grimes & Breidenbach, 1987; Lynes *et al.*, 1987). It is not possible to confirm this, however, without antibody characterisation or immunocytochemistry.

Apoplastic AGPs are proteoglycans, but plasma membrane AGPs are glycoproteins

Apoplastic AGPs have been known for many years. The ease with which these AGPs can be isolated and analysed has allowed many of them to become well characterised: for example, the *Acacia* AGP gum arabic

(Churms *et al.*, 1978), the AGP from root slime of *Vigna* (Moody *et al.*, 1988), and the extracellular AGPs purified from suspension cultures of *Lolium* endosperm cells (Anderson *et al.*, 1977). All of these AGPs are proteoglycans. Plasma membrane AGPs have only recently been characterised, however, with immunochemical (Pennell *et al.*, 1989) and biochemical (Norman *et al.*, 1990) techniques. The mobility and resolution in two-dimensional Western blots of the MAC 207-reactive *Daucus* plasma membrane AGPs, together with their detergent-partitioning properties, suggests that they are only moderately glycosylated (Pennell *et al.*, 1989). Chemical analysis of the PN 16.4B4-reactive *Nicotiana* AGPs confirmed the presence of a 50 kDa core protein in an AGP with a relative mobility in polyacrylamide gels of 160 kDa (Norman *et al.*, 1990). These AGPs are glycoproteins.

Although nothing is known directly about the sugar linkages in any glycoprotein AGP, epitope studies with MAC 207 suggest that, like many other AGPs, at least some contain 4-linked galactans (Pennell *et al.*, 1989). The classification of Aspinall (1969) therefore does not seem to relate to the structural AGP groupings now known. Groupings of AGPs based on the extent of glycosylation may now be more appropriate.

Developmentally regulated AGPs

Many AGPs are developmentally regulated, and it is the discovery of this regulation that is breathing new life into the study of these compounds (Table 5). The chemical basis of the developmental regulations – whether transcriptional or post-translational – is unknown, although differences between two extracellular AGPs in *Daucus* seem to be linked to glycosylation. The first descriptions of the spatial regulation of AGPs were from studies with β-glucosyl Yariv reagent. They are probably apoplastic proteoglycans. These studies demonstrated that extracellular AGPs are associated specifically with the hollow stylar canal of *Gladiolus* (Gleeson & Clarke, 1979) and, in combination with MAb PCBC3, with the solid transmitting tract and vascular bundles of *Nicotiana* carpels, in which they also accumulated during development (Sedgley *et al.*, 1985; Gell *et al.*, 1986). Concurrently, a derivation of crossed electrophoresis with β-glucosyl Yariv reagent in the second dimension demonstrated that different organs of *Lycopersicon* contain different and characteristic sets of AGPs (van Holst & Clarke, 1986) and also that the stylar AGPs of *Nicotiana* undergo developmental modulation (Gell *et al.*, 1986). In *Glycine* (soybean), β-glucosyl Yariv reagent has been used with crossed electrophoresis to demonstrate the accumulation of AGPs in the medulla of the root nodules, and also to show that tissue-specific sets of AGPs can be

Table 5. *Developmentally regulated AGPs*

System	Regulation	Probe	Reference
Nicotiana pistil	Stigma, style	β-Glucosyl Yariv reagent, PCBC3	Sedgely *et al.*, 1985; Gell *et al.*, 1986
Lycopersicon sporophyte	Vegetative tissues	β-Glucosyl Yariv reagent	van Holst & Clarke, 1986
Glycine sporophyte, root nodule	Vegetative tissues, nodule medulla	β-Glucosyl Yariv reagent	Cassab, 1986
Gladiolus pistil	Style	β-Glucosyl Yariv reagent	Gleeson & Clarke, 1979
Daucus root	Pericycle	JIM4	Knox *et al.*, 1989
Daucus suspension culture	Proembryogenic masses, post-globular embryos	JIM4	Stacey *et al.*, 1990
Pisum flower	Stamens, carpels, embryos	MAC 207	Pennell & Roberts, 1990
Zea hypocotyl	Hypocotyl epidermis	β-Glucosyl Yariv reagent, UEA 1, RCA 120	Schopfer, 1990
Brassica flower	Stamens, carpels, embryos	JIM8	Pennell *et al.*, 1991

resolved from the major organs of that species (Cassab, 1986). In *Zea*, β-glucosyl Yariv reagent has been used in conjunction with *Ulex* lectin (UEA 1) and *Ricinus* lectin (RCA 120), which bind fucose and galactose, respectively, both being present in certain AGPs, to localise AGP in the vascular cells and outer epidermis of coleoptiles (Schopfer, 1990). More recently, the subtle techniques of immunocytochemistry have revealed spatio-temporal expression patterns among related plasma membrane-associated AGPs. Plasma membrane AGPs are more precisely regulated than are apoplastic AGPs, and the expression patterns reported so far (which probably represent only a small fraction of those still to be defined) are interesting. Two developmental systems in particular have been studied: somatic embryogenesis and histogenesis in *Daucus*, and sexual determination and sexual embryogenesis in *Pisum*.

Somatic embryogenesis and histogenesis in *Daucus*

In the *Daucus* system, the MAbs MAC 207 and JIM4 have been used to demonstrate that a specific set of AGPs is associated with certain cells in undifferentiated tissues and with cells in differentiated tissues that span tissue boundaries. During *Daucus* somatic embryogenesis (Stacey *et al.*, 1990), the JIM4 epitope first appears in occasional surface cells in large, callus-like proembryogenic masses. The relationship between these cells and the adventive embryos that can be induced to develop from the cell clumps was not determined, but there are more JIM4-reactive cells than there are future embryos. The somatic embryo itself does not express the JIM4 epitope until it reaches the globular stage, when it appears in the developing epidermis and in the region of the future shoot apex and in small cell aggregates in the cotyledonary ridges. Later in development of *Daucus* somatic embryos and seedling roots (Knox *et al.*, 1989), the JIM4 epitope is expressed by two blocks of cells in the stele. Each block is centred on the poles of the protoxylem but contains only some of the protoxylem cells. The most reactive stele cells are those of the pericycle. The developmental expression pattern of the JIM4 epitope is quite novel in plant biology. Since the JIM4 epitope is first expressed in undifferentiated embryos, and since the spatial distribution of the JIM4 epitope cuts across tissue boundaries in differentiated roots, its expression does not relate to tissue identity. Also, since the JIM4 epitope is expressed within one or two cells of the root apical meristem, well before the architecture of the stele has been established, the JIM4 AGP epitope is not even a marker for future tissues. Rather, the JIM4 epitope appears to be a novel marker for certain cell aggregates in undifferentiated tissues, both embryogenic clumps in suspension cultures and in root apices. Since the JIM4 cell aggregates give rise to different tissues in which the JIM4 epitope is expressed differentially and in relation to the geometry of the vasculature, the JIM4 epitope seems to predict the fate of certain cells. In the sense that the cells in these aggregates appear to develop in accord not with particular tissues but in relation only to one another, they may be better termed cell collectives.

Sexual determination and sexual embryogenesis in *Pisum*

In *Pisum* flowers, the MAb MAC 207 has been used to obtain biochemical information on the formation of sexual cells from somatic cells in developing stamens and carpels (Pennell & Roberts, 1990). Serial frozen sectioning of *Pisum* plants at all stages of development has shown that the MAC 207 epitope, while present throughout the mature embryo and all vegetative parts of the mature sporophyte, is not expressed in the anther tapetum or the anther sporogenous tissue or in the ovule nucellus or the ovule sporogenous tissue. Morphologically, these compartments are the sporangia, giving rise by the process of meiosis to the spores and then to the gametophytes (tricellular pollen and mature embryo sac). Analysis of the gametophytes then demonstrated that all but one of the gametophytic plasma membranes also failed to express the MAC 207 epitope; the exception was that of the vegetative cell. Analysis of the developing embryo showed that the MAC 207 epitope is re-expressed only when meristems are formed and tissues differentiate in the heart-stage embryo. Like the JIM4-reactive AGPs, the significance of the regulation of those that bind MAC 207 is not obvious. The sporangia and the gametophytes form the sexual tissues, but the presence of the MAC 207 epitope of the vegetative cell plasma membrane indicates that the absence of the MAC 207 epitope is not a specific marker for sexual tissues. However, since the alteration in expression of the MAC 207 epitope does not take place in somatic embryogenesis, the function of the regulation of the MAC 207 epitope can not be concerned with the process of plant embryogenesis *per se*. Rather, it seems that its function is specific to sexual embryogenesis and therefore to sexual reproduction. At present, it seems most likely that the developmental window defined by the MAC 207 epitope relates to the expression in certain sexual cells of other plasma membrane determinants. Other AGPs – identified by the MAb JIM8 – are now known to be expressed within this developmental window, and demarcate specific cells in the sporophytic and gametophytic lineages (Pennell *et al.*, 1991).

Extracellular AGPs may control gelling in the apoplast, but plasma membrane AGPs seem to participate in cell fate determination

The functions of AGPs in plant growth and development are still unknown. Several hypotheses have been advanced, but without exception they lack critical data. Proteoglycan AGPs in apoplasts are very different from those in plasma membranes, and they presumably serve different functions. The apoplastic AGPs in styles are probably relatively homogeneous in distribution (there is no evidence for local AGP modulation within the transmitting tissue). They do not seem to be involved with recognition of foreign or self-incompatible pollen. Since apoplastic AGPs form hydrophilic gels, their presence in styles (and for that matter in many plant exudates such as gums) may relate more to water balance control. This is likely to be of peremptory importance for the growing pollen tube, since at its tip it is bounded by a particularly thin cell wall composed principally of hydrophilic pectin. The plasma membrane AGPs are more difficult to begin to understand. Unlike the apoplastic AGPs, those in plasma membranes are heterogeneous in distribution and their developmental modulations are complex. Some AGPs – probably including some plasma membrane AGPs – are certainly lectins, so they have the potential to bind cell wall ligands, but all the plasma membrane AGPs so far examined are peripheral glycoproteins (Pennell *et al.*, 1989; Norman *et al.*, 1990; R. Pennell & P. Kjellbom, unpublished results) and are therefore unlikely to bind cytoskeletal proteins in the cortical cytoplasm and participate in transduction of mechanical signals, at least directly. Recent research demonstrates that plasma membrane AGPs are organised into separate sets of closely related molecules. It is likely that the number, composition and spatial overlap (the modulation) of these sets underlies their function at the cell surface. For example, the MAC 207 AGP appears to define somatic cells; its combination with the JIM4 AGP identifies position-specific cell collectives close to meristems and during differentiation, but expression of the JIM4 AGP in the absence of that recognised by MAC 207 may specify ovule nucellus. Add to this simple pattern the expression in another cell set of a third plasma membrane AGP, and the number of additional AGP combinations increases. The potential for plasma membrane AGPs in cell fate determination in flowering plants seems to be great. Plasma membrane AGPs may therefore participate in the local control of histogenesis. This view receives some support from the apparent similarity between plasma membrane AGPs and cell and substrate adhesion molecules (Edelman, 1986).

Acknowledgements

I thank Peter Bell for criticism of the manuscript and Caroline Steeman-Clarke for typing the tables.

References

Akiyama, Y., Mori, M. & Kato, K. (1980). ^{13}C-NMR analysis of hydroxyproline arabinosides from *Nicotiana tabacum*. *Agricultural and Biological Chemistry* **44**, 2487–9.

Allen, A. K., Desai, N. N., Neuberger, A. & Creeth, J. M. (1978). Properties of potato lectin and the nature of its glycoprotein linkages. *Biochemical Journal* **171**, 665–74.

Allen, A. K. & Neuberger, A. (1973). The purification and properties of the lectin from potato tubers, a hydroxyproline-rich glycoprotein. *Biochemical Journal* **135**, 307–14.

Anderson, R. L., Clarke, A. E., Jermyn, M. A., Knox, R. B. & Stone, B. A. (1977). A carbohydrate-binding arabinogalactan-protein from liquid suspension cultures of endosperm from *Lolium multiflorum*. *Australian Journal of Plant Physiology* **4**, 143–58.

Anderson, M. A., Sandrin, M. S. & Clarke, A. E. (1984). A high proportion of hybridomas raised to a plant extract secrete antibody to arabinose or galactose. *Plant Physiology* **75**, 1013–16.

Ashford, D., Desai, N. N., Allen, A. K. & Neuberger, A. (1982). Structural studies of the carbohydrate moieties of lectins from potato (*Solanum tuberosum*) tubers and thorn-apple (*Datura stramonium*) seeds. *Biochemical Journal* **201**, 199–208.

Aspinall, G. O. (1969). Gums and mucilages. *Advances in Carbohydrate Chemistry and Biochemistry* **24**, 333–79.

Bradley, D. J., Wood, E. A., Larkins, A. P., Galfrè, G., Butcher, G. W. & Brewin, N. J. (1988). Isolation of monoclonal antibodies reacting with peribacteroid membranes and other components of pea root nodules containing *Rhizobium leguminosarum*. *Planta* **173**, 149–60.

Cassab, G. I. (1986). Arabinogalactan proteins during the development of soybean root nodules. *Planta* **168**, 441–6.

Cassab, G. I., Nieto-Sotelo, J., Cooper, J. D., van Holst, G. J. & Varner, J. E. (1985). A developmentally regulated hydroxyproline-rich glycoprotein from the cell walls of soybean seed coats. *Plant Physiology* **77**, 532–5.

Cassab, G. I. & Varner, J. E. (1988). Cell wall proteins. *Annual Review of Plant Physiology and Plant Molecular Biology* **39**, 321–53.

Catt, J. W., Hills, G. J. & Roberts, K. (1976). A structural glycoprotein containing hydroxyproline, isolated from the cell wall of *Chlamydomonas reinhardii*. *Planta* **131**, 165–71.

Churms, S. C., Merrifield, E. H. & Stephen, A. M. (1978). A comparative examination of two polysaccharide components from the gum of *Acacia mabellae*. *Carbohydrate Research* **63**, 337–41.

Clarke, A. E., Anderson, R. L. & Stone, B. A. (1979). Form and function of arabinogalactans and arabinogalactan proteins. *Phytochemistry* **18**, 521–40.

Coen, E. S., Romero, J. M., Doyle, S., Elliot, R., Murphy, G. & Carpenter, R. (1991). *floricaula*: a homeotic gene required for flower development in *Antirrhinum majus*. *Cell* **63**, 1311–22.

Edelman, G. M. (1986). Cell adhesion molecules in the regulation of animal form and tissue pattern. *Annual Review of Cell Biology* **2**, 81–116.

Fincher, G. B., Stone, B. A. & Clarke, A. E. (1983). Arabinogalactan-proteins: structure, biosynthesis and function. *Annual Review of Plant Physiology* **34**, 47–70.

Gell, A. C., Bacic, A. & Clarke, A. E. (1986). Arabinogalactan-proteins of the female sexual tissue of *Nicotiana alata* 1. Changes during flower development. *Plant Physiology* **82**, 885–9.

Gleeson, P. A. & Clarke, A. E. (1979). Structural studies on the arabinogalactan-protein from the style canal of *Gladiolus gandavensis*. *Biochemical Journal* **181**, 607–21.

Gleeson, P. A. & Clarke, A. E. (1980). Antigenic determinants of a plant proteoglycan, the *Gladiolus* arabinogalactan-protein. *Biochemical Journal* **191**, 437–47.

Gleeson, P. A., McNamara, M., Wettenhall, R. E. H., Stone, B. A. & Fincher, G. B. (1989). Characterization of the hydroxyproline-rich protein core of an arabinogalactan protein secreted from suspension-cultured *Lolium multiflorum* (Italian ryegrass) endosperm cells. *Biochemical Journal* **264**, 857–62.

Grimes, H. D. & Breidenbach, R. W. (1987). Plasma membrane proteins. Immunological characterization of a major 75 kilodalton protein group. *Plant Physiology* **85**, 1048–54.

Hillestead, A., Wold, J. K. & Engen, T. (1977). Water-soluble glycoprotein from *Cannabis sativa* (Thailand). *Phytochemistry* **16**, 1953–6.

Ishii, T., Thomas, J., Darvill, A. & Albersheim, P. (1989). Structure of plant cell walls. 26. The walls of suspension-cultured sycamore cells contain a family of rhamnogalacturonan-I-like polysaccharides. *Plant Physiology* **89**, 421–8.

Hong, J. C., Nagao, R. T. & Key, J. L. (1989). Developmentally regulated expression of soybean proline-rich cell wall protein genes. *Plant Cell* **1**, 937–43.

Keller, B., Templeton, M. D. & Lamb, C. J. (1989). Specific localization of a plant cell wall glycine-rich protein in protoxylem cells of the vascular system. *Proceedings of the National Academy of Sciences, USA* **86**, 1529–33.

Knox, J. P., Day, S. & Roberts, K. A. (1989). A set of cell surface glycoproteins forms an early marker of cell position, but not cell type, in the root apical meristem of *Daucus carota* L. *Development* **106**, 47–56.

Lamport, D. T. A. (1967). Hydroxyproline-*O*-glycosidic linkage in the plant cell wall glycoprotein extensin. *Nature* **216**, 1322–4.

Lamport, D. T. A., Katona, L. & Roerig, S. (1973). Galactosylserine in extensin. *Biochemical Journal* **133**, 125–31.

Lamport, D. T. A. & Miller, D. H. (1971). Hydroxyproline arabinosides in the plant kingdom. *Plant Physiology* **48**, 454–6.

Larkin, P. J. (1977). Plant protoplast agglutination and membrane-bound β-lectins. *Journal of Cell Science* **26**, 31–46.

Lynes, M., Lamb, C. A., Napolitano, L. A. & Stout, R. G. (1987). Antibodies to cell surface antigens of plant protoplasts. *Plant Science* **50**, 225–32.

Miller, D. H., Lamport, D. T. A. & Miller, M. (1972). Hydroxyproline heterooligosaccharides in *Chlamydomonas*. *Science* **176**, 918–20.

Moody, S. F., Clarke, A. F. & Bacic, A. (1988). Structural analysis of secreted slime from wheat and cowpea roots. *Phytochemistry* **27**, 2857–61.

Norman, P. M., Kjellbom, P., Bradley, D. J., Hahn, M. G. & Lamb, C. J. (1990). Immunoaffinity purification and biochemical characterization of plasma membrane arabino-galactan-rich glycoproteins of *Nicotiana glutinosa*. *Planta* **181**, 365–73.

Norman, P. M., Wingate, V. P. M., Fitter, M. S. & Lamb, C. J. (1986). Monoclonal antibodies to plant plasma-membrane antigens. *Planta* **167**, 452–9.

Northcote, D. H., Davey, R. & Lay, J. (1989). Use of antisera to localize callose, xylan and arabinogalactan in the cell plate, primary and secondary walls of plant cells. *Planta* **178**, 353–66.

Pennell, R. I., Janniche, L., Kjellbom, P., Scofield, G. N., Peart, J. M. & Roberts, K. (1991). Developmental regulation of a plasma membrane arabinogalactan protein epitope in oilseed rape flowers. *Plant Cell*, **3**, 1317–26.

Pennell, R. I., Knox, J. P., Scofield, G. N., Selvendran, R. R. & Roberts, K. (1989). A family of abundant plasma membrane-associated glycoproteins related to the arabinogalactan proteins is unique to flowering plants. *J. Cell Biol.* **108**, 1967–77.

Pennell, R. I. & Roberts, K. (1990). Sexual development in the pea is presaged by altered expression of arabinogalactan protein. *Nature* **344**, 547–9.

Roberts, K. (1974). Crystalline glycoprotein cell walls of algae: their structure, composition and assembly. *Proceedings of the Royal Society of London, Series B* **268**, 129–46.

Roberts, K. (1979). Hydroxyproline: its assymetric distribution in a cell wall glycoprotein. *Planta* **146**, 275–9.

Roberts, K. (1989). The plant extracellular matrix. *Current Opinion in Cell Biology* **1**, 1020–7.

Schopfer, P. (1990). Cytochemical identification of arabinogalactan protein in the outer epidermal wall of maize coleoptiles. *Planta* **183**, 139–42.

Sedgely, M., Blesing, M. A., Bonig, I., Anderson, M. A. & Clarke, A. E. (1985). Arabinogalactan-proteins are localized extracellularly in the transmitting tissue of *Nicotiana alata* Link and Otto, an ornamental tobacco. *Micron and Microscopica Acta* **16**, 247–54.

Showalter, A. M. & Varner, J. E. (1989). Plant hydroxyproline-rich glycoproteins. In *Biochemistry of Plants*, vol. 15, ed P. K. Stumpf & E. E. Conn, pp. 485–520. New York: Academic Press.

Smith, J. J., Muldoon, E. P., Willard, J. & Lamport, D. T. A. (1986). Tomato extensin precursors P1 and P2 are highly periodic structures. *Phytochemistry* **25**, 1021–30.

Stacey, N. J., Roberts, K. & Knox, J. P. (1990). Patterns of expression of the JIM4 arabinogalactan-protein epitope in cell cultures and during somatic embryogenesis in *Daucus carota* L. *Planta* **180**, 285–92.

Stafstrom, J. P. & Staehelin, A. (1988). Antibody localization of extensin in cell walls of carrot storage roots. *Planta* **174**, 321–32.

Strahm, A., Amado, R. & Neukom, H. (1981). Hydroxyproline-galactoside as a protein-polysaccharide linkage in a water soluble arabinogalactan peptide from wheat endosperm. *Biochemistry* **20**, 1061–3.

Stuart, D. A. & Varner, J. E. (1980). Purification and characterization of a salt-extractable hydroxyproline-rich glycoprotein from aerated carrot discs. *Plant Physiology* **66**, 787–92.

van Holst, J.-G. & Clarke, A. E. (1986). Organ-specific arabinogalactan-proteins of *Lycopersicon peruvianum* (Mill.) demonstrated by crossed electrophoresis. *Plant Physiology* **80**, 786–9.

Varma, R., Vercellotti, J. & Varma, R. S. (1977). On arabinose as a component of brain hyaluronate. Confirmation by chromatographic, enzymatic and chemical ionization-mass spectrometric analyses. *Biochimica et Biophysica Acta* **497**, 608–14.

Villanueva, M. A., Metcalf, T. N. III & Wang, J. L. (1986). Monoclonal antibodies directed against soybean cells. Generation of hybridomas and characterization of a monoclonal antibody reactive with the cell surface. *Planta* **168**, 503–11.

Yamagishi, T., Matsuda, K. & Watanabe, T. (1976). Characterization of the fragments obtained by enzymatic and alkaline degradation of rice-bran proteoglycans. *Carbohydrate Research* **50**, 63–74.

Yariv, J., Lis, H. & Katchalaski, E. (1967). Precipitation of arabic acid and some seed polysaccharides by glycosylphenylazo dyes. *Biochemical Journal* **105**, 1c.

Ye, Z.-H. & Varner, J. E. (1991). Tissue-specific expression of cell wall proteins in developing soybean tissues. *Plant Cell* **3**, 23–37.

D. BOWLES

Local and systemic signalling during a plant defence response

A principal feature of plant growth is the maximisation of surface area. This arises from the need of a sedentary organism to obtain the full spectrum of nutrients from the environment. One consequence of this survival strategy is increased vulnerability to pathogens and adverse conditions, since the subterranean and aerial boundaries of the organism with the external world will be immense.

Given this immensity and the lack of any specialised surveillance cells equivalent to the mammalian immune system, cells throughout the organism have evolved an ability to recognise foreign from self. The results of these molecular recognition events are reflected at the local site of stimulus perception but, importantly, are also transmitted to distant regions of the plant. There is now good evidence that defence gene expression and changes in the levels of defence-related products such as phytoalexins, callose and lignin are modulated by these local and systemic signalling events.

This chapter reviews research carried out on these topics at the University of Leeds, with particular reference to (1) plant defence responses to parasitic nematodes and (2) the molecular effectors of the wound-response. General literature to 1990, on defence-related proteins in higher plants has been reviewed by Bowles (1990a).

Plant–nematode interactions

One of the plant–nematode systems under study in my laboratory at Leeds University involves the response of potato (*Solanum tuberosum*) plants to potato cyst nematodes (*Globodera* spp.). The potato cultivar Maris Piper carries a single dominant gene for resistance (H1) that is effective against certain pathotypes of *Globodera rostochiensis* (for example, Ro1), but is ineffective against others (for example, Ro2) and against

Society for Experimental Biology Seminar Series 48: *Perspectives in Plant Cell Recognition*, ed. J. A. Callow & J. R. Green. © Cambridge University Press 1992, pp. 123–35.

the closely related *Globodera pallida* (Sidhu & Webster, 1981). We have used the interaction to determine molecular changes associated specifically with either resistance (an incompatible interaction, Maris Piper/Ro1) or susceptibility (a compatible interaction, Maris Piper/Ro2).

To determine stage-specific changes in gene expression, correlated with defined times in the infection process, we have optimised a method for synchronising the system (Bowles *et al.*, 1991). The synchrony of infection is maintained up to approximately 6 days following penetration of the root by the juvenile nematode. This has enabled us to study early local events during the determination of resistance or susceptibility, and also the time course of any systemic events reflected elsewhere in the plant in response to invasion of the root system by the parasite (Hammond-Kosack *et al.*, 1989, 1990; Gurr *et al.*, 1991; Bowles *et al.*, 1991).

Local signalling: establishment of the feeding cell

The cell biology of the interaction between the root and the nematode has been well characterised (Jones & Northcote, 1972; Jones, 1981; Rice *et al.*, 1985; Rumpenhorst, 1984). Juveniles of cyst nematodes (in contrast to those of root knot nematodes) penetrate the root behind the tip, by cutting their way from cell to cell of the epidermis and cortex. On reaching the stele, it is thought that the worm starts actively secreting products into the plant cells. Certainly, modification of the plant cells starts to occur and eventually within several days leads to the establishment of a functional feeding cell system.

For cyst nematodes, this system is called the syncytium and arises from the redifferentiation of existing plant cells into a transfer cell system that links the head of the nematode to the xylem and phloem of the vascular tissue. In a susceptible host plant, the syncytium becomes completed at 5–6 days following invasion and is used by the worm to draw-off nutrients from the plant, i.e. the nematode starts to represent a competing 'sink' for photosynthate. In resistant plants (expressing the H1 gene) the syncytium begins to form but then degenerates.

It would seem that sexual differentiation of cyst nematodes is regulated by nutrient supply: under conditions of ample nutrients, the juvenile differentiates into an egg-bearing female, whereas when nutrients are restricted, males develop. Yield loss arising from male growth and development is negligible relative to that of females.

Interestingly, the natural resistance mechanism involving H1 is particularly effective, since the syncytium degenerates after the juvenile worms lose their locomotory muscles and are committed to remain within the root system. In several other natural resistance mechanisms, resistance is expressed earlier during the invasion process, such as during the initial migration of the worms through the root. Under these circumstances the juveniles merely evacuate the plant tissue only to re-enter subsequently. The cycles of exit and re-entry from these resistance mechanisms can cause severe root damage, which in turn handicaps the plant and leads either directly to eventual yield loss or indirectly to loss caused by secondary infections by fungi and bacteria at the wound-sites.

Clearly, the effectiveness of H1 resistance relies on disruption of the feeding-site at a time when the juvenile nematode becomes committed to remaining within the root. Is it possible to mimic this natural resistance in the design of a novel strategy? To begin to investigate this possibility it is necessary to identify plant genes that are specific to the feeding-site, and in effect are regulated locally by signals originating from the nematode. In practice, the identification of these genes is complicated by the severe limitation of relevant biomass. The syncytium, even when fully functional and supporting the growth of an egg-bearing female, remains a comparatively minute group of cells within the plant root. In my laboratory, we have optimised a strategy involving the polymerase chain reaction (PCR) to overcome this problem, and have succeeded in identifying several plant genes that certainly would seem to be expressed highly specifically within the vicinity of the feeding nematode.

A cDNA library has been constructed from RNA extracted from potato roots infected with *G. rostochiensis* (Ro2), at 21 days following invasion. At this point, the female of cyst nematodes has expanded in size such that the body of the worm is visible on the outside of the root and can be readily removed. Root tissue was harvested at the site of the nematode, and from either side of the infection point: i.e. local and near-systemic tissue. The methods for construction and screening of the library and the characterisation of one plant gene identified by these techniques (PMR1) is described by Gurr *et al.* (1991).

To summarise these results: the expression of PMR1, as judged by Northern analysis, is specific to the local site of the infection, and is expressed only in a compatible interaction. Although the cDNA library was constructed from RNA derived from tissue comparatively late in the infection sequence (21 days), expression of the gene corresponding to PMR1 could be detected within 4 days of invasion; that is, at a very early time in syncytium formation. Expression remained at high levels throughout the colonisation period investigated (4–21 days).

The expression pattern is interesting, since it implies that highly localised changes in plant gene expression can be induced by the nematode. The signals perceived in a compatible interaction, whether they arise directly from the animal or indirectly from nematode-induced changes in the plant cells, can be distinguished from those of an incompatible interaction, or those perceived in a wound-response.

In situ hybridisation, and visualisation of promoter activity using GUS (β-glucuronidase) constructs in transgenics, will be necessary fully to define the cell specificity of expression of PMR1, or indeed that of any of the several additional plant genes that we have subsequently identified in screens of the cDNA library (S. J. Gurr *et al.*, unpublished results). These techniques have been established in other projects in my laboratory (L. Smith *et al.*, unpublished results) and can readily be applied to plant–nematode interactions. As yet, the expression of PMR1 does indeed seem to be highly localised at the infection zone of the root, since mRNA transcripts were not detected by Northern analysis of any other potato plant tissue investigated, including flowers. This contrasts with the expression pattern of an additional gene we have identified, that is found throughout the root system, is not modulated by nematode infection, but is again not found elsewhere in the plant.

Currently we are in the process of constructing promoter/reporter gene fusions to define precisely the cell-specificity of PMR promoter activity. The intention is to then develop novel resistance mechanisms to nematodes, using the PMR1 promoters (and those of the other genes identified) to drive the expression of other genes that we know are able to affect the colonisation of the root by the nematode. This transgenic solution for nematode control has a number of major advantages over classical plant breeding strategies: speed, universality and, importantly, a way around the current problems of natural resistance genes caused by pathotype-switching in the field. Since all cyst nematodes produce an identical structure for their feeding-sites, irrespective of the plant species colonised, it is highly probable that the genes activated in the potato/cyst nematode system will have homologues in other systems of economic importance such as soybean/cyst nematode, sugar beet/cyst nematode and cereal/cyst nematode.

Systemic signalling: a role in acquired resistance?

In the plant–nematode interaction, we have focused on systemic events within the leaves of the root-infected plant. This contrasts with our more detailed analysis of systemic signalling in the wound-response, described below. Using the synchronised method of root infection, we have the opportunity of studying very rapid systemic events: within 6 h of root invasion through to 6 days and then at subsequent times in the later stages of a fully susceptible or resistant response.

We have studied these systemic events in a number of ways, for example analysis of changes in the populations of translatable mRNAs (Hammond-Kosack *et al.*, 1990), and changes in defined defence genes (Hammond-Kosack *et al.*., 1989; S. J. Gurr *et al.*, unpublished results). The full details of each of these analyses can be gained from the cited literature and only certain aspects will be highlighted in this review.

The first point of interest is that the data clearly demonstrate that local infection of a plant root by a nematode leads to a massive and rapid systemic response. Systemic responses to pathogens are rarely discussed, since the vast majority of studies have focused on local events induced within the vicinity of the invading fungi or bacteria. An exception involves analysis of plant–virus interactions, when systemic responses have often been both demonstrated and discussed, particularly with respect to the activation of pathogenesis-related (PR) genes and the role of 'spread factor' proteins in the systemic infection of the plant by the virus (for a review, see Bowles, 1990*a*).

The systemic responses induced by nematodes occur over a wide time-frame: certain events are detectable within 6 h, whilst others occur only 10–12 days into the interaction. Certain systemic responses are highly specific; that is, they reflect (and can be used as markers of) a compatible versus incompatible interaction (Hammond-Kosack *et al.*, 1990). Other systemic responses, such as those involving the accumulation of PR proteins (Hammond-Kosack *et al.*, 1989) and proteinase inhibitor (PI) proteins (S. J. Gurr *et al.*, unpublished results) are clearly general defence responses, since the timing and level of the systemic response would seem to be identical, whether the potato plant is resistant or susceptible.

It would be interesting to determine whether broad-ranged general defences such as these are elicited with similar timing following root infection by other pathogens of potato, such as *Phytophora infestans*. In addition to the effects of nematode challenge, we demonstrated that the PR proteins of potato accumulate in the leaves in response to direct spray application of aspirin, and the PI proteins of potato accumulate in leaves in response to physical injury of the root. This suggests that it is highly probable that the signalling events initiated during the plant–nematode interaction in some way lock into more general signalling pathways that can be triggered by a range of stimuli and, perhaps, a range of pathogens.

Many of the systemic responses detected were transient. For example, the induction of specific translatable mRNAs during the interaction often showed clear differences in timing. Similarly, PI transcripts, detected in Northern analyses of leaf tissues, appeared within 6 h, peaked by 21 h, and had decreased to below detectable levels by 4 days. These data must reflect the dynamic quality of the interaction between the nematode and the plant root: different local signals are perceived at different stages of the invasion and colonisation process and, in turn, different systemic signals lead to a range of responses elsewhere in the plant. It should be noted, however, that these signalling events may have nothing directly to do with the pathogen per se but could equally reflect changes in metabolic balance within the whole organism, such as perturbations of sink–source relations or water deficit.

The wound-response

Physical injury to a plant is known to activate a wide range of genes at the local wound-site, such as those reflecting cellular damage, those involved in tissue repair, and those encoding defence products. However, the wound-response of tomato and potato plants has attracted considerable interest for a different reason: local injury is known to lead to effects of distant, unwounded regions of the plants. The systemic nature of the wound-response to potato was discovered in the early 1970s so that the system has been instrumental in providing working models of the molecular mechanisms underlying systemic signalling for more than 20 years. Although a number of mechanisms have been suggested, and in recent years suggested with increasing regularity, as yet the nature of the systemic signal is unknown.

Ryan and co-workers discovered the response (Green & Ryan, 1972) and their contributions have been extensively reviewed (e.g. Ryan et al., 1985; Ryan, 1987, 1988). The original finding centred on the wound-induced accumulation of PI proteins in leaf tissues of the potato plant. It had been recognised for a number of years prior to the observation that PI proteins are found at high levels in storage organs, such as potato tubers, and seeds of many plant species. Their presence in tissues associated with reproductive capacity is thought to reflect a defence strategy that has evolved to protect vulnerable stages in the life-cycle of the plant from predators, pathogens and pests. Certainly, more recently, PI proteins from a number of sources have been shown to have anti-nutritional qualities and when expressed in vegetative tissues of transformed plants can provide an effective transgenic solution to insect attack (for a review, see Bowles, 1990a).

The original observation showed that the leaves of the potato plant normally did not contain PI proteins corresponding to those known to exist in potato tubers, unless the leaf lamina was wounded. On wounding, PI proteins accumulated both in the wounded leaf and in other unwounded leaves of the plant. The systemic accumulation of PI proteins could be prevented by rapid removal of the wounded leaf, and this observation led Green & Ryan (1972) to coin the phrase 'proteinase inhibitor inducing factor' (PIIF) to describe the mobile signal that could be thought to be responsible for the induction of PIs in unwounded regions of the plants.

The problems incurred by this definition, and the way in which the notion of PIIF has determined the conceptual framework of experimental design for the last 20 years, have been reviewed recently (Bowles, 1990*a,b*; Bowles, 1991). In this review, I shall focus on the results arising from our recent work, and place these data within the context of the different molecular mechanisms suggested to give rise to the systemic wound-response of tomato plants. Our research has primarily used tomato plants as the experimental model for induction of systemic signalling by leaf injury, and this discussion will review these data, and not refer to the nematode induction of PIs described briefly above.

The range of genes responsive to the endogenous systemic signal

Until very recently, study of the wound-response in Ryan's laboratory and in Willmitzer's laboratory focused on the regulation of PI genes. Changes in expression of these genes has been studied in various ways, including promoter activity in transgenics, steady-state mRNA levels in Northern analyses, protein levels in radial immunodiffusion, and determination of PI activity using inhibition of serine proteinases as the assay (for a review, see Bowles, 1990*a*). Thus, the PIs have been used as 'markers' for a positive local or systemic response, and the full range of genes that may be affected, in particular by the systemic signal(s), have not been analysed.

In an attempt to gain some insight into the entire spectrum of genes affected, we compared the changes in translatable mRNAs in locally wounded leaves and systemically responding leaves over a time course from 2 to 24 h (Dalkin & Bowles, 1989). In summary: our data indicated that a number of novel ^{35}S-labelled translation products could be detected in regions distant to the local wound-site. Their appearance was transient, the time course being dependent on each particular gene product. More recently Willmitzer's group have carried out a much more extensive study and used a strategy involving differential screening of a wound-induced cDNA library to identify systemically responsive genes different from those for PIs (L. Willmitzer, personal communication).

Contrasting the wound-response and the bioassays

Comparison of the effects of wounding and elicitor treatment clearly showed that the pattern and timing of systemic changes induced by the leaf injury, differed to those changes induced by the application of PIIF to the plants (Dalkin & Bowles, 1989). This result highlights an inherent problem in current approaches to understanding systemic signalling and deserves some discussion.

In planta, injury to one leaf leads to increased levels of PIs in a different unwounded leaf. As a consequence of the damage to the leaf, a chain of events must be induced that ultimately affect gene expression elsewhere in the plant. Leaves above and below the wounded leaf are known to respond, so the events responsible for induction must (1) move out of the wounded leaf and petiole against the transpiration stream; (2) move up and down the stem; and (3) move into petioles of unwounded leaves with the transpiration stream. This transport route must be quite complex, since it involves movement with and against the transpiration stream, and the precise cellular path that the 'signal(s)' takes is completely unknown, e.g. is it apoplast, symplast or both?

However, when the chemical nature of PIIF is investigated, a bioassay is often used that involves excised plants. The stem is cut above the root system and PIIF (or more defined chemical agents) are applied to the cut surface. The difficulty with the bioassay is that it cannot be used to investigate triggers of the systemic response. The chemicals applied (depending on charge, size, mobility, etc.) are transported into the plant with the transpiration stream and may therefore exert their effects locally on arrival in the leaves. Certainly, the pattern and timing of changes induced in leaves by the application of PIIF much more closely resembled local changes at the wound-site, rather than those induced by the endogenous systemic signal (Dalkin & Bowles, 1989).

An alternative bioassay, again often used to study the molecular triggers of the wound-response (as assayed by PI induction), is the effect of chemicals applied to the leaves. These chemicals may be volatile agents sprayed on to the surfaces or vacuum-infiltrated into the leaf-spaces. Sometimes, spraying one leaf with an agent, e.g. abscisic acid, will lead to a systemic response, i.e. PIs can be detected in unsprayed leaves (Pera-Cortez *et al.*, 1989). Interpretation of this bioassay is again problematic, since it is unknown whether the agent acts locally in the sprayed leaf to set up an endogenous systematic signal or has access to internal transport routes and is carried to the distant leaf where it exerts an effect.

Using bioassays such as these, a number of compounds have been suggested to be the endogenous mobile signal that acts *in planta* during the wound-response. Compounds include, oligosaccharides, abscisic acid and methyl jasmonate (for a review, see Bowles, 1991). But the key question remains: does their ability to trigger the response in the bioassay necessarily mean that the same species function as the mobile signals in the plant during a wound-response? Surely their ability may equally reflect a capacity to mimic or induce events within an endogenous signalling chain that links wounding to gene activation.

Inhibitors of the wound-response

Although the bioassay provides no insight into the nature of endogenous systemic signals, there is no doubt that molecular species assayed as positive in the bioassays do lead to changes in PI gene expression. It has also been shown that chemical agents that inhibit the PI response of plants to elicitors in the bioassay also inhibit the PI response induced by wounding. This was first shown for the effect of aspirin (Doherty *et al.*, 1988) and has since been extended to include other inhibitors of ion transport (Doherty & Bowles, 1990; O'Donnell & Bowles, unpublished results).

Thus, induction of PI activity by wounding, pectic fragments or chitosan could be inhibited if the tomato plants were pretreated with aspirin, or with members of a related series of hydroxybenzoic acids. Interestingly, the same agents were also known to affect K^+ uptake and membrane potential in other systems, and the relative abilities of the different hydroxybenzoic acids to inhibit K^+ uptake closely resembled their ability to prevent the induction of PIs in tomato plants.

It should also be noted that precisely the same specificity is shown for the hydroxybenzoic acids to induce PR proteins in tobacco (for a review, see Bowles, 1990*a*) and most recently, to bind to the putative aspirin receptor in isolated plasma membrane vesicles from tobacco (Klessig, personal communication). The fact that aspirin prevents a plant response leading to PIs, whilst promoting a response leading to PRs, suggests that the two defence systems shared at least one common event, namely, one that can be modulated by aspirin. The event can be thought of as a switch: in one direction the consequences are PIs, in the other direction, the consequences are PRs.

Benzoic acids are thought to exert their effects on plant cells in a number of ways such as through their ability to act as protonophores, their weak acid character, and their inhibition of enzymes such as lipoxygenase. In turn, their effects on ion transport may lead to secondary effects, such as the known correlation between aspirin and ethylene.

This potential complexity arising from the use of aspirin led us to examine the effects of other weak acids and other inhibitors of ion transport (Doherty & Bowles, 1990). To summarise the data: other weak acids had little or no effect, whereas various agents known to influence ATPase activity and intracellular pH were able to inhibit PI induction. In particular, fusicoccin at 10 µM affected the response and at 100 µM completely inhibited PI induction. It should be remembered that all of these data on inhibitions were gained using the bioassay as the means to apply the chemical agent prior to testing its effect on wounding and/or subsequent application of elicitors. The data therefore provide no insights into the systemic signal or molecular events triggering the systemic response. But they do suggest that, at some level, ion transport phenomena are important in the induction of PIs, whether triggered by a wound stimulus or by an oligosaccharide elicitor.

Rapid effects of oligosaccharides on membrane potential

In order to gain insight into direct effects of oligosaccharides on membranes, cell wall fragments known to elicit PIs in the bioassay were studied for their effects on membrane potential, using an intracellular electrode system (Thain *et al.*, 1990). Two sizes of fragments were assayed with degrees of polymerization (d.p.) of 1–7 and 10–20. Both sizes induced substantial and rapid depolarisation of the membrane, which was fully reversible on removal of the elicitors. The rapidity of the depolarisation (i.e. immediate) suggests that the primary site of action of the fragments is the external surface of the plasma membrane. Effects on membrane potential had been shown previously with elicitors of plant–pathogen responses (Pelissier *et al.*, 1986; Low & Heinstein, 1986), although the characteristics of the responses differed, e.g. a considerable lag-phase preceding the effect (Low & Heinstein, 1986).

An alternative means of systemic signalling in the wound-response

A key feature of the systemic wound-response is amplification. This is rarely discussed, yet seems to me to provide an essential route to understanding the nature of the systemic signalling process. A small injury-site on the blade of one leaflet of a plant can lead to PIs throughout the aerial system: other unwounded leaflets of the same leaf, other leaves, petioles and stem.

Comparing size (area) of injury-site relative to leaf-blade area with distance achieved of a positive systemic PI response (amplification) leads to the intriguing observation that a correlation would appear to exist (O'Donnell & Bowles, unpublished results). If the injury-site is large compared to blade area of the leaflet the systemic response is observed in different leaves of the plant. In contrast, if the injury-site is small relative to leaflet area, the systemic response is restricted to other leaflets of the same leaf. These results could suggest that the response requires a 'kick-start'. A certain threshold must be overcome to get the signal out of the leaf, and this is higher than that required to achieve a systemic response amongst leaflets of the same leaf.

A second intriguing observation concerned with amplification is that *in situ* hybridisation using probes to detect PI genes indicates that the expression of these genes is throughout the cells of the systemically responding leaflet, with the exception of the epidermis (Doherty & Bowles, 1990). The epidermis is known to be a distinct symplastic domain. Similarly, both in tomato and in potato plants, workers in my laboratory and elsewhere have observed that no expression of PIs can be detected in the most basal stem internode, nor in the root system of a systemically responding plant (Pera-Cortez *et al.*, 1988; S. J. Gurr *et al.*, unpublished results). Internodes are also known to represent distinct symplastic domains, which have been established by considerable resistance to signal passage at each node.

Is it possible, therefore, that the spread of a systemic response does not involve transport of a mobile signal, but rather the use of a system analogous to that commonly used by primitive multicellular organisms, or epithelial sheets, when intercellular spread arises from electrical coupling of the cells (Mackie, 1965)? Certainly, many of the features of the systemic wound-response are entirely compatible with such a mechanism (first suggested by D. C. Wildon, personal communication; Bowles, 1991). Within this scenario, cellular damage sets up a depolarisation that is propagated through coupling of the cells within symplastic domains. A local heat stimulus is known to be propagated by an electrical induction system and, although it is also known to lead to systemic induction of PIs, is no proof that the signal leading to PIs is the same signal that is electrically propagated (Wildon *et al.*, 1989). However, more recent data do show that physical injury also leads to measurable propagated action potentials and that these are unaffected by cold block treatments that completely abolish phloem transport through a petiole (D. C. Wildon *et al.*, unpublished results).

If the systemic response to wounding were to involve these progressive changes in membrane potential spreading out from the local injury-site, the consequences in terms of gene activation would depend on the responsive capacity and signal transduction pathways of the individual cells that are coupled. This means of coordinating a defence response through the use of a common long-range conduction system such as the phloem parenchyma, with individual cell types determining specific localised events, was discussed in detail by Bowles (1990*a*).

Acknowledgements

Research described in this review has been supported by the AFRC (D.J.B.) and a grant-in-aid from Enichem to D.J.B. and H. J. Atkinson.

References

Bowles, D. J. (1990*a*). Defence-related proteins in higher plants. *Annual Review of Biochemistry* **59**, 873–907.

Bowles, D. J. (1990*b*). Signals in the wounded plant. *Nature* **343**, 314–15.

Bowles, D. J. (1991). Long-range signalling in the wound-response of plants. *Current Biology* **1**, 165–8.

Bowles, D. J., Hammond-Kosack, K., Gurr, S. J. & Atkinson, H. J. (1991). Local and systemic changes in plant gene expression following root infection with cyst nemotodes. In *Biochemistry and Molecular Biology of Plant Pathogen Interactions*, ed. C. J. Smith. Oxford University Press, in press.

Dalkin, K. & Bowles, D. J. (1989). Local and systemic changes in gene expression induced in tomato plants by wounding and by elicitor treatments. *Planta* **179**, 367–78.

Doherty, H. M. & Bowles, D. J. (1990). The role of pH and ion transport in oligosaccharide-induced PI accumulation in tomato plants. *Plant Cell and Environment* **13**, 851–5.

Doherty, H. M., Selvendran, R. R. & Bowles, D. J. (1988). The wound response of tomato plants can be inhibited by aspirin and related hydroxy-benzoic acids. *Physiological and Molecular Plant Pathology* **33**, 377–84.

Green, T. R. & Ryan, C. A. (1972). Wound-induced proteinase inhibitor in plant leaves: a possible defence mechanism against insects. *Science* **175**, 776–7.

Gurr, S. J. & McPherson, M. J. (1991). *PCR-directed cDNA Libraries.* In *Polymerase Chain Reaction: A Practical Approach.* Oxford: Oxford University Press.

Gurr, S. J., McPherson, M. J., Atkinson, H. J. & Bowles, D. J. (1991). Gene expression in nematode-infected plant roots. *Molecular and General Genetics* **226**, 361–6.

Hammond-Kosack, K. E., Atkinson, H. J. & Bowles, D. J. (1989). Systemic accumulation of novel proteins in the apoplast of the leaves of potato plant following root invasion by the cyst nematode *Globodera rostochiensis. Physiological and Molecular Plant Pathology* **35**, 495–506.

Hammond-Kosack, K. E., Atkinson, H. J. & Bowles, D.J. (1990). Changes in abundance of translatable mRNA species in potato roots and leaves following root invasion by cyst nematode *G. rostochiensis* pathotypes. *Physiological and Molecular Plant Pathology* **37**, 339–54.

Jones, M G. K. (1981). Host cell responses to endoparasitic nematode attack: structure and function of giant cells and syncytia. *Annals of Applied Biology* **97**, 353–72.

Jones, M. G. K. & Northcote, D. H. (1972). Nematode induced syncytium – a multinucleate transfer cell. *Journal of Cell Science* **10**, 789–809.

Low, P. S. & Heinstein, P. F. (1986). Elicitor stimulation of the defence response in cultured plant cells monitored by fluorescent dyes. *Archives of Biochemistry and Biophysics* **249**, 472–9.

Mackie, G. O. (1965). Conduction in the nerve-free epithelia of siphonophores. *American Journal of Zoology* **5**, 439–53.

Pelissier, B., Thiband, J. B., Grignon, C. & Esquerre-Tugaye, M. T. (1986). Cell surfaces in plant-microorganism interactions. VII. Elicitor preparations from two fungal pathogens depolarize plant membranes. *Plant Science* **46**, 103–9.

Pera-Cortez, H., Sanchez-Serrano, J. J., Mertens, R., Willmitzer, L. & Prat, S. (1989). Abscisic acid is involved in the wound-induced expression of the PI 2 gene in potato and tomato. *Proceedings of the National Academy of Sciences, USA* **86**, 9851–5.

Pera-Cortez, H., Serrano-Sanchez, J., Rocha-Sosa, M. & Willmitzer, L. (1988). Systemic induction of PI2 gene expression in potato plants by wounding. *Planta* **174**, 84–9.

Rice, S. L., Leadbeater, B. S. C. & Stone, A. R. (1985). Changes in cell structure in roots of resistant potatoes parasitised by potato cyst-nematodes. I. Potatoes with resistance gene H_1 derived from *Solanum ruberosum* ssp. *andigena*. *Physiological Plant Pathology* **27**, 219–34.

Rumpenhorst, H. J. (1984). Intracellular feeding tubes associated with sedentary plant parasitic nematodes. *Nematologica* **30**, 77–85.

Ryan, C. A. (1987). Oligosaccharide signalling in plants. *Annual Review of Cell Biology* **3**, 295–317.

Ryan, C. A. (1988). In *Plant Gene Research, Temporal and Spatial Regulation of Plant Genes*, ed. D. P. S. Verma & R. B. Goldberg, pp. 22–33. Berlin, Heidelberg, New York: Springer-Verlag.

Ryan, C. A., Bishop, P. D., Walker-Simmons, M., Brown, W. E. & Graham, J. S. (1985). Pectic fragments regulate the expression of proteinase inhibitor genes in plants. In *Cellular and Molecular Biology of Plant Stress*, ed. J. L. Key & T. Kosuge, pp. 319–34. New York: Alan R. Liss Inc.

Sidhu, G. S. & Webster, J. M. (1981). The genetics of plant nematode parasitic systems. *Botanical Review* **47**, 387–419.

Thain, J. F., Doherty, H. M., Bowles, D. J. & Wildon, D. C. (1990). Oligosaccharides that induce PI activity in tomato plants cause depolarization of tomato leaf cells. *Plant Cell and Environment* **13**, 569–74.

Wildon, D. C., Doherty, H. M., Eagles, G., Bowles, D. J. & Thain, J. F. (1989). Systemic responses arising from localized heat stimuli in tomato plants. *Annals of Botany* **64**, 691–5.

N. D. READ*, L. J. KELLOCK,
H. KNIGHT AND A. J. TREWAVAS

Contact sensing during infection by fungal pathogens

Summary

Many fungal pathogens use contact sensing of the host surface to achieve successful infection. Contact-mediated responses can be induced and experimentally examined on artificial substrata which mimic physical characteristics of the host surface. The necessary physical and spatial characteristics of surfaces which induce contact-mediated responses are described and classified as either topographical or non-topographical. Topographical signals may be in the form of steps, ridges or furrows and these sometimes require very precise dimensions and/or spacings to induce a fungal response. Non-topographical signals lack this specificity because the fungus recognises only that it is in contact with a surface possessing suitable physical features. Examples of contact-mediated responses include the asymmetric organisation of a cell relative to its substratum, adhesion of a cell to the contact surface, changes in spore surface morphology and accompanying release of enzymes, directional growth of hyphae, and induction of appressorium differentiation. Evidence of these processes being contact-mediated is assessed and possible mechanisms by which fungal contact sensing may operate are discussed. Finally, strategies for controlling plant diseases through an understanding of contact sensing are defined.

Introduction

A large number of eukaryotic cells are sensitive to contact (touch) stimuli which affect their behaviour, growth and morphogenesis in a variety of ways. Numerous examples of contact sensing by animal and plant cells have been reported (see e.g. Braam & Davis, 1990; Curtis & Clark, 1990;

*Author to whom correspondence should be sent.
Society for Experimental Biology Seminar Series 48: *Perspectives in Plant Cell Recognition*, ed. J. A. Callow & J. R. Green. © Cambridge University Press 1992, pp. 137–72.

and references cited therein). Contact sensing also plays an important role in the infection of plants by many fungal pathogens.

A contact-mediated response by a fungal pathogen is a physiological or developmental response to physical features of a host surface. Evidence for a response being contact-mediated requires it to be induced on artificial substrata which mimic physical features of the natural substratum. Reasonably well-characterised examples of contact-mediated responses by filamentous fungal pathogens include: (1) the asymmetric organisation of a cell relative to its substratum; (2) cell adhesion to a contact surface; (3) changes in spore surface morphology and accompanying release of enzymes; (4) directional growth of germ tubes; and (5) induction of appressorium differentiation.

Contact signals identified as being important for fungal infection can be classified as either topographical or non-topographical. Topographical signals may be in the form of steps, ridges or furrows and these sometimes require very precise dimensions and/or spacings to induce a response. Non-topographical signals lack this specificity because topography is not sensed. The fungus recognises only that it is in contact with a surface, although physical features of the latter are important for this response.

In this review, the various artificial substrata used in contact sensing studies are described, evidence for different responses by fungal pathogens being contact-mediated is assessed, possible mechanisms by which these processes may operate are considered, and, finally, strategies for controlling plant diseases through an understanding of fungal contact sensing are outlined.

Artificial substrata for *in vitro* studies of contact sensing

A wide range of artificial substrata providing topographical and non-topographical signals have been used in the investigation of contact sensing by fungal pathogens (Table 1). One might expect that the ideal artifical substratum for these *in vitro* studies should be inert in order to remove the possibility of chemical signals complicating interpretation of the fungus–substratum interaction. However, we know very little about the precise interaction between cells and the surfaces which they are sensing. Close adherence is usually important for the perception of the substratum but the chemical and physical basis of these adhesion phenomena is little understood (Nicholson & Epstein, 1991). Factors such as the hydrophobicity and chemistry of the substratum surface may be very significant in facilitating adhesion and therefore the sensing of surface signals by the fungus (Wynn & Staples, 1981; see also p. 154 below). Thus, cell adhesion may be tied up in a complex way with the various reactions of cells to contact (see Possible mechanisms of contact sensing, below).

Artificial substrata providing topographical signals

The gross surface topography of a leaf is best reproduced artificially as a leaf replica, although fine detail (e.g. of wax ornamentation) is often lost (Wynn, 1976). Making leaf replicas commonly involves making a negative replica of part of the leaf surface with silicone rubber, and then using this replica to form a positive replica of transparent polystyrene, nail varnish or another material (Table 1). The negative replica can be used for the production of multiple positive replicas without apparent damage. A good practice is to discard the first set of positive replicas in order to eliminate any contaminants or chemicals carried over from the leaf surface. Other safeguards can involve making further negative and positive replicas from the first positive replica. Appressorium differentiation in some species (e.g. *Uromyces phaseoli* var. *typica* = *U. appendiculatus*) on leaf replicas can be just as high as on the host leaves themselves (Wynn, 1976).

Oil-collodion membranes possess a highly cratered surface morphology which can be very inductive for appressorium formation in rusts (Dickinson, 1949*b*; Maheshwari *et al.*, 1967*a*; Wynn, 1976; Heath, 1989). They are produced by dissolving collodion in an organic solvent in the presence of a hydrocarbon, such as paraffin oil, and then evaporating off the solvent to produce a thin film which can be floated off on water. The oil droplets produce cratering in the collodion. Wrinkled collodion membranes in the absence of oil, or collodion lightly atomised with a solvent (to produce cratering by partially dissolving the collodion membrane) also induce rust appressoria, but flat collodion membranes do not (Wynn, 1976).

Dickinson (1969) reported that stretching nitrocellulose, rubber or gelatine membranes induced sufficient structural changes to cause rust germ tubes to exhibit directional growth over them. Subsequent experiments by Dickinson (1971, 1972) used stretched nitrocellulose membranes to study a variety of thigmotropic responses. The method of producing these membranes, which possess an oriented polymeric structure, has been described in detail by Dickinson (1974). It involves forming a gelated nitrocellulose membrane on glass from a complex chemical cocktail containing dissolved nitrocellulose. The membrane is then floated off on water and traversed by a jet of vapour from an acetone–ether solvent mixture provided by a 'molecular iron'. This causes strain lines of expansion and contraction in the membrane, with

Table 1. *Artificial substrata used for* in vitro *studies of contact sensing by fungal pathogens*

Non-topographical substrata	Topographical substrata
Polytetrafluoroethylene[39,50,53] (PTFE, Teflon)	Leaf replicas[18,21,27,58] (e.g. polystyrene, nail varnish, Araldite, Araldite coated with carbon-gold)
Polyparaphenylene[48] (Kevlar)	
Polystyrene[29,31,48,50,57,58,63] (e.g. some plastic Petri dishes)	Oil-collodion[2,4-7,11,17-20,24,34,35,43,45,46,52,54,55,60]
Primaria[48,50] (polystyrene with +ve charge)	Stretched membranes[9,10,13,15] (e.g. nitrocellulose, gelatin, rubber)
Polyvinyl chloride[50] (PVC)	Scratched surfaces (e.g. glass[28,37], polystyrene[28,32,58], polyethylene[18,28,59], cellophane[27], cellulose acetate[51], Parafilm[38])
Polyvinyl chloride[47] + polyvinyl acetate (e.g. some plastic coverslips)	Polystyrene replicas of microfabricated silicon wafers[40,41,58]
Paraffin wax[14]	
Polypropylene[48,50] (e.g. some centrifuge tubes)	
Polymethyl methacrylate (PMMA, Acrylic or Perspex)[48]	
Polyethylene[4,11,24,59] (e.g. many plastic bags and plastic gloves)	
Epoxy resin[12,48]	
Aluminium foil[24,51]	
Teflon-coated aluminium[24]	
Polycarbonate[24,38,61] (e.g. nucleopore membranes)	
Polyesters (e.g. Gelbond[51], Mylar[37,48])	

Agarose-coated Gelbond[50]

Polyvinyl formal (Formvar)[29, 56, 30, 62]

Polyvinylpyrrolidone (PVP)-coated Mylar[24]

PVP-coated polycarbonate[38]

Glass[3, 4, 14, 26, 39, 42, 50]

Cellulose[16, 22, 25, 26, 33, 36, 44, 48, 49, 53] (e.g. Cellophane, Visking tubing, dialysis membrane)

Cellulose acetate[51]

Nitrocellulose[1, 4, 8, 11, 18, 22–25, 42, 44, 52] (e.g. collodion)

Agar[41, 44, 50, 51]

Superscript numbers denote references as: 1, Dickinson, 1949a; 2, Dickinson, 1949b; 3, Purdy, 1958; 4, Parberry, 1963; 5, Endo & Amacher, 1964; 6, Maheshwari et al., 1967a; 7, Maheshwari et al., 1967b; 8, Hunt, 1968; 9, Dickinson, 1969; 10, Dickinson, 1970; 11, Woodbury & Stahmann, 1970; 12, Mercer et al., 1971; 13, Dickinson, 1972; 14, Purkayastha & Gupta, 1973; 15, Dickinson, 1974; 16, Abawi et al., 1975; 17, Staples et al., 1975; 18, Wynm, 1976; 19, Heath, 1977; 20, Heath & Heath, 1978; 21, Lapp & Skoropad, 1978; 22, Suzuki et al., 1981; 23, Herr & Heath, 1982; 24, Staples & Hoch, 1982; 25 Suzuki et al., 1982; 26, Kubo et al., 1983; 27, Sotomayor et al., 1983; 28, Staples et al., 1983a; 29, Wolkow et al., 1983; 30, Woloshuk et al., 1983; 31, Kubo et al., 1984; 32, Epstein et al., 1985; 33, Kubo et al., 1985; 34, Hoch et al., 1986; 35, Kubo & Furusawa, 1986; 36, Kubo et al., 1986; 37, Bourett et al., 1987; 38, Epstein et al., 1987; 39, Hamer et al., 1988; 40, Hoch et al., 1987a; 41, Hoch et al., 1987b; 42, Freytag et al., 1988; 43, Heath & Perumalla, 1988; 44, Koch & Hoppe, 1988; 45, Bhairi et al., 1989; 46, Heath, 1989; 47, Howard & Ferrari, 1989; 48, R.C. Howard, personal communication (used in studies on Magnaporthe grisea); 49, Mims & Richardson, 1989; 50, St Leger et al., 1989a; 51, Beckett et al., 1990; 52, Bhairi et al., 1990; 53, Bourett & Howard, 1990; 54, Heath, 1990a; 55, Heath, 1990b; 56, O'Connell & Ride, 1990; 57, St Leger et al., 1990; 58, Allen et al., 1991; 59, Deising et al., 1991; 60, Freytag & Mendgen, 1991; 61, Nicholson & Epstein, 1991; 62, O'Connell & Bailey, 1991; 63, St Leger et al., 1991.

the result that polymers in the nitrocellulose become oriented perpendicular to the jet's path. The methodology for producing these membranes is complicated and has not been employed by other workers.

A range of scratched surfaces has been used to provide topographical cues (Table 1). The scratches can be rather coarse and non-uniform in the dimensions of the ridges and furrows they possess. However, Mylar can be scratched with a microabrasive film to yield uniform submicron ridges and furrows (Bourett *et al.*, 1987).

Perhaps the most elegant, versatile and reproducible approach to mimicking topographical cues of the plant host surface is to use polystyrene replicas of microfabricated silicon wafers (Figs 2, 3, 6, and 7). In this way topographies of precise and predetermined dimensions can be produced in a substratum of known chemistry. Microfabrication can be achieved either by electron beam lithography (Hoch *et al.*, 1987*b*; Allen *et al.*, 1991) or photolithography (Allen *et al.*, 1991; the method used in our laboratory). Depending on which technique is employed, silicon wafers are initially coated with a resist, which is either electron- or photosensitive. As a result of its chemistry, the resist is either cross-linked or depolymerised by the action of the electron beam or light. Electron beam lithography is a maskless technique which involves selectively exposing the resist to an electron beam under computer control in a scanning

---

Fig. 1. Scanning electron micrograph of germ tubes of *Puccinia hordei* (barley brown rust) growing on the surface of a barley leaf. Note directional growth of primary germ tubes perpendicular to cell junctions (g), initiation of short branches at cell junctions (b), and directional growth of these branches along the cell junctions. Also note the formation of appressoria (a1, a2 and a3) over stomata which are arranged in rows. Two of these appressoria (a2 and a3) have differentiated terminally from primary germ tubes, whilst the other (a1) has formed from a side branch. Partially freeze-dried sample prepared for low temperature scanning electron microscopy according to methods described by Beckett & Read (1986) and Jeffree & Read (1991). The bar represents 100 µm.

Fig. 2. Scanning electron micrograph of a germ tube of *Puccinia hordei* growing on a polystyrene replica of a microfabricated silicon wafer. The replica has ridges with a height of 2.0 µm, width of 2.5 µm and spacing of 1.2 µm. Note that the germ tube is growing more-or-less perpendicular to the orientation of the ridges and that the narrow ridge spacing has induced it to form abnormal repetitive swellings along its length. Partially freeze-dried. The bar represents 20 µm.

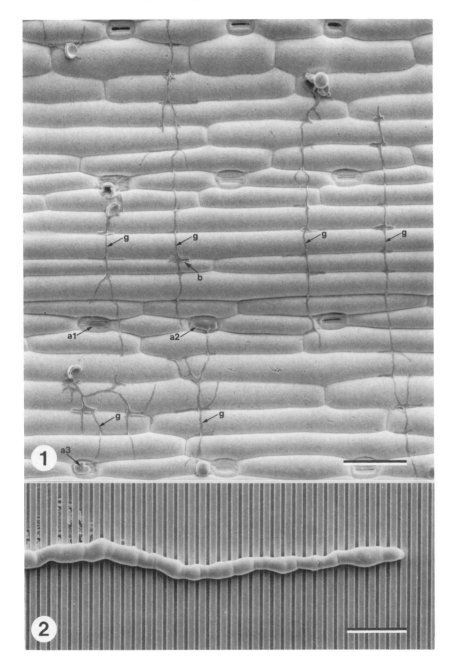

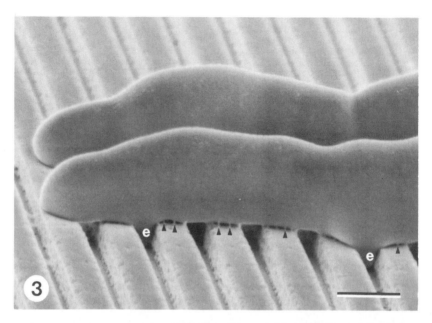

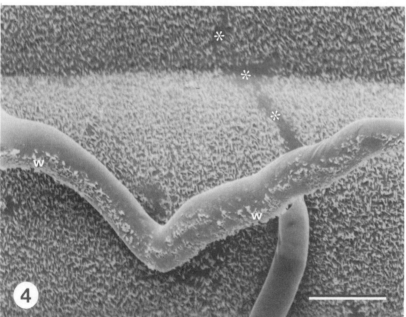

electron microscope to create specific patterns in the resist. Photolithography is basically a contact-printing process in which light is shone through a pattern which has been printed on a metal mask. After exposure to the electron beam or light, the less polymerised resist material is removed from the wafer with suitable solvents. The exposed areas of the wafer are then usually etched (e.g. by ion beam etching) to produce the microfabricated pattern of required depth. Finally, the remaining resist material is removed with suitable solvents.

Electron beam lithography is a high resolution technique allowing details as fine as 3 nm to be defined in the *xy* plane. Photolithography, on the other hand, can achieve a resolution of only 1–2 μm in this plane (e.g. as evident in Figs 2 and 3). Whether the higher resolution possible with electron lithography will be advantageous in studies of fungal contact sensing has not been determined. Both techniques allow for similar resolution in the *z* axis because this is dependent on the extent of etching. Troughs of depths of 0.03–6.7 μm have been obtained as substrata for rust germlings (Hoch *et al.*, 1987*a*; Allen *et al.*, 1991). The microfabricated silicon wafers are usually used repeatedly as templates to produce negative, transparent polystyrene replicas, thus allowing fungi growing on them to be visualised by light microscopy (Figs 5–7).

This account of methods used for wafer microfabrication is very simplified and various other techniques as well as other substrata can be used. For more details of microfabrication methods applied in contact sensing studies the reader is referred to papers by Dow *et al.* (1987) and Hoch *et al.* (1987*b*).

Fig. 3. Scanning electron micrograph of two germ tubes of *Puccinia hordei* growing on a polystyrene replica of a silicon wafer with a topography similar to that in Fig. 2. Note the asymmetric ('nose-down') shape of the hyphae relative to the substratum, extensions of the hypha/cell wall (e) down into some of the troughs and presence of strands of extracellular material (arrowheads) connecting the underside of the hypha with the polystyrene substratum. Partially freeze-dried sample. The bar represents 5 μm.

Fig. 4. Scanning electron micrograph of a germ tube of *Puccinia striiformis* (barley yellow rust) on a barley leaf. Part of the germ tube has been fractured away from the leaf surface and fallen back across the top of the germ tube which it was joined to and which has remained *in situ*. Note that wax has been removed from the leaf surface (*) and can be seen tightly adhered to the underside of the germ tube (w). Partially freeze-dried sample. The bar represents 10 μm.

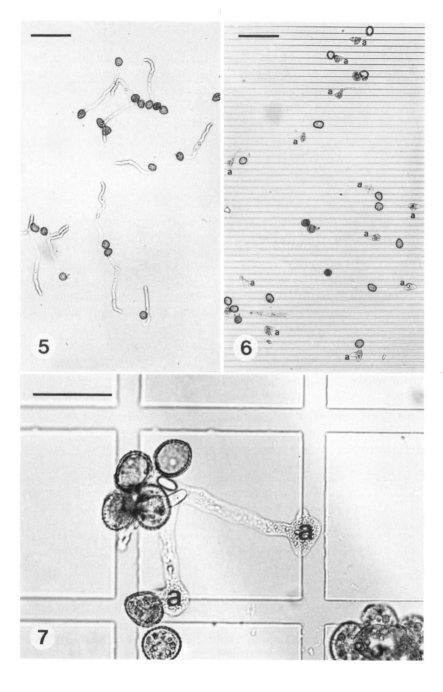

Substrata providing non-topographical contact signals

A large range of substrata lacking topography has been used in contact sensing studies (Table 1) but with variable success. Factors correlated with the efficiency of a substratum to be inductive for a contact-mediated response are hydrophobicity and hardness, although the responses of different species and sometimes even different strains of the same species vary (see p. 154 below).

Controls for *in vitro* studies

Two types of control are commonly used for studies of contact-mediated responses induced by topographical signals on artificial substrata. Either flat surfaces (Fig. 5) or substrata with topographical features which are non-inductive for the contact-mediated response can be employed.

In some applications it can be useful to have one or more controls where the cells are prevented or at least restricted from coming into contact with hard surfaces. One method is to keep the cells in liquid suspension in a 'hanging drop' culture (see e.g. Purkayastha & Gupta, 1973). Alternatively, they can be kept suspended in liquid shake culture, although there can be problems with cells (e.g. spores) sticking to surfaces of the culture vessel. A second approach is to float cells on the surface of water or liquid medium. A third technique is to grow the cells on a soft surface to which the fungus does not respond (e.g. soft agar, St Leger *et al.*, 1989*a*). A fourth method is to simulate the airborne state by trapping cells in microthreads produced by spiders (May & Druett, 1968). This novel approach has been used very effectively by Carver & Ingerson (1987) in studies of contact sensing by germlings of *Erysiphe graminis*.

Figs 5–7. Germlings of *Uromyces vignae* (cowpea rust) on different polystyrene substrata. Fig. 5. Undifferentiated uredospore germlings growing in random directions on a flat surface. The bar represents 100 μm. Fig. 6. Differentiated uredospore germlings growing on a replica of a silicon wafer. The replica has ridges with a height of 0.5 μm, width of 2.5 μm and spacing of 15 μm. Note the pigmented uredospores and that appressoria (a) are formed on the first or second ridge encountered by each germ tube. The bar represents 100 μm. Fig. 7. Young appressoria (a) differentiating over a single step down from plateau. The plateau is 100 μm×100 μm and 0.5 μm high. The plateaus are separated by troughs which are 15 μm in width. The bar represents 50 μm.

Contact-mediated responses

Asymmetric cell organisation and adhesion

A feature of cells which are contact sensing is that they exhibit an asymmetry in their organisation relative to the substratum upon which they reside. Contact sensing cells are often flattened against the substratum. This is particularly apparent in germ tubes of rusts which commonly exhibit a 'nose-down' asymmetry (Fig. 3; Dickinson, 1969; Wynn, 1976; Epstein *et al.*, 1987; Hoch *et al.*, 1987*b*; Koch & Hoppe, 1988). The substratum is frequently closely appressed to the underside of the cell, even to the extent of the cell moulding itself into the fissures and folds of its underlying surface (Fig. 3; Mendgen, 1973; Bourett *et al.*, 1987). Germ tubes are typically tightly adherent to waxes on the leaf surface. This is indicated by wax crystals remaining stuck to the undersides of detached germ tubes (Fig. 4; Lewis & Day, 1972). Intimate contact with the substratum is usually brought about by an extracellular matrix with adhesive properties which is locally secreted at the cell–substratum interface (Figs 3 and 4; Nicholson & Epstein, 1991). The fungal cell wall is sometimes thinner in this region (Littlefield & Heath, 1979; Bourett *et al.*, 1987) or even non-existent (Howard & Ferrari, 1989; Bourett & Howard, 1990; Howard *et al.*, 1991). Within the cell, vesicles, microtubules and F-actin microfilaments are often concentrated at the inner surface of the plasma membrane adjacent to the substratum (Hoch & Staples, 1983*a*,*b*; Hoch *et al.*, 1987*b*; Bourett *et al.*, 1987). Furthermore, microtubules and microfilaments in *U. appendiculatus* have been shown to be predominately orientated parallel with topographies (scratches or ridges) inductive for appressorium formation (Hoch *et al.*, 1986; Bourett *et al.*, 1987).

Changes in spore surface morphology and accompanying enzyme release

Changes in the morphology of the conidium surface of *Erysiphe graminis* occur 10 to 30 min after contact with a barley leaf or a cellophane membrane. The spore surface is initially ornamented with a reticulate network containing spine-like protrusions but then converts to having prominent globose bodies. A thin film flows out from the spore onto the leaf or substratum surface and this occurs concurrently with the loss of the crystal integrity of leaf waxes. The release of esterases accompanies these morphological changes. These processes are believed to be necessary for the successful preparation of the infection court (Kunoh *et al.*, 1988, 1990, 1991; Nicholson *et al.*, 1988).

Directional growth of germ tubes

Johnson (1934) first described the directional growth of germ tubes of *Puccinia graminis* f. sp. *tritici* across wheat leaves and suggested that it might be a thigmotropic response to the parallel arrangement of epidermal cell junctions (anticlinal cell walls). The directional growth of germ tubes at right angles to cell junctions, or possibly other host surface features (e.g. the parallel cuticular ridges of some species) has since been demonstrated in many rust species (Fig. 1; Maheshwari & Hildebrandt, 1967; Lewis & Day, 1972; Wynn, 1976; Pring, 1980; Wynn & Staples, 1981; Mendgen, 1982; Hoch *et al.*, 1987*b*; Lennox & Rijkenberg, 1989; Allen *et al.*, 1991). In many, if not all, cases, directional growth perpendicular to cell junctions reduces wandering and maximises the probability of the rust germ tube making contact with a stoma. This is best exemplified with rusts growing on graminaceous leaves in which stomata are arranged in staggered longitudinal rows (Fig. 1; Johnson, 1934; Lewis & Day, 1972). On dicotyledonous leaves, such as *Phaseolus vulgaris*, the advantage of this type of directional growth is less obvious. The loose concentric circles of epidermal cells surrounding guard cells may provide guidance cues towards stomata (Wynn, 1976). However, the advantages of directional growth on dicotyledonous leaves needs more critical assessment.

As an alternative explanation of the leaf topographical cues which induce directional growth, Lewis & Day (1972) suggested that a regular wax lattice structure on the wheat leaf surface was responsible for causing oriented growth by *P. graminis* f. sp. *tritici*. However, these authors did not produce convincing evidence to support their proposal.

Maheshwari & Hildebrandt (1967) first demonstrated the directional growth of rust germ tubes on artificial substrata. This was achieved using *Puccinia antirrhini* grown on collodion or cellulose acetate replicas of snapdragon leaves. Dickinson (1969) later showed that the germ tubes of six different rust species grew at right angles to straight and repetitive structural features on five chemically different artificial membranes. Using polystyrene replicas of silicon wafers, Hoch *et al.* (1987*b*) demonstrated that bean rust germlings grew perpendicular to ridges spaced between 0.5 and 30 μm apart but germ tube straightness diminished as the spacing of the ridges increased. However, wandering germling growth was usually corrected once the germ tube grew over a ridge. Furthermore, ridges or grooves were equally effective in promoting directional growth. Interestingly, an artificially close spacing of ridges and grooves causes germ tubes of *Puccinia hordei* to produce an abnormal, repetitive series of slight swellings along its length (Fig. 2). This further indicates the intimate signal–response coupling of rust germlings to topographical features.

The primary germ tubes of a number of pathogens have been reported to grow along the anticlinal cell walls of their plant hosts (Flentje, 1957; Flentje *et al.*, 1963; Dodman & Flentje, 1970; Pring, 1980; Gold & Mendgen, 1984; Daniels *et al.*, 1991), although, to our knowledge, this has not been mimicked on artificial substrata.

Some rusts, such as *P. hordei* and *P. graminis* f. sp. *tritici*, show directional growth of branches of the primary germ tube and these branches usually only grow along cell junctions for a short distance (Fig. 1; Johnson, 1934; Lewis & Day, 1972; Lennox & Rijkenberg, 1989). We have found that the branches of germ tubes of *P. hordei* will also exhibit directional growth along 1.2 µm wide and 0.5 µm deep grooves on polystyrene replicas of silicon wafers (N. D. Read & L. J. Kellock, unpublished results). It therefore seems that the primary germ tube and its branches in *P. hordei* exhibit different types of directional growth.

Appressorium formation

For many pathogens, the appressorium is the first infection structure formed as a prerequisite for host colonisation. The appressorium provides a platform, anchored onto the host surface, from which the fungal pathogen can penetrate into the host plant. Depending on the species, or even strain within a species, appressoria may be precisely positioned over stomata, form preferentially over anticlinal walls, or develop at random over the host surface. Fungi forming appressoria can be divided into those which enter via stomata and those which do not (e.g. Wynn & Staples, 1981; Staples & Hoch, 1987). Alternatively, they can be classified according to whether under natural conditions their appressoria are induced by topographical or non-topographical contact signals. The latter classification will be used here, although it is somewhat artificial because appressorium differentiation can involve a complex interaction of thigmotropic, chemotropic and environmental stimuli (Allen *et al.*, 1991).

Appressoria induced by topographical contact signals

Appressoria which form over stomata or anticlinal walls *in vivo* are typically formed in response to specific topographical cues provided by surface features of the host plant. The purpose of this recognition phenomenon is to place the appressorium in an optimal position for penetration.

The production by rust uredospore germlings of appressoria specifically over host stomata can be very efficient (e.g. over 99% in *U. appendi-*

culatus; Wynn, 1976). Dickinson (1949*b*) was the first to show that the formation of these appressoria might be contact-mediated by demonstrating that they could be produced *in vitro* on an oil–collodion membrane. Dickinson's original pioneering study has been much expanded and appressoria of numerous rust species have been induced on a wide range of artificial substrata bearing topographical signals. These include oil–collodion membranes, stretched membranes, leaf replicas, various scratched surfaces, and polystyrene replicas of microfabricated silicon wafers (Figs 6 and 7; Table 1).

As a result of the observation that just a single scratch can induce appressorium formation in bean rust (Staples *et al.*, 1985; Tucker *et al.*, 1986; Bourett *et al.*, 1987), the response of germ tubes to single ridges or steps in polystyrene replicas of silicon wafer templates was investigated. Hoch *et al.* (1987*b*) found that *U. appendiculatus* is able to distinguish minute differences in these artificial topographies. They demonstrated that a ridge 0.5 µm high is optimal for appressorium formation whereas on ridges above (1.0 µm) or below (0.25 µm) this height, appressorium development was significantly reduced. Furthermore, they showed that two abrupt changes in topography were sufficient as an effective inductive stimulus. Thus, appressoria were induced by germ tubes growing onto or off a plateau. A similar phenomenon is illustrated by *Uromyces vignae* (Fig. 7). Hoch *et al.* (1987*b*) also demonstrated that the height and form of the guard cell lip on the leaf of the host (*Phaseolus vulgaris*) was similar to the polystyrene ridge which was optimal for appressorium induction *in vitro*. This supported Wynn's (1976) earlier conclusion that the guard cell lip was probably the specific stimulus for the bean rust appressorium. However, more recently Terhune *et al.* (1991) concluded that other components of *Phaseolus* stomatal architecture, notably the guard cell 'ledge', may also stimulate appressorial formation.

The most comprehensive survey of the precise ridge heights which induce rust appressoria has recently been made by Allen *et al.* (1991) using polystyrene replicas of microfabricated silicon wafers. They analysed the responses of 27 rust species to single ridges spaced 60 µm apart and of specific heights between 0.11 and 6.7 µm. Four broad categories of response were observed: (1) no differentiation on ridges and low differentiation (<3%) on scratched membranes (this category included a number of cereal rusts); (2) differentiation on ridge heights with a distinct optimum range and varied differentiation (13–94%) on scratched membranes (this group included *U. appendiculatus* and *U. vignae*); (3) differentiation on a broader range of ridge heights, even on the highest ridges tested, and high differentiation (66–100%) on scratched membranes; (4) differentiation in loose association with ridges and high differentiation (100%) on scratched membranes. Some species showed a remarkable ability to sense very minute ridge heights. *Puccinia hieracii* (the dandelion rust), for example, formed appressoria over minor flaws or cracks approximately 0.01 μm high or deep in the replica surface.

In some rusts (e.g. the cereal rusts), it has often proved very difficult to induce appressoria on artificial substrata. With *P. graminis* f. sp. *tritici*, for example, leaf and wafer replicas have failed to induce appressoria, although scratched polystyrene has given results varying from <3% to 68% differentiation (Maheshwari *et al.*, 1967a; Woodbury & Stahmann, 1970; Staples *et al.*, 1983a; Allen *et al.*, 1991). Reasons for the largely negative results with wafer replicas may have been because: (1) the stomatal complexes of the wheat leaves were imprecisely reproduced in the polystyrene replicas; (2) the signalling topographies of the leaf surface are poorly mimicked by the simple microfabricated topographies used by Allen *et al.* (1991); or (3) chemotropic or other environmental stimuli are additionally, or even primarily, required for appressorium formation (Staples *et al.*, 1983a; Allen *et al.*, 1991).

A feature of the stomatal complexes of cereal leaves is that they do not possess prominent guard cell lips (Wynn, 1976), suggesting that the latter may not provide the sole signal for appressorium induction. We have evidence that a close spacing of multiple, topographical signals may be important for inducing appressorium formation in *P. hordei*, a cereal rust which was not analysed by Allen *et al.* (1991). We found that this rust will not form appressoria on single ridges spaced 15 or 50 μm apart but will differentiate over ridges with a narrow (1.2 μm) spacing (L. J. Kellock & N. D. Read, unpublished results). This suggests that a repetitive, close spacing of topographical features might be the inductive stimulus *in vivo* such as is characteristically exhibited by the lateral cell junctions associated with the dumb bell-shaped guard cells of cereal leaves (Fig. 1).

A number of rust and non-rust species form appressoria preferentially in the 'valleys' over anticlinal cell walls of leaves (Preece *et al.*, 1967; Wood, 1967; Hunt, 1968; Mercer *et al.*, 1971; Bonde *et al.*, 1976, 1982; Clark & Lorbeer, 1976; Lapp & Skoropad, 1978; Gold & Mendgen, 1984; Allen *et al.*, 1991; Daniels *et al.*, 1991). In some of these species, appressoria have been induced to form in similar locations on leaf replicas (Clark & Lorbeer, 1976; Lapp & Skoropad, 1978; Allen *et al.*, 1991) indicating a response to surface topography. This is further supported by the observation that two rust species which form appressoria over anticlinal walls also develop them in association with ridges on polystyrene replicas of microfabricated wafers (Allen *et al.*, 1991).

The formation of appressoria over anticlinal cell walls suggests that germ tube contact sensing is at a less sophisticated level than in the majority of rusts in which appressoria are formed more specifically over stomata (Staples & Macko, 1980). Overall, the wide range of responses of different pathogens to different topographies also suggests that individual rust contact sensing mechanisms may have evolved to respond to the specific topographical signals offered by the specific host plant(s) which they infect. This in turn may influence the host range of some rust pathogens. Such speculation, however, still requires determination of the precise identity and physical nature of these signals on host plants.

There are a number of cytological, biochemical and genetic consequences of appressorial induction by topographical signals. The timescale of events involved in appressorium thigmodifferentiation by *U. appendiculatus* has been summarised by Staples & Hoch (1988). Morphological changes involve: (1) cessation of forward growth of the germ tube and start of appressorial expansion (<1 min); (2) nuclear migration into the developing appressorium (36 min); (3) initiation of mitosis (44 min); (4) termination of mitosis (56 min); and (5) completion of appressorial septum formation (83 min). DNA replication and differentiation-related gene expression and protein synthesis accompany these morphogenetic changes in *U. appendiculatus* and other rusts (Staples, 1974; Staples *et al.*, 1975, 1984*a*, 1986; Huang & Staples, 1982; Bhairi *et al.*, 1989; Kwon & Hoch, 1989; Deising *et al.*, 1991). Changes in the pattern of wall synthesis must also occur in order to effect the change in shape during the transition from a germ tube to an appressorium.

A detailed discussion of appressorium differentiation is beyond the scope of this review because many of the processes involved are indirect rather than direct responses to contact sensing. Nevertheless, some aspects of appressorium differentiation clearly involve the continued perception of the topographical stimulus. The shape and polarity of the appressorium is frequently influenced by the underlying topography and limited bipolar growth of the differentiating germling often occurs along the inductive feature (Figs 1, 6, 7; Maheshwari *et al.*, 1967*a*; Staples *et al.*, 1983*b*; Bourett *et al.*, 1987; Hoch *et al.*, 1987*b*,*c*; Allen *et al.*, 1991). This is also accompanied by the re-orientation of many microtubules and F-actin microfilaments parallel to the topographic feature (Hoch *et al.*, 1986; Bourett *et al.*, 1987). Thus there remains a close stimulus–response coupling during appressorial formation.

Appressoria induced by non-topographical contact signals

In species which form appressoria on natural and artificial surfaces in the absence of topographical signals, a non-specific contact stimulus is recognised by the fungus. Fungi which respond in this way include

conidial germlings of *Magnaporthe grisea* (Woloshuk *et al.*, 1983; Bourett
& Howard, 1990; Howard *et al.*, 1991), *Colletotrichum* spp. (Staples &
Macko, 1980; Kubo & Furusawa, 1986; Staples & Hoch, 1987; Bhairi *et
al.*, 1990; O'Connell & Bailey, 1991) and *Metarhizium anisopliae* (St
Leger *et al.*, 1989*a*, 1990, 1991). Furthermore, species which tend to form
appressoria in association with cell junctions (see above) commonly also
form them on surfaces lacking topography (e.g. Lapp & Skoropad, 1978;
Gold & Mendgen, 1984; Freytag *et al.*, 1988; Allen *et al.*, 1991). This
further supports the notion put forward above that germ tube contact
sensing in these fungi is less sophisticated than in the stomatal penetrating
rusts.

Two aspects of the substratum which often seem to be important in the
induction of appressoria by non-topographical contact signals are
hydrophobicity and hardness. With some species (e.g. *Met. anisopliae*)
there is a clear correlation between increased appressorium formation
and increased substratum hydrophobicity (St Leger *et al.*, 1989*a*). In
others the correlation is not clearcut. In *Mag. grisea*, for example, appressoria readily form on hydrophobic surfaces such as Teflon or Mylar but
also on cellophane which is hydrophilic (Hamer *et al.*, 1988; Bourett &
Howard, 1990; Howard *et al.*, 1991). However, the same strain does not
develop appressoria on glass which has an intermediate hydrophobicity
(Hamer *et al.*, 1988). That surface hardness might be an important
requirement for contact sensing was recognised by Dey (1933). It is
supported by the observation that several direct penetrating species will
not form appressoria on soft agar but will form them on hard agar or
another hard surface (Staples & Macko, 1980; Freytag *et al.*, 1988; St
Leger *et al.*, 1989*a*). It is possible that different responses to variations in
substratum hydrophobicity and hardness may reflect the nature of the
adhesion process about which, as indicated earlier (see Artificial
substrata for *in vitro* studies of contact sensing, above), we know very
little.

Other possible contact-mediated responses

It is likely that a number of other fungal responses during the infection
process involve contact sensing. Some evidence suggests that the following processes may be wholly or partially contact-mediated:

1. *Directional emergence of germ tubes from uredospores.*
   Uredospore germ tubes commonly emerge from germ pores
   close to, rather than away from, an underlying leaf or artificial surface (N. D. Read, unpublished results).
2. *Directional emergence from spores of germ tubes towards cell
   junctions.* Conidia of *Botrytis squamosa* germinate on the

side closest to anticlinal wall junctions on both onion leaves and onion leaf replicas (Clark & Lorbeer, 1976; Wynn & Staples, 1981).

3. *Directional emergence of infection peg from appressorium.* The infection pegs of stomatal penetrating rusts grow down through stomatal pores whilst pegs of directly penetrating fungi grow specifically into the plant or artificial substratum (Wynn & Staples, 1981).

4. *Haustorial mother cell formation.* In *U. appendiculatus* it has been shown that the single stimulus which induces the appressorium to form also results in the subsequent development of three other infection structures (infection peg, vesicle and infection hyphae; Staples & Hoch, 1988). A fourth infection structure, the haustorial mother cell, forms *in vivo* only when an infection hypha makes contact with a mesophyll cell (Wynn & Staples, 1981; Mendgen, 1982). Whether the signal for this is provided by contact or by a chemical, or even a combination of the two, is not clear. However, haustorial mother cells of several rust species have been produced *in vitro* on oil–collodion membranes (Maheshwari *et al.*, 1967*b*; Heath, 1977, 1989, 1990*a,b*; Heath & Perumalla, 1988) and scratched polystyrene (Deising *et al.*, 1991), suggesting that contact stimuli may be involved.

Although not within the remit of this review, it is noteworthy that zoospores of the oomycete *Pseudoperonospora humuli* have also been reported to recognise stomatal complexes in leaf replicas (Royle & Thomas, 1973).

Possible mechanisms for contact sensing

Basic characteristics of contact sensing

The mechanism of contact sensing by fungal pathogens is unknown. However, it is reasonable to assume that it involves, firstly, the initial reception of a physical signal provided by the inductive surface and, secondly, the mediation of the signal to the biochemical machinery which results in the contact-mediated response. Any proposed mechanism(s) to explain contact sensing need(s) to take account of a number of its features. These include: (1) the ability of a single cell to respond to, and distinguish between, different types of contact signals; (2) rapid signal–response coupling; and (3) a necessity for close adhesion with the contact surface.

Because a range of contact signals can elicit different but very specific responses, even within a single cell, it seems likely that a number of different contact signal transduction pathways exist. For example, there is evidence that rusts can respond to non-topographical and different topographical signals by exhibiting (1) directional emergence of germ tubes from uredospores, (2) asymmetric cell organisation, (3) localised adhesion, (4) directional growth of primary germ tubes, (5) directional growth of germ tube branches, (6) appressorium induction, (7) polarised appressorial expansion, (8) directional emergence of infection pegs, and (9) induction of haustorial mother cells. Some of these diverse responses may be related to the stage of growth or differentiation of the fungus and thus the competence of a cell to transduce a signal at a particular time. However, other contact responses clearly involve the independent yet simultaneous sensing of different signals. From the above list it is clear that rust germ tubes can exhibit this complex behaviour and, to date, it appears that no other fungal pathogen, or possibly even eukaryotic cell, has evolved such sophisticated contact sensing.

A feature of many of these responses is that they can occur very quickly. For example, Staples & Hoch (1988) have shown that a germ tube responds to a topographical stimulus in less than a minute. Precisely how rapidly different types of contact-mediated responses occur needs to be determined precisely.

For contact signal reception to operate, close and intimate adhesion with the inductive surface is required in most, if not all, cases (Fig. 4; pp. 138–9). It has been shown, for example, that adding proteases reduces the adhesion of germ tubes of *U. appendiculatus* to a polystyrene surface and prevent appressorium induction by a thigmo-stimulus. Germination or germ tube growth was unaffected by this treatment (Epstein *et al.*, 1985, 1987). We believe that localised adhesion may have a critical role in contact sensing, because it will provide localised stress at the cell–substratum interface which might be sensed by a mechanoreceptor (or 'touch' receptor).

Nature of the mechanoreceptor

So what could the mechanoreceptor be? There are a number of possibilities but we will speculate about only one. The most appealing idea is that changes in membrane tension are sensed by stretch-activated ion channels (Sachs, 1989) located in the plasma membrane adjacent to the contact surface. The resultant membrane depolarisation may then have a number of rapid effects including activation of other voltage-dependent ion channels and second-messenger systems. Preliminary evidence for the involvement of stretch-activated channels in contact-sensing has recently been obtained from inhibitor studies. St Leger *et al.* (1991) have shown that appressorium formation in *M. anisopliae* is blocked by gadolinium, the most potent blocker known for stretch-activated channels (Yang & Sachs, 1989). To date, the only published reports of these channels in fungi have been in *Saccharomyces cerevisiae* (Gustin *et al.*, 1988) but their existence in filamentous fungi has also been proposed (Wessels, 1990). How such a mechanoreceptor could discriminate between, and effect, a variety of specific responses to different contact signals is not clear. It may be that different stretch-activated channels respond to different types and degrees of stretch. Thus, localised, and sometimes very precise, deformation of the plasma membrane by topographical features may be distinguished from a more general membrane stress by a flat surface to which the cell had adhered. If stretch-activated channels are involved it will be important to know their ion selectivity and whether they are located and/or active in only those regions of the cell involved in contact sensing.

Induction of appressorium formation by chemicals or heat

In vivo, contact signals seem to play the primary role in triggering appressorium differentiation in many plant pathogens. This mode of development has been described as 'thigmodifferentiation' (Staples *et al.*, 1983c). However, *in vitro* certain chemicals and heat shock can also induce appressoria and the terms 'chemodifferentiation' (Staples *et al.*, 1983b) and 'thermodifferentiation' (Deising *et al.*, 1991) have been used to describe these processes.

Chemicals which can, when added exogenously, induce rusts to differentiate appressoria include: various cations, especially K^+, Mg^{2+} and Ca^{2+} (Staples *et al.*, 1983a, 1984b, 1985; Kaminskyj & Day, 1984b; Hoch *et al.*, 1986, 1987a; N. D. Read & K. Taylor, unpublished results); cAMP and cGMP (Hoch & Staples, 1984; Epstein *et al.*, 1989); simple sugars (Kaminskyj & Day, 1984a; Hoch *et al.*, 1987a); GTP (Staples & Hoch, 1988); the dipalmitoyl derivative of phosphatidic acid and several diacylglycerols (Staples & Hoch, 1988); and acrolein (Macko *et al.*, 1978; Staples *et al.*, 1983b). It seems likely that these chemical treatments may, in many cases, have bypassed the initial step of sensing the contact stimulus. As a result they may provide a clue to the second-messenger signalling systems which operate in contact sensing (see below). However, although chemodifferentiation and thigmodifferentiation may share a number of common signal steps, there are also clear differences between the two processes. Appressoria induced by chemicals are often initiated after a considerable lag phase (e.g. 6 h), in contrast to thigmodifferentiation, which results in the formation of a recognisable appressorium within 1.5 h (Hoch & Staples, 1984; Hoch *et al.*, 1987*a*). Secondly, in *U. appendiculatus* at least, appressoria induced by K^+, Ca^{2+} or sucrose are formed aerially away from the substratum (Hoch *et al.*, 1987*a*). The microtubule cytoskeleton in these aerial germlings is symmetrically arranged whilst in germlings grown on a firm substratum it is asymmetrical (Hoch *et al.*, 1987*a*). Other differences have also been noted (Hoch *et al.*, 1986). In addition, not all rust species respond to chemicals in a similar manner (e.g. *P. graminis* f. sp. *tritici* does not differentiate after the addition of K^+; Kaminskyj & Day, 1984*b*). The significance of chemicals for inducing rust appressoria under natural conditions is not clear. There is some evidence that the chemical environment around stomata, including that from volatiles, may provide important cues or modulate contact-mediated appressorium induction in some rusts (Grambow & Reidel, 1977; Staples & Macko, 1980). In other fungi (e.g. various insect pathogens) both chemical and physical stimuli are important for appressorium formation (Magalhaes *et al.*, 1990; St Leger *et al.*, 1989*a*).

Heat shock will usually induce rust germ tubes to differentiate appressoria (Maheshwari *et al.*, 1967*a*; Dunkle *et al.*, 1969; Dunkle & Allen, 1971; Kim *et al.*, 1982; Mendgen, 1982; Staples & Hoch, 1982; Mendgen & Dressler, 1983; Wanner *et al.*, 1985; Hoch *et al.*, 1986; Staples *et al.*, 1989; Bhairi *et al.*, 1990), although apparently not in all species (Boasson & Shaw, 1984; Shaw *et al.*, 1985). The heat shock treatments used have typically taken the form of germ tube growth at 20 °C followed by exposure to 30 °C for 1.5 h and then a return to growth at 20 °C (Staples *et al.*, 1989). The time course for appressorium development in response to heat shock is similar to that for thigmodifferentiation. However, marked differences occur in the pattern of proteins synthesised following these two modes of induction: the contact signal results in the synthesis of a unique set of proteins (the dr-proteins) which are not evident after heat shock; heat shock causes the formation of six unique heat shock proteins (Staples *et al.*, 1989). However, some of the genes for the dr-proteins are expressed after heat shock (Bhairi *et al.*, 1990).

In conclusion, the mechanisms by which the germlings of different rust species are induced to form appressoria in response to contact, chemicals and heat appear to differ, although they exhibit many common features. It seems that rusts have a number of different inducer-responsive pathways (Staples *et al.*, 1989) or a network of interconnecting signal transduction pathways which can be entered from a number of different points.

Signal transduction components involved in contact sensing

There is growing evidence for a multiplicity of signal transduction components and processes operating during contact sensing by plant pathogens. The evidence can be summarised as follows:

1. *Membrane depolarization.* Appressorium initiation in *Metarhizium anisopliae* and *Zoophthora radicans* can be blocked by using inhibitors of Ca^{2+} channels (Magalhaes *et al.*, 1991; St Leger *et al.*, 1991), stretch-activated channels (St Leger *et al.*, 1991) and the plasma membrane ATPase (St Leger *et al.*, 1991). A role for ion transport, and thus possibly membrane depolarisation, during this process is therefore indicated. However, electrophysiological or dye measurements of membrane potential during appressorium induction have not yet been published.

2. *G-proteins.* These guanine nucleotide-binding proteins act as intermediaries in transmembrane signalling by coupling membrane receptors to effector proteins which, in many cases, regulate second messengers (Iyengar & Birnbaumer, 1990). A multiplicity of G-proteins have been isolated from plasma membranes of *M. anisopliae* germ tubes, suggesting that G-proteins may have a role in contact-mediated appressorium formation (St Leger *et al.*, 1989c). In addition, GTP is an effective inducer of chemodifferentiation in *U. appendiculatus* (Staples & Hoch, 1988).

3. *Calcium.* Localised elevations or spatial gradients of free cytosolic Ca^{2+} are important in regulating a wide range of eukaryotic cell processes including polarity, secretion, cytoskeletal organisation and mitosis (Campbell, 1983; Hepler & Wayne, 1985). Some, or all, of these processes feature in contact-mediated responses. External Ca^{2+} is required for appressorium formation in *Z. radicans* (Magalhaes *et al.*, 1991) but not in *M. anisopliae* (St Leger *et al.*, 1990). In *Uromyces*, appressorium differentiation can be induced by exogenous Ca^{2+} (Hoch *et al.*, 1987a; N. D. Read & K. Taylor, unpublished results) and can increase the frequency of chemodifferentiation induced by low levels of K^+ (Staples *et al.*, 1983a; Kaminskyj & Day, 1984b). We are currently imaging and measuring Ca^{2+} in germ tubes of *U. vignae* using fluorescence ratio imaging and photometry

(Gilroy *et al.*, 1991), and laser scanning confocal microscopy (Shotton, 1989). A major problem at present is that all the Ca^{2+}-selective fluorescent dyes which we have loaded into germ tubes using either the ester- (Tsien, 1981) or acid-loading (Bush & Jones, 1988) procedures become compartmentalised into organelles. Dyes we have tested include Fura-2, Indo-1, Fluo-3, Calcium Green and Calcium Orange (P. H. Knight, W. T. G. Allan, A. J. Trewavas & N. D. Read, unpublished results). A major breakthrough will be achieved when it is possible to image and measure cytosolic Ca^{2+} and other ions during the induction of contact-mediated responses.

Calmodulin is generally considered to be the primary Ca^{2+} receptor in eukaryotic cells although this universal view has recently been questioned (Geiser *et al.*, 1991). Nevertheless, in many eukaryotic cells calmodulin, after combination with Ca^{2+}, activates numerous enzymes central to cell regulation. Calmodulin has been identified in a number of contact sensing fungi. It has been isolated and purified from uredospores of *U. appendiculatus* and has been substituted for by bovine calmodulin as an activator of Ca^{2+}-dependent cyclic nucleotide phosphodiesterase (Laccetti *et al.*, 1987). Calmodulin, other calcium-binding proteins and calmodulin target proteins have also been detected in conidia and germ tubes of *M. anisopliae* (St Leger *et al.*, 1989*b*, 1990). A possible role in contact sensing is indicated by the finding that calmodulin antagonists inhibit appressorium formation in *Z. radicans* at concentrations which have no effect on spore germination (Magalhaes *et al.*, 1991). The distribution of calmodulin in germlings and appressoria of *M. anisopliae* seems to be more-or-less uniform, as revealed by localisation with the fluorescent calmodulin probe W-5 (St Leger *et al.*, 1990).

4.   *Inositol phosphates and diacylglyerol.* Inositol 1,4,5-triphosphate ($InsP_3$) and diacylglycerol are produced from the hydrolysis of plasma membrane inositol lipids and act as second messengers in mammalian cells. $InsP_3$ is important in mobilising Ca^{2+} from internal stores, whilst diacylglycerol activates protein kinase C, which phosphorylates specific proteins (see below). Other inositol lipids may also be involved in cell signalling (Berridge & Irvine, 1989). There is preliminary evidence that these second messengers operate

during contact sensing because chemodifferentiation is induced by the addition of either diacylglycerols or the dipalmitoyl derivative of phosphatidic acid (a component of inositol lipid metabolism; Staples & Hoch, 1988).

5. *cAMP and cGMP.* cAMP is a second messenger involved in the control of numerous different processes within fungal and animal cells by regulating cAMP-dependent protein kinases. When added exogenously, cAMP (or cGMP) induces mitosis and septum formation in germ tubes of *U. appendiculatus*. These processes are also induced by phosphodiesterase inhibitors (which block the breakdown of cAMP) and stimulators of adenylate cyclase (which synthesises cAMP; Hoch & Staples, 1984). Three cyclic nucleotide binding proteins which bind either cAMP or cGMP have been detected and one peptide is phosphorylated by either cyclic nucleotide (Epstein *et al.*, 1989). Roles for cyclic nucleotides in appressorium formation are thus suggested.

6. *Cytoskeleton.* It has been proposed that the cytoskeleton in *U. appendiculatus* plays a role in the transmission of a contact signal to the nucleus. So far the evidence for this is only circumstantial. The concentration of microtubules and microfilaments next to the plasma membrane at the fungus–substratum interface, and their reorganisation during appressorium differentiation, suggests a role for the cytoskeleton during this process (Staples & Hoch, 1988). However, these observations may just reflect the localisation of certain cellular activities (e.g. vesicle transport, cell wall synthesis and extracellular secretion) in this region. Nevertheless it has been shown that microtubules must be intact for germ tubes to respond to an inductive thigmostimulus (Hoch *et al.*, 1987*c*).

Disease control through an understanding of contact sensing

Contact sensing has undoubtedly played an extremely important, if not vital, role in the success of many fungal plant pathogens. In particular, the sensing of topographical signals during the prepenetration phase of some pathogens provides the fungus with the ability to discriminate between different surface features of the host in order efficiently to acquire a suitable penetration site without unnecessary energy expenditure. The prepenetration phase represents the stage when the fungus is at its most vulnerable to various control measures which could disrupt fungal contact sensing. Two strategies may be used to reduce or prevent disease:

1. *Targeting specific aspects of contact sensing with fungicides.* Here a molecular understanding of the fundamental processes of signal transduction and adhesion will be particularly important.
2. *Producing resistant plants which lack or form non-optimal contact signals.* An understanding of the nature of the physical signals which stimulate contact-mediated responses may provide a means to modify these features of plant surfaces by host gene manipulation or traditional plant breeding techniques. This will be aided by a knowledge of the precise ways in which plants produce these stimulatory signals.

Acknowledgements

We thank Dr Rick Howard for helpful discussions during the preparation of the manuscript. This work was funded by a grant from the AFRC. L.J.K. was supported by the 1969 Fund from the University of Edinburgh.

References

Abawi, G. S., Polach, F. J. & Molin, W. T. (1975). Infection of bean by ascospores of *Whetzelinia sclerotiorum*. *Phytopathology* **65**, 673–8.

Allen, E. A., Hazen, B. E., Hoch, H. C., Kwon, Y., Leinhos, G. M. E., Staples, R. C., Stumpf, M. A. & Terhune, B. T. (1991). Appressorium formation in response to topographical signals by 27 rust species. *Phytopathology* **81**, 323–31.

Beckett, A. & Read, N. D. (1986). Low-temperature scanning electron microscopy. In *Ultrastructural Techniques for Microorganisms*, ed. H. C. Aldrich & W. J. Todd, pp. 45–86. New York: Plenum Press.

Beckett, A., Tatnell, J. A. & Taylor, N. (1990). Adhesion and pre-invasion behaviour of uredospores of *Uromyces viciae-fabae* during germination on host and synthetic surfaces. *Mycological Research* **94**, 865–75.

Berridge, M. J. & Irvine, R. F. (1989). Inositol phosphates and cell signalling. *Nature* **341**, 197–205.

Bhairi, S. M., Laccetti, L. & Staples, R. C. (1990). Effect of heat shock on expression of thigmo-specific genes from a rust fungus. *Experimental Mycology* **14**, 94–8.

Bhairi, S. M., Staples, R. C., Freve, P. & Yoder, O. C. (1989). Characterization of an infection structure-specific gene from the rust fungus *Uromyces appendiculatus. Gene* **81**, 237–43.

Boasson, R. & Shaw, M. (1984). Further observations on the growth of flax rust fungus in axenic culture. *Canadian Journal of Botany* **62**, 2175–80.

Bonde, M. R., Bromfield, K. R. & Melching, J. S. (1982). Morphological development of *Physopella zeae* on corn. *Phytopathology* **72**, 1489–91.

Bonde, M. R., Melching, J. S. & Bromfield, K. R. (1976). Histology of the suscept–pathogen relationship between *Glycine max* and *Phakopsora pachyrhizi*, the cause of soybean rust. *Phytopathology* **66**, 1290–4.

Bourett, T., Hoch, H. C. & Staples, R. C. (1987). Association of the microtubule cytoskeleton with the thigmotropic signal for appressorium formation in *Uromyces. Mycologia* **79**, 540–5.

Bourett, T. & Howard, R. C. (1990). *In vitro* development of penetration structures in the rice blast fungus *Magnaporthe grisea. Canadian Journal of Botany* **68**, 329–42.

Braam, J. & Davis, R. W. (1990). Rain-, wind-, and touch-induced expression of calmodulin and calmodulin-related genes in *Arabidopsis. Cell* **60**, 357–64.

Bush, D. S. & Jones, R. L. (1988). Measurement of cytoplasmic calcium in aleurone protoplasts using Indo-1 and Fura-2. *Cell Calcium* **8**, 455–72.

Campbell, A. K. (1983). *Intracellular Calcium: Its Universal Role as Regulator*. New York: Wiley.

Carver, T. L. W. & Ingerson, S. M. (1987). Responses of *Erysiphe graminis* germlings to contact with artificial and host surfaces. *Physiological and Molecular Plant Pathology* **30**, 359–72.

Clark, C. A. & Lorbeer, J. W. (1976). Comparative histopathology of *Botrytis squamosa* and *B. cinerea* on onion leaves. *Phytopathology* **66**, 1279–89.

Curtis, A. S. G. & Clark, P. (1990). The effects of topographic and mechanical properties of materials on cell behaviour. *Critical Reviews in Biocompatibility* **5**, 343–62.

Daniels, A., Lucas, J. A. & Peberdy, J. F. (1991). Morphology and ultrastructure of W and R pathotypes of *Pseudocercosporella herpotrichoides* on wheat seedlings. *Mycological Research* **95**, 385–97.

Deising, H., Jungblut, P. R. & Mendgen, K. (1991). Differentiation-related proteins of the broad bean rust fungus *Uromyces viciae-fabae*, as revealed by high resolution two-dimensional polyacrylamide gel electrophoresis. *Archives of Microbiology* **155**, 191–8.

Dey, P. K. (1933). Studies in the physiology of the appressorium of *Colletotrichum gloesporioides. Annals of Botany* **47**, 305–12.

Dickinson, S. (1949a). Studies in the physiology of obligate parasitism.

I. The stimuli determining the direction of growth of the germ tubes of rust and mildew spores. *Annals of Botany* **13**, 89–104.

Dickinson, S. (1949*b*). Studies in the physiology of obligate parasitism. II. The behaviour of the germ-tubes of certain rusts in contact with various membranes. *Annals of Botany* **13**, 219–30.

Dickinson, S. (1969). Studies in the physiology of obligate parasitism. Directed growth. *Phytopathologische Zeitschrift* **66**, 38–49.

Dickinson, S. (1970). Studies in the physiology of obligate parasitism. VII. The effect of a curved thigmotropic stimulus. *Phytopathologische Zeitschrift* **69**, 115–24.

Dickinson, S. (1971). Studies in the physiology of obligate parasitism. VIII. An analysis of responses to a thigmotropic stimulus. *Phytopathologische Zeitschrift* **70**, 62–70.

Dickinson, S. (1972). Studies in the physiology of obligate parasitism. IX. Measurement of the thigmotropic stimulus. *Phytopathologische Zeitschrift* **73**, 347–58.

Dickinson, S. (1974). The production of nitrocellulose membranes as used in the study of thigmotropism. *Physiological Plant Pathology* **4**, 373–7.

Dodman, R. L. & Flentje, N. T. (1970). The mechanism and physiology of plant penetration by *Rhizoctonia solani*. In *Rhizoctonia solani: Biology and Pathology*, ed. J. R. Parmeter, pp. 149–60. Berkeley: University of California Press.

Dow, J. A. T., Clark, P., Connolly, P., Curtis, A. S. G. & Wilkinson, C. D. W. (1987). Novel methods for the guidance and monitoring of single cells and simple networks in culture. *Journal of Cell Science Supplement* **8**, 55–79.

Dunkle, L. D. & Allen, P. J. (1971). Infection structure differentiation by wheat stem rust uredospores in suspension. *Phytopathology* **61**, 649–52.

Dunkle, L. D., Maheshwari, R. & Allen, P. J. (1969). Infection structures from rust uredospores: effect of RNA and protein synthesis inhibitors. *Science* **163**, 481–2.

Endo, R. M. & Amacher, R. H. (1964). Influence of guttation fluid on infection structures of *Helminthosporium sorokineanum*. *Phytopathology* **54**, 1327–34.

Epstein, L., Laccetti, L. & Staples, R. C. (1987). Cell–substratum adhesive protein involved in surface contact responses of the bean rust fungus. *Physiology and Molecular Plant Pathology* **30**, 373–88.

Epstein, L., Laccetti, L., Staples, R. C., Hoch, H. C. & Hoose, W. A. (1985). Extracellular proteins associated with induction of differentiation in bean rust uredospore germlings. *Phytopathology* **75**, 1073–6.

Epstein, L., Staples, R. C. & Hoch, H. C. (1989). Cyclic AMP, cyclic GMP, and bean rust uredospore germlings. *Experimental Mycology* **13**, 100–4.

Flentje, N. T. (1957). Studies on *Pellicularia filamentosa* (Pat.) Rogers.

III. Host penetration and resistance, and strain specialization. *Transaction of the British Mycological Society* **40**, 322–36.

Flentje, N. T., Dodman, R. L. & Kerr, A. (1963). The mechanism of host penetration by *Thanatephorus cumeris*. *Australian Journal of Biological Sciences* **16**, 784–99.

Freytag, S., Bruscaglioni, L., Gold, R. E. & Mendgen, K. (1988). Basidiospores of rust fungi (*Uromyces* species) differentiate infection structures *in vitro*. *Experimental Mycology* **12**, 275–83.

Freytag, S. & Mendgen, K. (1991). Surface carbohydrates and cell wall structure of in vitro-induced uredospore infection structures of *Uromyces viciae-fabae* before and after treatment. *Protoplasma* **161**, 94–103.

Geiser, J. R., van Tuinen, D., Brockerhoff, S. E., Neff, M. M. & Davis, T. N. (1991). Can calmodulin function without binding calcium? *Cell* **65**, 949–59.

Gilroy, S., Fricker, M. D., Read, N. D. & Trewavas, A. J. (1991). Role of calcium in signal transduction of *Commelina* guard cells. *Plant Cell* **3**, 333–44.

Gold, R. E. & Mendgen, K. (1984). Cytology of basidiospore germination, penetration, and early colonization of *Phaseolus vulgaris* by *Uromyces appendiculatus*. *Canadian Journal of Botany* **62**, 1989–2002.

Grambow, H. J. & Reidel, S. (1977). The effect of morphogenically active factors from host and nonhost plants on the in vitro differentiation of infection structures of *Puccinia graminis* f. sp. *tritici*. *Physiological Plant Pathology* **11**, 213–24.

Gustin, M. C., Zhou, X.-L., Martinac, B. & Kung, C. (1988). A mechanosensitive ion channel in the yeast plasma membrane. *Science* **242**, 762–5.

Hamer, J. E., Howard, R. J., Chumley, F. G. & Valent, B. (1988). A mechanism for surface attachment in spores of a plant pathogenic fungus. *Science* **239**, 288–90.

Heath, M. C. (1977). A comparative study of non-host interactions with rust fungi. *Physiological Plant Pathology* **10**, 73–8.

Heath, M. C. (1989). *In vitro* formation of haustoria of the cowpea rust fungus, *Uromyces vignae*, in the absence of a living plant cell. I. Light microscopy. *Physiological and Molecular Plant Pathology* **35**, 357–66.

Heath, M. C. (1990*a*). *In vitro* formation of haustoria of the cowpea rust fungus *Uromyces vignae* in the absence of a living plant cell. II. Electron microscopy. *Canadian Journal of Botany* **68**, 278–87.

Heath, M. C. (1990*b*). Influence of carbohydrates on the induction of haustoria of the cowpea rust fungus *in vitro*. *Experimental Mycology* **14**, 84–8.

Heath, M. C. & Heath, I. B. (1978). Structural studies of the development of infection structures of cowpea rust, *Uromyces phaseoli* var. *vignae*. I. Nucleoli and nuclei. *Canadian Journal of Botany* **56**, 648–61.

Heath, M. C. & Perumalla, C. J. (1988). Haustorial mother cell development by *Uromyces vignae* on collodion membranes. *Canadian Journal of Botany* **66**, 736–41.

Hepler, P. K. & Wayne, R. O. (1985). Calcium and plant development. *Annual Review of Plant Physiology* **36**, 397–439.

Herr, F. B. & Heath, M. C. (1982). The effects of antimicrotubule agents on organelle positioning in the cowpea rust fungus, *Uromyces phaseoli* var. *vignae*. *Experimental Mycology* **6**, 15–24.

Hoch, H. C., Bourett, T. & Staples, R. C. (1986). Inhibition of cell differentiation in *Uromyces* with D_2O and taxol. *European Journal of Cell Biology* **41**, 290–7.

Hoch, H. C. & Staples, R. C. (1983*a*). Ultrastructural organization of the non-differentiated uredospore germling of *Uromyces phaseoli* variety *typica*. *Mycologia* **75**, 795–824.

Hoch, H. C. & Staples, R. C. (1983*b*). Visualization of actin *in situ* by rhodamine-conjugated phalloin in the fungus *Uromyces phaseoli*. *European Journal of Cell Biology*. **32**, 52–8.

Hoch, H. C. & Staples, R. C. (1984). Evidence that cAMP initiates nuclear division and infection structure formation in the bean rust fungus, *Uromyces phaseoli*. *Experimental Mycology* **8**, 37–46.

Hoch, H. C., Staples, R. C. & Bourett, T. (1987*a*). Chemically induced appressoria in *Uromyces appendiculatus* are formed aerially, apart from the substrate. *Mycologia* **79**, 418–24.

Hoch, H. C., Staples, R. C., Whitehead, B., Comeau, J. & Wolf, E. D. (1987*b*). Signaling for growth orientation and cell differentiation by surface topography in *Uromyces*. *Science* **235**, 1659–62.

Hoch, H. C., Tucker, B. E. & Staples, R. C. (1987*c*). An intact microtubule cytoskeleton is necessary for mediation of the signal for cell differentiation in *Uromyces*. *European Journal of Cell Biology* **45**, 209–18.

Howard, R. J., Bourett, T. M. & Ferrari, M. A. (1991). Infection by *Magnaporthe*: an *in vitro* analysis. In *Electron Microscopy of Plant Pathogens*, ed. K. Mendgen & D.-E. Lesemann, pp. 251–64. Berlin, Heidelberg, New York: Springer-Verlag.

Howard, R. J. & Ferrari, M. A. (1989). Role of melanin in appressorium function. *Experimental Mycology* **13**, 403–18.

Huang, B.-F. & Staples, R. C. (1982). Synthesis of proteins during differentiation of the bean rust fungus. *Experimental Mycology* **6**, 7–14.

Hunt, P. (1968). Cuticular penetration by germinating uredospores. *Transactions of the British Mycological Society* **51**, 103–12.

Iyengar, R. & Birnbaumer, L. (1990). *G-proteins*. San Diego: Academic Press.

Jeffree, C. E. & Read, N. D. (1991). Ambient- and low-temperature scanning electron microscopy. In *Electron Microscopy of Plant Cells*, ed. J. Hall & C. Hawes, pp. 313–413. New York: Academic Press.

Johnson, T. (1934). A tropic response in germ tubes of uredospores of *Puccinia graminis tritici*. *Phytopathology* **24**, 80–2.

Kaminskyj, S. G. W. & Day, A. W. (1984*a*). Chemical induction of infection structures in rust fungi. I. Sugars and complex media. *Experimental Mycology* **8**, 63–72.

Kaminskyj, S. G. W. & Day, A. W. (1984*b*). Chemical induction of infection structures in rust fungi. II. Inorganic ions. *Experimental Mycology* **8**, 193–201.

Kim, W. K., Howes, N. K. & Rohringer, R. (1982). Detergent-soluble polypeptides in germinated uredospores and differentiated uredosporelings of wheat stem rust. *Canadian Journal of Plant Pathology* **4**, 328–33.

Koch, E. & Hoppe, H. H. (1988). Development of infection structures by the direct-penetrating soybean rust fungus (*Phakopsora pachyrhizi* Syd.) on artificial membranes. *Journal of Phytopathology* **122**, 232–44.

Kubo, Y. & Furusawa, I. (1986). Localization of melanin in appressoria of *Colletotrichum lagenarium*. *Canadian Journal of Microbiology* **32**, 280–2.

Kubo, Y., Furusawa, I. & Yamamoto, M. (1984). Regulation of melanin biosynthesis during appressorium formation in *Colletotrichum lagenarium*. *Experimental Mycology* **8**, 364–9.

Kubo, Y., Katoh, M., Furusawa, I. & Shishiyama, J. (1986). Inhibition of melanin biosynthesis by ceulenin in appressoria of *Colletrotrichum lagenarium*. *Experimental Mycology* **10**, 301–6.

Kubo, Y., Suzuki, K., Furusawa, I. & Yamamoto, M. (1983). Scytalone as a natural intermediate of melanin biosynthesis in appressoria of *Colletotrichum lagenarium*. *Experimental Mycology* **7**, 208–15.

Kubo, Y., Suzuki, K., Furusawa, I. & Yamamoto, M. (1985). Melanin biosynthesis as a prerequisite for penetration by appressoria of *Colletotrichum lagenarium*: site of inhibition by melanin-inhibiting fungicides and their action on appressoria. *Pesticide Biochemistry and Physiology* **23**, 47–55.

Kunoh, H., Nicholson, R. L. & Kobayashi, I. (1991). Extracellular materials of fungal structures: their significance at prepenetration stages of infection. In *Electron Microscopy of Plant Pathogens*, ed. K. Mendgen & D.-E. Lesemann, pp. 223–34. Berlin, Heidelberg, New York: Springer-Verlag.

Kunoh, H., Nicholson, R. L., Yoshioka, H., Yamaoka, N. & Kobayashi, I. (1990). Preparation of the infection court by *Erysiphe graminis*: Degradation of the host cuticle. *Physiological and Molecular Plant Pathology* **36**, 397–407.

Kunoh, H., Yamaoka, N., Yoshioka, H. & Nicholson, R. L. (1988). Preparation of the infection court by *Erysiphe graminis*. I. Contact-mediated changes in morphology of the conidium surface. *Experimental Mycology* **12**, 325–35.

Kwon, Y. H. & Hoch, H. C. (1989). Nuclear DNA synthesis; time course during appressorium formation in *Uromyces*. *Mycological Society of America Newsletter* **40**, 35 (abstract).

Laccetti, L., Staples, R. C. & Hoch, H. C. (1987). Purification of calmodulin from bean rust uredospores. *Experimental Mycology* **11**, 231–5.

Lapp, M. S. & Skoropad, W. P. (1978). Location of appressoria of *Colletotrichum graminicola* on natural and artificial barley leaf surfaces. *Transactions of the British Mycological Society* **70**, 225–8.

Lennox, C. L. & Rijkenberg, F. H. J. (1989). Scanning electron microscopy study of infection structure formation of *Puccinia graminis* f. sp. *tritici* in host and non-host cereal species. *Plant Pathology* **38**, 547–56.

Lewis, B. G. & Day, J. R. (1972). Behaviour of uredospore germ-tubes of *Puccinia graminis tritici* in relation to the fine structure of wheat leaf surfaces. *Transactions of the British Mycological Society* **58**, 139–45.

Littlefield, L.J. & Heath, M. C. (1979). *Ultrastructure of rust fungi*. New York: Academic Press.

Macko, V., Renwick, J. A. & Rissler, J. F. (1978). Acrolein induces differentiation of infection structures in the wheat stem rust. *Science* **199**, 442–3.

Magalhaes, B. P., Butt, T. M., Humber, R. A., Shields, E. J. & Roberts, D. W. (1990). Formation of appressoria in vitro by the entomopathogenic fungus *Zoophthora radicans* (Zygomycetes: Entomophthorales). *Journal of Invertebrate Pathology* **55**, 284–8.

Magalhaes, B. P., Wayne, R., Humber, R. A., Shields, E. J. & Roberts, D. W. (1991). Calcium-regulated appressorium formation of the entomopathogenic fungus *Zoophthora radicans*. *Protoplasma* **160**, 77–88.

Maheshwari, R., Allen, P. J. & Hildebrandt, A. C. (1967a). Physical and chemical factors controlling the development of infection structures from uredospore germ tubes of rust fungi. *Phytopathology* **57**, 855–62.

Maheshwari, R. & Hildebrandt, A. C. (1967). Directional growth of the uredospore germ tubes and stomatal penetration. *Nature* **214**, 1145–6.

Maheshwari, R., Hildebrandt, A. C. & Allen, P. J. (1967b). The cytology of infection structure development in uredospore germ tubes of *Uromyces phaseoli* var. *typica* (Pers.) Wint. *Canadian Journal of Botany* **45**, 447–50.

May, K. R. & Druett, H. A. (1968). A microthread technique for studying the viability of microbes in a simulated airborne state. *Journal of General Microbiology* **51**, 353–66.

Mendgen, K. (1973). Feinbau der Infektionsstrukturen von *Uromyces phaseoli*. *Phytopathologische Zeischrift* **78**, 109–20.

Mendgen, K. (1982). Differential recognition of the outer and inner walls of epidermal cells by a rust fungus. *Naturwissenschaften* **69**, 502–3.

Mendgen, K. & Dressler, E. (1983). Culturing *Puccinia coronata* on a cell monolayer of the *Avena sativa* coleoptile. *Phytopathologische Zeitschrift* **108**, 226–34.

Mercer, P. C., Wood, R. K. S. & Greenwood, A. D. (1971). Initial infection of *Phaseolus vulgaris* by *Colletotrichum lindemuthianum*. In *Ecology of Leaf Surface Micro-organisms*, ed. T. F. Preece & C. H. Dickinson, pp. 381–9. New York: Academic Press.

Mims, C. W. & Richardson, E. A. (1989). Ultrastructure of appressorium development by basidiospore germlings of the rust fungus *Gymnosporangium junerperi-virginianae*. *Protoplasma* **148**, 111–19.

Nicholson, R. L. & Epstein, L. (1991). Adhesion of fungi to the plant surface: prerequisite for pathogenesis. In *The Fungal Spore and Disease Initiation in Plants and Animals*, ed G. T. Cole & H. C. Hoch, pp. 3–23. New York: Plenum Press.

Nicholson, R. L., Yoshioka, H., Yamaoka, N. & Kunoh, H. (1988). Preparation of the infection court by *Erysiphe graminis*. II. Release of esterase enzymes from conidia in response to a contact stimulus. *Experimental Mycology* **12**, 336–49.

O'Connell, R. J. & Bailey, J. A. (1991). Hemibiotrophy in *Colletotrichum lindemuthianum*. In *Electron Microscopy of Plant Pathogens*, ed K. Mendgen & D.-E. Lesemann, pp. 211–22. Berlin, Heidelberg, New York: Springer-Verlag.

O'Connell, R. J. & Ride, J. P. (1990). Chemical detection and ultrastructural localization of chitin in cell walls of *Colletotrichum lindemuthianum*. *Physiological and Molecular Plant Pathology* **37**, 39–53.

Parberry, D. G. (1963). Studies on graminicolous species of *Phyllachora* Fckl. I. Ascospores – their liberation and germination. *Australian Journal of Botany* **11**, 117–30.

Preece, T. F., Barnes, G. & Bayley, J. M. (1967). Junction between epidermal cells as sites of appressorium formation by plant pathogenic fungi. *Plant Pathology* **16**, 117–18.

Pring, R. J. (1980). A fine structural study of the infection of leaves of *Phaseolus vulgaris* by uredospores of *Uromyces phaseoli*. *Physiological Plant Pathology* **17**, 269–76.

Purdy, L. H. (1958). Some factors affecting penetration and infection by *Sclerotinia sclerotiorum*. *Phytopathology* **48**, 605–9.

Purkayastha, R. P. & Gupta, M. S. (1973). Studies on conidial germination and appressoria formation in *Colletotrichum gloesosporioides* Penz. causing anthracnose of jute (*Corchorus olitorius* L.). *Zeitschrift für Pflanzenkrankeiten und Pflanzenschutz* **80**, 718–24.

Royle, D. J. & Thomas, G. G. (1973). Factors affecting zoospore responses towards stomata in hop downy mildew (*Pseudoperonospora humuli*) including some comparisons with grapevine downy mildew (*Plasmopara viticola*). *Physiological Plant Pathology* **3**, 405–17.

Sachs, F. (1989). Ion channels as mechanical transducers. In *Cell Shape: Determinants, Regulation and Regulatory Role*, ed. W. D. Stein & F. Bonner, pp. 63–92. San Diego: Academic Press.

St Leger, R. J., Butt, T. M., Goettel, M. S., Staples, R. C. & Roberts, D. W. (1989*a*). Production of *in vitro* appressoria by the entomopathogenic fungus *Metarhizium anisopliae*. *Experimental Mycology* **13**, 174–288.

St Leger, R. J., Butt, T. M., Staples, R. C. & Roberts, D. W. (1990). Second messenger involvement in differentiation of the entomopathogenic fungus *Metarhizium anisopliae*. *Journal of General Microbiology* **136**, 1779–89.

St Leger, R. J., Roberts, D. W. & Staples, R. C. (1989*b*). Calcium- and calmodulin-mediated protein synthesis and protein phosphorylation during germination, growth and protease production by *Metarhizium anisopliae*. *Journal of General Microbiology* **135**, 2141–54.

St Leger, R. J., Roberts, D. W. & Staples, R. C. (1989*c*). Novel GTP-binding proteins in plasma membranes of *Metarhizium anisopliae*. *Biochemical and Biophysical Research Communications* **164**, 562–6.

St Leger, R. J., Roberts, D. W. & Staples, R. C. (1991). A model to explain differentiation of appressoria by germlings of *Metarhizium anisopliae*. *Journal of Invertebrate Pathology* **57**, 299–310.

Shaw, M., Boasson, R. & Scrubb, L. (1985). Effect of heat on protein synthesis in flax rust uredosporelings. *Canadian Journal of Botany* **63**, 2069–76.

Shotton, D. (1989). Confocal scanning optical microscopy and its application for biological specimens. *Journal of Cell Science* **94**, 175–206.

Sotomayor, I. A., Purdy, L. H. & Trese, A. T. (1983). Infection of sugarcane leaves by *Puccinia melanocephala*. *Phytopathology* **73**, 695–9.

Staples, R. C. (1974). Synthesis of DNA during differentiation of bean rust uredospores. *Physiological Plant Pathology* **4**, 415–24.

Staples, R. C., App, A. A. & Ricci, P. (1975). DNA synthesis and nuclear division during formation of infection structures by bean rust uredospore germlings. *Archives of Microbiology* **104**, 123–7.

Staples, R. C., Grambow, H.-J. & Hoch, H. C. (1983*a*). Potassium induces rust fungi to develop infection structures. *Experimental Mycology* **7**, 40–6.

Staples, R. C., Grambow, H.-J., Hoch, H. C. & Wynn, W. K. (1983*b*). Contact with membrane grooves induces wheat stem rust uredospore germlings to differentiate appressoria but not vesicles. *Phytopathology* **73**, 1436–9.

Staples, R. C., Gross, D., Tiburzy, R., Hoch, H. C. & Webb, W. W. (1984*a*). Changes in DNA content of nuclei in rust uredospore germlings during the start of differentiation. *Experimental Mycology* **8**, 245–55.

Staples, R. C., Hassouna, S. & Hoch, H. C. (1985). Effect of potassium on sugar uptake and assimilation by bean rust germlings. *Mycologia* **77**, 248–52.

Staples, R. C., Hassouna, S., Laccetti, L. & Hoch, H. C. (1984*b*). Metabolic alterations in bean rust germlings during differentiation by the potassium ion. *Experimental Mycology* **8**, 183–92.

Staples, R. C. & Hoch, H. C. (1982). A possible role for microtubules and microfilaments in the induction of nuclear division in bean rust uredospore germlings. *Experimental Mycology* **6**, 293–302.

Staples, R. C. & Hoch, H. C. (1987). Infection structures – form and function. *Experimental Mycology* **11**, 163–9.

Staples, R. C. & Hoch, H. C. (1988). Preinfection changes in germlings of a rust fungus induced by host contact. In *Biotechnology for Crop Protection*, ed. P. A. Hedin, J. J. Menn & R. M. Hollingworth, pp. 82–93. Washington, DC: American Chemical Society.

Staples, R. C., Hoch, H. C., Freve, P. & Bourett, T. M. (1989). Heat shock-induced development of infection structures by bean rust uredospore germlings. *Experimental Mycology* **13**, 149–57.

Staples, R. C. & Macko, V. (1980). Formation of infection structures as a recognition response in fungi. *Experimental Mycology* **4**, 2–16.

Staples, R. C., Macko, V., Wynn, W. K. & Hoch, H. C. (1983*c*). Terminology to describe the differentiation response by germlings of fungal spores. *Phytopathology* **74**, 380.

Staples, R. C., Yoder, O. C., Hoch, H. C., Epstein, L. & Bhairi, S. (1986). Gene expression during infection structure development by germlings of the rust fungi. In *Biology and Molecular Biology of the Plant–Pathogen Interactions*, ed. J. Bailey, pp. 331–41. London: Springer-Verlag.

Suzuki, K., Furusawa, I., Ishida, N. & Yamamoto, M. (1981). Protein synthesis during germination and appressorium formation of *Colletotrichum lagenarium* spores. *Journal of General Microbiology* **124**, 61–9.

Suzuki, K., Furusawa, I., Ishida, N. & Yamamoto, M. (1982). Chemical dissolution of cellulose membranes as a prerequisite for penetration from appressoria of *Colletotrichum lagenarium*. *Journal of General Microbiology* **128**, 1035–9.

Terhune, B. T., Allen, E. A., Hoch, H. C., Wergin, W. P. & Erbe, E F. (1991). Stomatal ontogeny and morphology in *Phaseolus vulgaris* in relation to infection structure initiation by *Uromyces appendiculatus*. *Canadian Journal of Botany* **69**, 477–84.

Tsien, R. Y. (1981). A non-disruptive technique for loading calcium buffers and indicators into cells. *Nature* **290**, 527–8.

Tucker, B. E., Hoch, H. C. & Staples, R. C. (1986). The involvement of

F-actin in *Uromyces* cell differentiation: the effects of cytochalasin E and phalloidin. *Protoplasma* **135**, 88–101.

Wanner, R., Forster, H., Mendgen, K. & Staples, R. C. (1985). Synthesis of differentiation-specific proteins in germlings of the wheat stem rust fungus after heat shock. *Experimental Mycology* **9**, 279–83.

Wessels, J. G. H. (1990). Role of cell wall architecture in fungal tip growth generation. In *Tip Growth in Plant and Fungal Cells*, ed. I. B. Heath, pp. 1–29. San Diego: Academic Press.

Wolkow, P. M., Sisler, H. D. & Vigil, E. L. (1983). Effects of inhibitors of melanin biosynthesis on structure and function of appressoria of *Colletotrichum lindemuthianum*. *Physiological Plant Pathology* **22**, 55–71.

Woloshuk, C. P., Sisler, H. D. & Vigil, E. L. (1983). Action of the antipenetrant, tricyclazole, on appressoria of *Pyricularia oryzae*. *Physiological Plant Pathology* **22**, 55–72.

Wood, R. K. S. (1967). *Physiological Plant Pathology*. Oxford: Blackwell Scientific Publications.

Woodbury, W. & Stahmann, M. A. (1970). Role of surface films in the germination of rust uredospores. *Canadian Journal of Botany* **48**, 499–511.

Wynn, W. K. (1976). Appressorium formation over stomates by the bean rust fungus: response to a surface contact stimulus. *Phytopathology* **66**, 136–46.

Wynn, W. K. & Staples, R. C. (1981). Tropisms of fungi in host recognition. In *Plant Disease Control: Resistance and Susceptibility*, ed. R. C. Staples & G. H. Toenniessen, pp. 45–69. New York: Wiley-Interscience.

Yang, X.-E. & Sachs, F. (1989). Block of stretch-activated ion channels in *Xenopus* oocytes by gadolinium and calcium ions. *Science* **243**, 1068–71.

N. A. R. GOW, B. M. MORRIS AND B. REID

The electrophysiology of root–zoospore interactions

Introduction

Plant roots generate electrical currents and voltages in the rhizosphere that may influence the behaviour of the many pathogenic, symbiotic or commensal microorganisms that live in association with them. These electrical currents represent circulations of protons and other ions. Consequently, they also lead to the creation of substantial ionic and pH gradients whose affect on the infection and colonisation of the root is only now being explored. Here, we summarise briefly what is known about the electrical currents of plant roots and discuss the ways in which they may influence the root microflora. In particular, we focus on the swimming zoospores of *Phytophthora* species, which are exquisitely sensitive to electrical fields and may target their host roots using a combination of chemotaxis and electrotaxis.

Growth and electricity

It has been known for many years that plants generate electrical currents (Müller-Hettlingen, 1883; Lund & Kenyon, 1927; Lund, 1947). These were first measured using microelectrodes inserted into cells at different regions of a root or tissue. Voltage differences were found between different sites and electrical current was presumed to flow through the cells between the electrodes and through the extracellular medium to complete the circuit. Endogenous currents in the extracellular loop of the circuit can now be measured directly, without invading cells with intracellular microelectrodes. In the 1970s ultrasensitive, voltage-sensing vibrating electrodes were devised that are capable of detecting the minute electrical fields generated by individual cells or tissues (Jaffe & Nuccitelli, 1974). These have since been used to map electrical fields and currents

Society for Experimental Biology Seminar Series 48: *Perspectives in Plant Cell Recognition*, ed. J. A. Callow & J. R. Green. © Cambridge University Press 1992, pp. 173–92.

around a wide variety of microbial, animal and plant cells (Jaffe, 1981; Gow, 1989; Nuccitelli, 1990). These vibrating electrodes are moved around a specimen with a micromanipulator to build an electrochemical profile, of high spatial precision, of the undisturbed cells. Primary interest in these bio-electric currents has centred on their physiological roles, principally on the hypothesis that they serve to establish the polar axis for growth of a cell or tissue such as a root (Jaffe, 1981; Nuccitelli, 1990). However, they have also been shown to be important in wound healing (Rajnicek *et al.*, 1988; Hush & Overall, 1989) and are frequently part of the normal homeostatic or maintenance activities of cells with no special consequence to morphogenesis (Waaland & Lucas, 1984; Schreurs & Harold, 1988; Youatt *et al.*, 1988). Here, we describe another aspect of bio-electrical currents, with an ecological rather than physiological perspective; namely, that the electrical current of an organism may affect the growth, movement or activities of others that live in close physical association with it.

Plant roots as batteries

Vibrating microelectrodes and other microelectrode systems have been used to detect and map electrical currents around the roots of at least 19 species of plant (Table 1). In most cases the current pattern is remarkably similar, with positive electrical current entering the meristematic tissue and region of cell elongation at the root tip and exiting at the mature tissues at the rear (Fig. 1). In a few cases outward current is found around the root cap (e.g. Miller & Gow, 1989*b*; Iwabuchi *et al.*, 1989) and inward electrical current is found at sites of root wounds (Miller *et al.*, 1988; Hush & Overall, 1989). There are, however, reports of outward currents along most of the region of cell elongation of roots of *Zea mays* and *Lepidium sativum* (Behrens *et al.*, 1982; Bjorkman & Leopold, 1987*a,b*; Iwabuchi *et al.*, 1989). The regions of cell division and expansion have only been measured directly in a few cases and some of these differences may reflect difficulties in defining accurately the borders between zones of growth and maturity.

The current pattern is a dynamic and labile property of roots that varies according to the plant species and individual root. External pH (Fig. 2; Miller & Gow, 1989*b*), the source of combined nitrogen (Miller *et al.*, 1991), and the presence of growth regulators and toxins (Miller & Gow, 1989*b*; Rathore *et al.*, 1990) also modulate the pattern and magnitude of the current.

In *Z. mays* and *Hordeum vulgare* the current is carried mainly by protons (Weisenseel *et al.*, 1979; Miller & Gow, 1989*b*). In these species

Table 1. *Plant species for which root currents have been measured*

Plant species	Reference
Arachis hypogaea	Miller & Gow, 1989a
Avena sativa	Miller & Gow, 1989a
Dalbergia nigra	Miller & Gow, 1989a
Fragaria vesca	Miller & Gow, 1989a
Hordeum vulgare	Weisenseel et al., 1979; Miller et al., 1991
Lepidium sativum	Behrens et al., 1982; Iwabuchi et al., 1989
Lolium perenne	Miller & Gow, 1989a
Lonchocarpus leucanthus	Miller & Gow, 1989a
Neptunia plena	Miller & Gow, 1989a
Nicotiana tabacum	Miller et al., 1988
Phaseolus chrysanthus	Iiyama et al., 1985[a]; Toko et al., 1987[a]; Ezaki et al., 1988[a]; Toko & Yamafuji, 1988[a]; Toko et al., 1989[a]
Picea abies	Miller & Gow, 1989a
Pisum sativum	Miller & Gow, 1989a; Hush & Overall, 1989
Rathanus sativus	Rathore et al., 1990
Solanum tuberosum	Miller et al., 1989a
Trifolium repens	Miller et al., 1986; Miller et al., 1991
Triticum aestivum	Miller & Gow, 1989a
Vigna radiata	Miller & Gow, 1989a
Zea mays	Bjorkman & Leopold, 1987a, b; Bjorkman, 1989; Miller & Gow, 1989b; Miller, 1989

[a]These studies made use of a parallel series of microelectrodes. All other studies used vibrating microelectrodes.

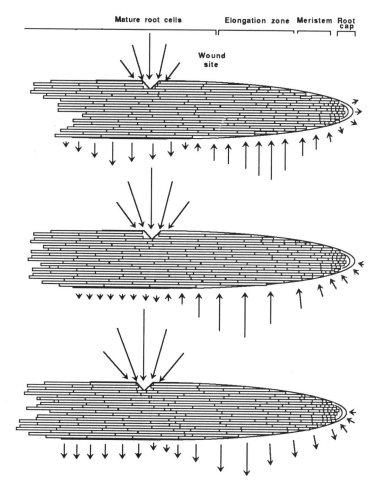

Fig. 1. Patterns of electrical current flow around plant roots as described in the text. In the upper two roots the meristematic region and zone of cell elongation are sites of inward current (see e.g. Weisenseel *et al.*, 1979; Miller & Gow, 1989*b*; Rathore *et al.*, 1990), while in the lowest root this is a site of outward current (see e.g. Bjorkman & Leopold, 1987*a*). The current at the root cap can vary from outward (upper root) to inward (lower two roots) for different root species or the same root under different conditions (see Miller & Gow, 1989*b*). Currents at the sites of wounds are inward (Miller *et al.*, 1988; Hush & Overall, 1989).

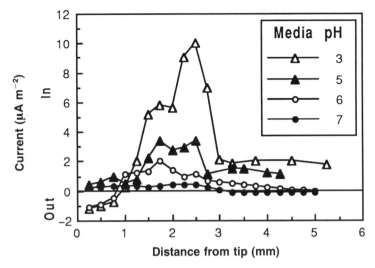

Fig. 2. Effect of medium pH on the electrical current density of a root of *Zea mays*. A single root was used to generate all four current profiles. (Adapted from Miller & Gow, 1989*b*.)

and in *Trifolium repens* (Miller *et al.*, 1991) the proton current has also been shown to establish spatial gradients of pH along the root (Fig. 3). The extracellular medium at zones of proton entry is rendered relatively alkaline, while regions of proton exit are made acidic.

The physiological functions of the ionic currents of roots have not been established unequivocally. The first possible role is that they serve to orient the axis of cell polarity. The evidence for this is essentially correlative and it has proved difficult to separate the causes and consequences of ionic currents from polarized development. It is, however, interesting that ionic currents have also been measured in developing plant embryos when the polar axis is forming (Brawley *et al.*, 1984; Overall & Wernicke, 1986; Rathore & Robinson, 1989) and that applied electrical fields can induce root/shoot differentiation and stimulate growth of plant callus cells (Rathore & Goldsworthy, 1985*a,b*; Rathore *et al.*, 1988). It has been suggested that the electrical currents may cause growth regulators such as indole-3-acetic acid to be transported basipetally, resulting in stimulated excretion of protons from the basal cells (Raven, 1979). In this scheme of things the effects of currents and growth are linked together in a positive feedback loop. Whatever the mechanism, it is clear that there is a tight correlation between the morphological and electrical polarity of the root.

A second physiological role for electrical currents is in gravity sensing.

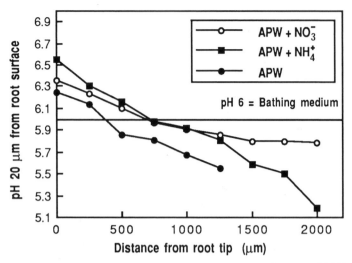

Fig. 3. Profiles of extracellular pH mapped around roots of *Trifolium repens* using a pH microelectrode. The roots were grown in artificial pond water (APW), or in APW supplemented with 1 mM KNO₃ or APW plus 1 mM NH₄Cl. (Adapted from Miller *et al.*, 1991.)

When roots are placed horizontally the cells on the upper side of the root expand preferentially to reorient the root downwards. The mechanism of signal transduction is complex and multifactorial, but it is clear that the outward current at the columella and root cap regions of a horizontally oriented root is displaced to the upper side prior to the initial gravitropic response (Behrens *et al.*, 1982; Bjorkman & Leopold, 1987a,b). Again, there is a correlation between the pattern of current flow and the pattern of growth.

Electrochemistry of the rhizosphere

The growing root generates electrical currents, voltage and pH gradients within the rhizosphere as a consequence of spatial differences in the arrangement of electrogenic proton transport systems in the cells of the various root tissues. Electrical current is defined as the direction of flow of positive charge, so sites of current entry are cathodic and regions around a root where current is outward are anodic. From the discussion above it is clear that the region of elongation can be considered cathodic, anodic or both, according to the root type and the growth conditions.

The following data describe quantitatively the electrochemical profile of roots. Current densities measured around roots (Table 1) vary from 0.1

to 10 µA cm^{-2}, with a mode of around 1 µA cm^{-2} at the region of cell elongation. Values at the upper end of this range are found at sites of wounds.

Gradients of pH are generated in the rhizosphere as a consequence of the electrochemical circulation of protons. Microelectrode experiments have revealed longitudinal gradients of more than 0.5 pH units (mm root)$^{-1}$ (tip alkaline; Fig. 3).

The resistivity of soils is proportional to the salt content and this varies enormously (Hesse, 1971). However, assuming a resistivity of 5000 Ω cm, typical of many garden soils, loams and clays, one can calculate an expected voltage gradient of between 0.5 and 50 mV cm^{-1}, with a modal field of 5 mV cm^{-1}, at the elongation zone and 20–50 mV cm^{-1} at wound sites. The voltage gradients could easily be an order of magnitude less in a salty soil on a soft rock formation, or more, in waterlogged soil on a hard-rock bed. The dryness of a soil will also influence the voltage gradient in the rhizosphere by affecting the thickness of the water films around the roots and soil particles. For any given soil–water conductivity and assuming root current density to be constant, a dry soil, where the conductive pathways are constrained, would support a larger voltage gradient than a wet one.

It should be noted that these values are recorded at the root surfaces of intact, uncolonised roots in artificial media of resistivities of several thousand Ω cm. We have little information on the extent to which the current density and voltage fall-off at increasing distances from the root, the consequences of mycorrhizal or other root infections, etc. However, if microorganisms are influenced by root-generated electrical fields, they need to be sensitive enough to respond to electrochemical gradients of the order of those described above. Cellular responses to electric fields are for the most part voltage- rather than current-dependent (see discussion by Jaffe, 1986), so the size of the voltage gradient, rather than the current density, is paramount when considering root–microbe interactions. These points will be returned to when we discuss the design of experiments to investigate zoospore electrotaxis.

Zoospore targeting to the root

A wide range of organisms including bacteria, fungi, protozoa and nematodes live in the rhizosphere of plant roots. Some, such as the mycorrhizal fungi, form symbiotic associations with roots. Others, such as the members of the genus *Phytophthora*, tend to be obligate, or near obligate pathogens. The swimming zoospores of *Phytophthora* spp. frequently invade the plant via the root tips, at the base of root hairs, sites of wounds and emerging lateral roots or through the stomata of leaves.

There are considerable advantages to being able to find these sites. The cells in the meristem and zones of elongation and wounding exude nutrients that allow the fungus to build up their inoculum potential. They are also more readily penetrated because their outer walls are thinner and more labile to enzymic attack or because wounding has broken their integrity. These sites are the preferred infection court of many pathogens and symbionts of plant roots and it is, therefore, important to gain an understanding of the local cues that enable microoganisms to detect and swim towards them.

Zoospores are short-lived cells with only limited energy reserves to enable them to swim to their target. They have therefore evolved a range of tactic responses to potentiate their site-selection capabilities. They are chemotactic to the nutrients that are exuded from root tips and wounds. In addition, some can respond to light (phototaxis), water currents (rheotaxis) and negatively to gravity (geotaxis) (Carlile, 1983, 1986). Rheotaxis and geotaxis are the physical consequence of the geometry and centre of gravity of the zoospore and do not involve sensing the environment (Cameron & Carlile, 1977). The mechanism of chemotaxis is not known, but it is clear that this phenomenon is non-specific with respect to the plant species. Zoospores of *Pythium* and *Phytophthora* spp. often are attracted equally to host and non-host roots and zoospores of saprophytic oomycetes and chytridiomycetes often accumulate around plant roots as readily as those of pathogenic zoosporic fungi (for a review, see Deacon, 1988). The specific chemotaxis of zoospores of *Phytophthora cinnamomi* to the roots of its host, avocado, is a notable exception to this rule (Zentmyer, 1961). However, for most zoosporic pathogens, the specificity of the fungus–plant interaction does not seem to occur at the swimming stage. Instead, tactic responses seem to be concerned with the finding of roots, not the identification of their nature.

Zoospore electrotaxis and pH taxis *in vitro*

First reports of electrotaxis

An additional zoospore guidance mechanism to those discussed above is electrotaxis. Since plant roots generate electrical fields, it has been suggested that they may use electrotaxis as well as chemotaxis in locating them (Troutman & Wills, 1964; Miller *et al.*, 1986, 1988).

The influence of applied electrical fields on the movements of zoospores has been reported several times (Table 2; Troutman & Wills, 1964; Katsura *et al.*, 1966; Ho & Hickman, 1967; Khew & Zentmyer, 1974). Two of these reports demonstrated zoospore accumulation at the cathode and one at the anode (Table 2). Two demonstrated immobilisa-

Table 2. *Published studies of Phytophthora zoospore electrotaxis*

Zoospore species	Field strength	Current or current density	Response	Electrode type	Reference
P. parasitica	~10 V cm^{-1a}	10–40 μA cm^{-2a}	Cathodic accumulation	Paper wicks	Troutman & Wills, 1964
P. capsici	2 V cm^{-1}	nd	Cathodic accumulation; anodic repulsion	Platinum wire	Katsura et al., 1966
P. megasperma	nd	0.1–0.15 μA	No preferential accumulation; cathodic immobilisation	Platinum wire	Ho & Hickman, 1967
P. cactorum P. capsici P. cinnamomi P. citrophthora P. megasperma P. palmivora P. parasitica	≤1.2 V cm^{-1}	(i) <0.5 μA (ii) >0.5 μA (iii) >0.5 μA	(i) anodic accumulation (ii) anodic repulsion (iii) cathodic immobilization	Platinum wire	Khew & Zentmyer, 1974

nd, note determined.
[a]Calculated from data in text.

tion and trapping of the zoospores around the cathode and active repulsion from the anode. These studies preceded the invention of vibrating microelectrodes that have allowed quantification of the magnitude of root electrical fields. The chosen field strengths in these studies are consequently far greater than the natural fields around roots.

Troutman & Wills (1964) mention that the swimming behaviour observed was not due to passive electrophoresis, but they provided no data or explanation of their conclusion. Khew & Zentmyer (1974) demonstrated that swimming and encysted zoospores were negatively charged, but they were not able to quantify the electromobility or the contribution that electrophoresis may have made to the electrotactic behaviour that they observed.

The most serious difficulties in interpreting these data relate to (1) the use, in three cases, of naked wire electrodes that would expose the zoospores to the products of electrolysis, (2) application of electric field strengths that are one or two orders of magnitude greater than those found around roots, and (3) the absence of quantitative data relating to the passive electromobility of the zoospore.

Electrotaxis under physiological conditions

We have undertaken a careful examination of zoospore electrotaxis using *Phytophthora palmivora* as a model system. In preliminary experiments we reproduced the protocols employed above and used platinum wire electrodes to generate electrical fields of between 0.5 and 6 V cm^{-1} (current 14 μA). Using pH indicators and pH-sensitive microelectrodes, the pH around the cathode and anode increased and decreased, respectively, by over 2 pH units (δpH = 4) within 10 s in unbuffered media (Fig. 4). Since the cathodic pH rose to over pH 9.5, and the anodic pH fell to below 3.5 by 30 s, reports of zoospore immobilisation, repulsion or lysis in the immediate vicinity of bare electrodes are readily understood. We therefore designed and fabricated a chamber in which zoospore suspensions were completely isolated from electrode products by agarose bridges (Fig. 5). Further controls were made in which microelectrodes were used to record pH changes at both ends and in the centre of the chamber. No detectable pH changes were observed during 2-h tests at any of the fields employed in electrotaxis experiments.

In electrotaxis experiments, zoospores of *P. palmivora* were placed in the chamber and an electrical field was applied for 60 min. A divider was then placed in the chamber that segregated the cells into separate compartments. CaCl$_2$ was added to encyst any swimming cells and the distribution of zoospores across the chamber was determined by counting

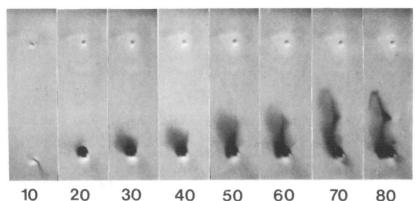

10   20   30   40   50   60   70   80

Fig. 4. Changes in pH around platinum-wire electrodes generating a field of 0.5 V cm^{-1}, 8.3 μA, as revealed by changes in coloration of a 1% solution of methyl red indicator at an initial pH of 6.1. The electrodes were 2 cm apart with the anode at the bottom of the photographs, which were taken at 10 s intervals. The darkening around the anode is due to the titration of the indicator to a red colour below pH 4.2. The rapid electrolysis of the medium highlights the dangers of using bare wire electrodes to apply electrical fields in electrotaxis experiments. (From Morris *et al.*, 1992.)

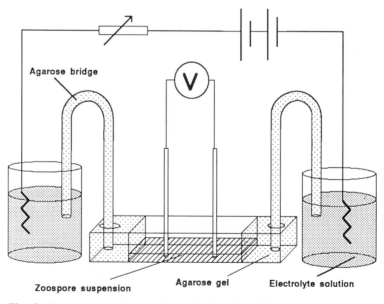

Fig. 5. Apparatus used to apply weak electrical fields to populations of zoospores. The agarose bridges protect the zoospores from the products of electrolysis. The electrical field in the chamber is measured directly with a multimeter. (After Morris *et al.*, 1992.)

the average numbers in a fixed area of the chamber base. The distribution was quantified as a tactic response quotient (TRQ),

$$TRQ = (A-C)/(A+C+2M)$$

where A, C and M are the mean densities of zoospores per mm^2 at the anode, cathode and centre of the chamber. The TRQ values vary between 1 (anodic) and -1 (cathodic) and indicate the anode–cathode bias relative to the numbers of zoospores in the centre of the chamber.

Under all conditions examined zoospores of *P. palmivora* accumulated at the anode (Fig. 6). The electrotactic response to fields of 0.1 V cm^{-1} was not significant in distilled water; however, marked anodic accumulation could be observed in electrical fields when weak osmotica were present. The chemical nature of the osmoticum did not seem to be critical to the promotion of electrotaxis. Another important variable was the zoospore density, with the electrotaxis quotient increasing with increasing zoospore densities. The basis of the potentiation of electrotaxis by osmotica and high zoospore densities might be by extending the swimming time and range of the zoospore. For example, weak osmotica may reduce the energy costs of osmoregulation via the contractile vacuole and therefore allow more of the internal energy reserves to be used for swimming. This may allow more time for anodic accumulation to occur.

The electrotactic response increased with increasing field strength, apparently saturating at fields at or above 100 mV cm^{-1} (Fig. 7). Since

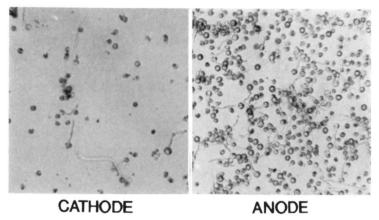

CATHODE   ANODE

Fig. 6. Density of encysted zoospores of *Phytophthora palmivora* at the anodic and cathodic ends of the electrotaxis chamber (see Fig. 5) after 60 min in an electrical field of 0.5 V cm^{-1}.

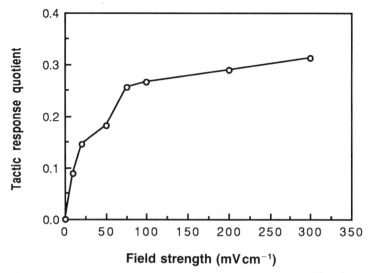

Fig. 7. The effect of the magnitude of an electrical field on the electro-tactic response of zoospores of *Phytophthora palmivora* as measured by the tactic response quotient (described in the text). (After Morris *et al.*, 1992.)

electrical fields around plant roots are normally about 5–50 mV cm^{-1} this represents the first report of electrotaxis at physiological field strengths.

Charged particles are subjected to electrophoretic forces when in free solution and electroosmotic currents when on, or close to, a surface. Electroosmosis is caused by fluid currents that result from the electro-phoretic displacement of counter-ions around fixed charges on a surface. For example, the flow of positively charged counter-ions and their associ-ated shells of hydration around negative charges on a glass plate will cause an electroosmotic flow towards the negative pole. Electrophoresis and electroosmosis are therefore forces that act in opposite directions. In order to assess the contribution of electrophoresis and electroosmosis to the observed zoospore distributions in our experiments, video-recordings of encysted zoospores were made and the velocity of the spores in free solution and along surfaces was measured in electrical fields of different strengths (Fig. 8). Encysted zoospores had electrophoretic and electro-osmotic mobilities of 3.97 μm s^{-1} V cm^{-1} (to the anode) and 1.60 μm s^{-1} V cm^{-1} (to the cathode), respectively. Assuming a comparable electro-mobility for the swimming zoospores and considering the field strengths (<50 mV cm^{-1}) and duration (60 min) of our experiments, the length of the chamber (8 cm) and the rate of swimming of the zoospores (~100 μm

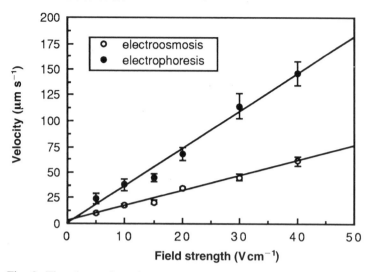

Fig. 8. The electrophoretic and electroosmotic movement of encysted zoospores of *Phytophthora palmivora*, at different field strengths at pH 6.0. (After Morris *et al.*, 1992.)

s^{-1}), it is clear that these physical forces did not account for the observed accumulation of zoospores at the anodic end of the chamber. Polystyrene beads of the same size and twice the electromobility of the encysted zoospores were also placed in the electrotaxis chamber and exposed to electrical fields. The distribution of these beads was not affected significantly (TRQ<0.02) after 60 min, even in strong fields up to 0.5 V cm^{-1}. A full report of these data will be given elsewhere (Morris *et al.*, 1992).

Under these conditions we have therefore observed electrotaxis that is not due to the presence of pH, oxygen, chemical or inhibitor gradients and shown that the tactic response was dependent on the electrical field strength, medium osmotic pressure and zoospore density. The data also suggest that electrotaxis is sensitive enough to have a physiological role, at least when a zoospore is in close proximity to the rhizoplane.

pH taxis

Plant roots generate a longitudinal pH gradient that may be a manifestation of the circulating proton traffic already described (Fig. 3). For example, the medium external to the *Hordeum* root tip region may have 20–30 times the proton concentration (δpH = 1.3–1.5) of the root 2 mm to the rear (Miller *et al.*, 1991). Zoospores of *P. palmivora* have been shown to be negatively chemotactic to H$^+$ (Cameron & Carlile, 1980), i.e. tending to move away from acidic areas. It is therefore possible that the accumulation of zoospores around roots is also affected by the local pH, which is in turn modulated by the root's electrochemical activity. Interestingly, Edwards & Bowling (1986) found that germ tubes of uredospores of *Uromyces vicea-fabae* grew tropically in the pH gradients that were generated by closed stomata. Gradients of pH may therefore effect both tactic and tropic movements of fungi.

Electrotaxis *in vivo*

In agreement with Khew & Zentmyer (1974), we find anode-directed electrotaxis of zoospores of *P. palmivora*. Artefacts due to electrolysis complicate interpretation of data from experiments *in vitro* using wire electrodes. Therefore it is not prudent to make broad generalisations about the behaviour of zoospores of other fungi until further adequately controlled experiments have been performed. The zone of root elongation, wound sites and sites of emerging lateral roots have been variously described as cathodic or anodic (see above). Until a detailed comparison of the root electrochemical profile (in terms of both the external electrical and pH gradients) and zones of zoospore accumulation are carried out on individual roots it will be difficult to assess whether electrotaxis is significant in the attraction of a zoospore to any one site of infection. Such a study would be facilitated by the fact that electrochemical maps can be measured and manipulated with high spatial precision. At the very least it should be easy to correlate accurately the electrochemical fields with the pattern of accumulation and encystment. At this stage it would seem unlikely that electrotaxis of zoospores of *P. palmivora* features strongly, for example, in the direct attraction to wound sites, since the two reported electrophysiological studies of root wounding concur that these too are cathodic in nature (Miller *et al.*, 1988; Hush & Overall, 1989). However, the requisite study of both pathogen and its host has yet to be performed. It may be that electrotaxis merely aids in bringing the zoospore to the root and that further cues dictate where they settle and encyst. Alternatively, electrical fields may have active roles in the processes of docking of the zoospore on the root, and on the triggering of encystment and germ tube outgrowth.

It should also be noted that consideration of electrotaxis should not be limited to infections by zoospores or infections of the root. The vibrating electrode studies of the last 15 years insist that we regard bio-electricity as an innate property of most cells and most organisms rather than a curious aspect of a few (Gow, 1989; Nuccitelli, 1990). Nematodes exhibit electrotaxis but preliminary experiments indicate that they are not sufficiently sensitive to detect root-sized electrical fields (L. MacCulloch, W. Robertson & N. A. R. Gow, unpublished results). Open stomata of *Communila communis* generate electrical currents of up to 4 μA cm⁻² (electrical field 27 mV cm⁻¹, for a resistivity of 5000 Ω cm; Bowling *et al.*, 1986) and this may be of significance when considering zoospores of *Phytophthora* species that gain access to the leaves through these natural openings. It is also possible that natural electrical fields of plants elicit electrotropism of germ tubes of plant pathogens or symbionts such as mycorrhizal fungi. However, studies indicate that the threshold fields that affect germ tube and hyphal orientation of fungi *in vitro* are around 1 V cm⁻¹ (McGillivray & Gow, 1986; Van Laere, 1988; Crombie *et al.*, 1990), which is above the magnitude of natural fields around plant tissues.

At least two other questions remain to be addressed. It will be important to assess to what extent chemotaxis and electrotaxis compete or support each other as targeting mechanisms. Also, the mechanism by which zoospores (and other cells) can sense electrical fields of the order of 5 mV cm⁻¹ (equivalent to 5 μV (cell diameter)⁻¹) is of fundamental importance.

Conclusions

Zoospores of *P. palmivora* exhibit electrotaxis in electrical fields that are comparable to those generated by growing roots. Electrotaxis may therefore have a physiological role in root–zoospore interactions. Chemotaxis is likely to remain the dominant behavioural response facilitating root detection; however, electrotaxis may augment chemotaxis or act synergistically with it. There may be additional roles for electrotaxis in the orientation and docking of zoospores at the infection court. Roots must be viable, have intact membranes and be actively engaged in ion transport to generate electrical currents. Electrotaxis may therefore be a useful adjunct to root targeting, since the presence of an electrical field provides an indication of the viability of the host root. We would highlight the need for future studies on root–microbe interactions to incorporate an electrophysiological perspective into cytological, biochemical and genetic investigations in this area.

Acknowledgements

We thank the AFRC for financial support and the NERC for a studentship in this area.

References

Behrens, H. M., Weisenseel, M. H. & Sievers, A. (1982). Rapid changes in the pattern of electric current around the root tip of *Lepidium sativum* L. following gravistimulation. *Plant Physiology* **70**, 1079–83.

Bjorkman, T. (1989). The use of bioelectric currents to study gravity sensing in roots. *Biological Bulletin* **176S**, 49–55.

Bjorkman, T. & Leopold, A. C. (1987*a*). An electric current associated with gravity sensing in maize roots. *Plant Physiology* **84**, 841–6.

Bjorkman, T. & Leopold, A. C. (1987*b*). Effect of inhibitors of auxin transport and of calmodulin on a gravisensing-dependent current in maize roots. *Plant Physiology* **84**, 847–50.

Bowling, D. J. F., Edwards, M. C. & Gow, N. A. R. (1986). Electrical currents at the leaf surface of *Commelina communis* and their relationship to stomatal activity. *Journal of Experimental Botany* **37**, 876–82.

Brawley, S. H., Wetherell, D. F. & Robinson, K. R. (1984). Electrical polarity in embryos of wild carrot precedes cotyledon differentiation. *Proceedings of the National Academy of Sciences, USA* **81**, 6064–7.

Cameron, J. N. & Carlile, M. J. (1977). Negative geotaxis of zoospores of the fungus *Phytophthora cinnamomi*. *Journal of General Microbiology* **94**, 23–38.

Cameron, J. N. & Carlile, M. J. (1980). Negative chemotaxis of the fungus *Phytophthora palmivora*. *Journal of General Microbiology* **120**, 347–53.

Carlile, M. J. (1983). Motility, taxis and tropism in *Phytophthora*. In *Phytophthora: Its Biology, Ecology and Pathology*, ed. D. C. Erwin, S. Bartnicki-Garcia & P. H. Tsao, pp. 95–107. St Paul, MN: American Phytopathological Society.

Carlile, M. J.(1986). The zoospore and its problems. In *Water, Fungi and Plants*, ed. P. G. Ayres & L. Boddy, *British Mycological Society Symposium* vol. 11, pp. 105–18. Cambridge: Cambridge University Press.

Crombie, T., Gow, N. A. R. & Gooday, G. W. (1990). Influence of applied electrical fields on yeast and hyphal growth of *Candida albicans*. *Journal of General Microbiology* **136**, 311–17.

Deacon, J. W. (1988). Behavioural responses of fungal zoospores. *Microbiological Reviews* **5**, 249–52.

Edwards, M. C. & Bowling, D. J. F. (1986). The growth of rust germ tubes towards stomata in relation to pH gradients. *Physiological and Molecular Plant Pathology* **29**, 185–96.

Ezaki, S., Toko, K., Yamafugi, K. & Irie, F. (1988). Electric potential patterns around a root of the higher plant. *Transactions of the IEICE* **E71**, 965–7.

Gow, N. A. R. (1989). Circulating ionic currents in micro-organisms. *Advances in Microbial Physiology* **30**, 89–123.

Hesse, P. R. (1971). *A Textbook of Soil Analysis*. London: Murray Ltd.

Ho, H. H. & Hickman, C. J. (1967). Factors governing zoospore responses of *Phytophthora megasperma* var. *sojae* to plants. *Canadian Journal of Botany* **45**, 1983–94.

Hush, J. M. & Overall, R. J. (1989). Steady ionic currents around pea (*Pisum sativum* L.) root tips; the effects of tissue wounding. *Biological Bulletin* **176S**, 56–64.

Iiyama, S., Toko, K. & Yamafuji, K. (1985). Band structure of surface electric potential in growing roots. *Biophysical Chemistry* **21**, 285–93.

Iwabuchi, A., Yano, M. & Shimizu, H. (1989). Development of extracellular electric pattern around *Lepidium* roots: its possible role in root growth and gravistimulation. *Protoplasma* **148**, 94–100.

Jaffe, L. F. (1981). The role of ionic currents in establishing developmental pattern. *Philosophical Transactions of the Royal Society of London* **295**, 553–6.

Jaffe, L. F. (1986). Ionic currents in development: an overview. In *Ionic Currents in Development*, ed. R. Nuccitelli, pp. 351–7. New York: Alan R. Liss, Inc.

Jaffe, L. F. & Nuccitelli, R. (1974). An ultrasensitive vibrating probe for measuring steady extracellular currents. *Journal of Cell Biology* **63**, 614–28.

Katsura, K., Masago, H. & Miyata, Y. (1966). Movements of zoospores of *Phytophthora capsici*. I. Electrotaxis in some organic solutions. *Annals of the Phytopathological Society of Japan* **32**, 215–20.

Khew, K. L. & Zentmeyer, G. A. (1974). Electrotactic response of zoospores of seven species of *Phytophthora*. *Phytopathology* **64**, 500–7.

Lund, E. J. (1947). *Bioelectric Fields and Growth*. Austin, TX: University of Texas Press.

Lund, E. L. & Kenyon, W. A. (1927). Relation between continuous bioelectrical currents and respiration. *Journal of Experimental Zoology* **44**, 333–57.

McGillivray, A. M. & Gow, N. A. R. (1986). Applied electrical fields polarize the growth of mycelial fungi. *Journal of General Microbiology* **132**, 2515–25.

Miller, A. L. (1989). Ion currents and growth regulators in plant root development. *Biological Bulletin* **176S**, 65–70.

Miller, A. L. & Gow, N. A. R. (1989*a*). Correlation between profile of ion-current circulation and root development. *Physiologia Plantarum* **75**, 102–8.

Miller, A. L. & Gow, N. A. R. (1989*b*). Correlation between root-generated ionic currents, pH, fusicoccin, indoleacetic acid, and growth of the primary root of *Zea mays*. *Plant Physiology* **89**, 1198–206.

Miller, A. L., Raven, J. A., Sprent, J. I. & Weisenseel, M. H. (1986). Endogenous ion currents traverse growing roots and root hairs of *Trifolium repens. Plant Cell and Environment* **9**, 79–83.

Miller, A. L., Shand, E. & Gow, N. A. R. (1988). Ion currents associated with root tips, emerging laterals and induced wound sites in *Nicotiana tabacum*: spatial relationship proposed between resulting electrical fields and phytophthoran zoospore infection. *Plant Cell and Environment* **11**, 21–5.

Miller, A. L., Smith, G. N., Raven, J. A. & Gow, N. A. R. (1991). Ion currents and the nitrogen status of roots of *Hordeum vulgare* and non-nodulated *Trifolium repens. Plant Cell and Environment*, **14**, 559–67.

Morris, B. M., Reid, B. & Gow, N. A. R. (1992). Electrotaxis of zoospores of *Phytophthora palmivora* at physiologically relevant field strengths. *Plant Cell and Environment* (in press).

Müller-Hettlingen, J. (1883). Über galvanische ersheinungen an keimenden samen. *Pfügers archiv für die Geamte Physiologie des Menschen under der Tiere* **31**, 193–214.

Nuccitelli, R. (1990). Vibrating probe technique for studies of ion transport. In *Noninvasive Techniques in Cell Biology*, pp. 272–310. New York: Wiley-Liss, Inc.

Overall, R. & Wernicke, W. (1986). Steady ionic currents around haploid embryos formed from tobacco pollen in culture. *Progress in Clinical Biological Research* **210**, 139–46.

Rajnicek, A. M., Stump, R. F. & Robinson, K. R. (1988). An endogenous sodium current may mediate wound healing in *Xenopus* nurulae. *Developmental Biology* **128**, 290–9.

Rathore, K. S. & Goldsworthy, A. (1985a). Electrical control of growth in plant tissue cultures. *Biotechnology* **3**, 253–4.

Rathore, K. S. & Goldsworthy, A. (1985b). The electrical control of growth in plant tissue cultures: the polar transport of auxin. *Journal of Experimental Botany* **36**, 1134–41.

Rathore, K. S., Hodges, T. K. & Robinson, K. R. (1988). A refined technique to apply electrical currents to callus cultures. *Plant Physiology* **88**, 515–17.

Rathore, K. S., Hotary, K. B. & Robinson, K. R. (1990). A two-dimensional vibrating probe study of currents around lateral roots of *Raphanus sativus* developing in culture. *Plant Physiology* **92**, 543–6.

Rathore, K. S. & Robinson, K. R. (1989). Ionic currents around developing embryos of higher plants in culture. *Biological Bulletin* **176S**, 46–8.

Raven, J. A. (1979). The possible role of membrane electrophoresis in the polar transport of IAA and other solutes in plant tissues. *New Phytologist* **82**, 285–91.

Schreurs, W. J. A. & Harold, F. M. (1988). Transcellular proton current in *Achlya bisexualis* hyphae: relationship to polarized growth. *Proceedings of the National Academy of Sciences, USA* **85**, 1534–8.

Toko, K., Fujiyoshi, T., Tanaka, C., Iiyama, S., Yoshida, T., Hayashi, K. & Yamafuji, K. (1989). Growth and electric current loops in plants. *Biophysical Chemistry* **33**, 161–76.

Toko, K., Iiyama, S., Tanaka, C., Hayashi, K. & Yamafuji, K. (1987). Relation of growth to spatial patterns of electric potential and enzyme activity in bean roots. *Biophysical Chemistry* **27**, 39–58.

Toko, K. & Yamafuji, K. (1988). Spontaneous formation of the spatial pattern of electric potential in biological systems. *Ferroelectrics* **86**, 269–79.

Troutman, J. T. & Wills, W. H. (1964). Electrotaxis of *Phytophthora parasitica* zoospores and its possible role in infection of tobacco by the fungus. *Phytopathology* **54**, 225–8.

Van Laere, A. (1988). Effects of electrical fields on polar growth of *Phycomyces blakesleeanus*. *FEMS Microbiology Letters* **49**, 111–16.

Waaland, S. D. & Lucas, W. J. (1984). An investigation of the role of transcellular ion currents in morphogenesis of *Griffithsia pacifica* Kylin. *Protoplasma* **123**, 184–91.

Weisenseel, M. H., Dorn, A. & Jaffe, L. F. (1979). Natural H^+ currents traverse growing roots and root hairs of barley (*Hordeum vulgare* L.) *Plant Physiology* **64**, 512–18.

Youatt, J., Gow, N. A. R. & Gooday, G. W. (1988). Bioelectric and biosynthetic aspects of cell polarity in *Allomyces macrogynus*. *Protoplasma* **146**, 118–26.

Zentmyer, G. A. (1961). Chemotaxis of zoospores for root exudates. *Science* **133**, 1595–6.

J. R. GREEN, A. J. MACKIE,
A. M. ROBERTS AND J. A. CALLOW

Molecular differentiation and development of the host–parasite interface in powdery mildew of pea

Introduction

The host–parasite interface formed during infection of plants by biotrophic powdery mildew fungi is a specialised structure involved in the
transfer of host nutrients to the fungus (Manners & Gay, 1983; Manners,
1989) and the efficiency with which such pathogens do this makes them
highly damaging to crops (Singh *et al.*, 1982). The haustorium is the
structure by which the fungus absorbs nutrients from its host. It forms
within the epidermal cells, enclosed by an invagination of the host plasma
membrane termed the extrahaustorial membrane (ehm); this is separated
from the haustorium by the polysaccharide-rich extrahaustorial matrix.
All these components collectively make up the haustorial complex (HC;
see Fig. 1.).

Powdery mildew fungi are obligate pathogens and exhibit a high degree
of host species and cultivar specificity. In certain incompatible combinations of host and pathogen genotype, haustoria are produced but appear
to be functionally inactive and so fail to produce elongating secondary
hyphae or secondary haustoria (e.g. *Erysiphe pisi* (Singh & Singh, 1983;
Manners & Gay, 1983; Manners, 1989); *Erysiphe betae* (Dickey & Levy,
1979)). Work on the *E. pisi*/*Pisum sativum* system using various
cytochemical reagents such as fluorescent lectins, as well as protein and
Ca^{2+}-specific fluorochromes has shown that there may be molecular differences between the ehm of haustoria from resistant compared with
susceptible interactions (Chard & Gay, 1984). It is therefore a possibility
that the molecular recognition events, mediated by the primary products
of host genes for resistance and pathogen genes for avirulence (Callow,
1984, 1987) and which are regarded as the first step in resistance triggering through intracellular signalling, are located at this interface. Thus the
HC, as well as having an important role in nutrient transfer, is probably

Society for Experimental Biology Seminar Series 48: *Perspectives in Plant Cell Recognition*,
ed. J. A. Callow & J. R. Green. © Cambridge University Press 1992, pp. 193–212.

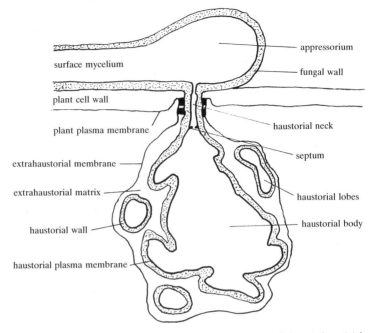

surface mycelium

appressorium

fungal wall

plant cell wall

plant plasma membrane

haustorial neck

septum

extrahaustorial membrane

extrahaustorial matrix

haustorial lobes

haustorial wall

haustorial body

haustorial plasma membrane

Fig. 1. Diagram showing the haustorial apparatus of *Erysiphe pisi* in transverse section, and its association with a leaf epidermal cell of *Pisum sativum*.

involved in recognition and signalling between the plant and fungal partners. However, little is known about the detailed molecular architecture and functions of the various parts of the HC.

Powdery mildews in general are good systems with which to work in investigating molecular aspects of biotrophy, since the interface with the living host is restricted to one cell type, the epidermal cell. However, though there are some clear advantages with these systems, the highly discrete nature of the host–pathogen interface in powdery mildews precludes gross biochemical studies to explore its structure in detail. The specificity and resolution of monoclonal antibodies (MAbs) at the level of the individual cell offers a viable approach to the study of this plant–pathogen interface and we have used this technology to study the *E. pisi/P. sativum* interaction. Pea leaves infected with *E. pisi* have been chosen as it is possible to isolate HCs from them (Gil & Gay, 1977; Manners & Gay, 1982). In addition, this pathosystem has previously been used extensively for structural and metabolic studies (Gil & Gay, 1977; Manners & Gay, 1980; Spencer-Phillips & Gay, 1981; Manners & Gay,

1983). Resistant and susceptible culivars of pea are available, and compatible and incompatible disease reactions have been well studied at the ultrastructural and physiological levels in addition to the specific studies on their haustoria (Manners, 1989; Stumpf & Gay, 1989).

In this review the main structural features of the HC are described as well as the current knowledge on the molecular components of some of its key regions. Some of the results obtained with MAbs raised to HCs isolated from the *E. pisi/P. sativum* interaction are then described, including evidence for molecular differentiation within this infection structure. Developmental aspects of HC formation are also described.

Formation and structure of haustoria

Powdery mildew fungi produce conidia which land on aerial parts of plants and germinate to form short germ tubes, usually within 3 h (Singh & Singh, 1983). The hyphae, in contact with the outer layer of the host cell wall, become firmly attached to it by adhesion and then usually begin to swell. This attached and swollen hyphal tip, the appressorium, serves as a reservoir for turgor pressure and is essentially a 'drilling platform' from which an extremely fine hyphal peg grows directly through the cell wall of the host cell, producing the specialised absorption structure, termed the haustorium (Fig. 1). A collar which surrounds the haustorial neck forms a papilla at the time of penetration. The penetration peg grows through the papilla forming the haustorial neck, which is a small tube connecting the haustorium to the surface mycelium. Although a septum forms between the haustorial body and the neck, it is perforated, allowing protoplasmic continuity between neck and body. Two annular structures develop in the neck region and these are distinguished as the A- and B-neckbands. The A-neckband adjoins the inner face of the host cell wall and it is usually at this level that the neck breaks during isolation of HCs (Bushnell & Gay, 1978). The B-neckband is separate and nearer to the haustorial body, and the ehm of *E. pisi* is firmly attached to it (Bushnell & Gay, 1978; Gil & Gay, 1977). The haustorium has distinct lobes which surround the body in most species but these are elongated in *Erysiphe graminis* in cereals (Bushnell, 1971; Manners & Gay, 1983).

Molecular components of haustoria

The haustorial plasma membrane

The haustorial cytoplasm contains a single nucleus and is surrounded by the haustorial plasma membrane (hpm) which has a typical membrane structure. It is a semipermeable membrane (Gay & Manners, 1987) and is continuous around the protoplast, extending through the septal pore to the haustorial neck (Bracker, 1968) and it is closely adherent to the neck wall (Spencer-Phillips & Gay, 1981). The hpm shows high cytochemical phosphatase activity using both ATP and β-glycerophosphate as substrates; this is in contrast to the surface hyphae, appressoria and adjoining region of the haustorial neck. The transition is abrupt and occurs between the A- and B-neckbands. This activity has been implicated in the transport of nutrients into the fungal haustoria (Spencer-Phillips & Gay, 1981; Gay et al., 1987).

The haustorial wall

This is continuous around the neck, body and lobes of the haustorium (Gil & Gay, 1977), and although structural linkages between the wall and ehm are rarely seen by electron microscopy, the lobe walls do adjoin the ehm in a few places. Although detailed analysis of the haustorial wall has not been reported it is likely to be similar in composition to that of hyphal walls which contain mainly α- and β-linked glucans, chitin and some undefined protein components. Labelling of the haustorial wall with Calcofluor and the fluorescently labelled lectin fluorescein isothiocyanate-wheatgerm agglutinin (FITC-WGA) suggests the presence of β-linked polysaccharides and chitin, respectively, and binding of the lectins FITC-ConA (concanavalin A) and FITC-RCA (Ricinus communis agglutinin) suggests the occurrence of terminal α-linked mannose or glucose residues and α- or β-linked galactose residues, respectively (Chard & Gay, 1984). These sugar residues could be part of glycoproteins or polysaccharides.

The extrahaustorial matrix

The extrahaustorial matrix is non-cytoplasmic, and leakage out between the ehm and the hpm is prevented by the seal-forming B-neckband, an annular ring in the neck region that is impermeable to matrical solutes. Bushnell (1971) and Gil & Gay (1977) have shown that, when complexes are placed in hypotonic solutions, the extrahaustorial matrix swells without rupturing of the HCs and have concluded that the matrix is fluid in nature. In media hypertonic to the host, the ehm of E. graminis contracts to a position contiguous with the haustorial wall indicating the absence of large amounts of insoluble material in the matrix (Bushnell, 1971). In addition, Chard & Gay (1984) could not demonstrate the presence of any protein, or protein-containing macromolecules in the extrahaustorial matrix of HCs of E. pisi. The lectin pokeweed mitogen, which binds to chitobiose, was found to bind to the extrahaustorial matrix specifically, though WGA did not bind, suggesting the absence of chitin. Gay & Manners (1987) have found that after ehm disruption the extrahaustorial matrix is permeable to uranyl ions but impermeable to the 40 kDa protein horseradish peroxidase. They suggest that the matrix is therefore probably a gel rather than a solution and that the structural organisation of β-linked polysaccharides results in mechanical blocking of horseradish peroxidase entry (Gil & Gay, 1977; Chard & Gay, 1984). Results of silver-methenamine staining, before and after treatment of HCs with cell wall degrading enzymes, suggested that the matrix in pea mildew HCs contains pectins and/or hemicelluloses (Gil & Gay, 1977). The idea of a gel would be compatible with the presence of pectic substances, since those with polyanionic galacturonan backbones are capable of binding Ca^{2+}, resulting in their aggregation and the formation of gels. The origin and function of the extrahaustorial matrix is unknown.

The extrahaustorial membrane

The ehm is structurally and functionally different from the plant plasma membrane. It is about twice as thick as the host plasma membrane, is convoluted in healthy HCs, and exhibits extraordinary resistance to some treatments which normally dissolve cell membranes (Gil & Gay, 1977). This strength allows the isolation of HCs, as they are not easily grossly disrupted. Gil & Gay (1977) found that with *E. pisi* in some instances the ehm appeared to be symmetrical (using electron microscopy (EM)), while in others the lamella joining the matrix was again slightly thicker. Most of the extra thickness of the ehm can be removed by pectinase and cellulase treatments and therefore is presumably polysaccharide material (Bushnell & Gay, 1978). This is probably mainly β1–4-linked as it is also stained by Tinopal BOPT, which is reported to bind to cellulosic material (Chard & Gay, 1984). Small amounts of α-linked sugars (possibly glucose, mannose and galactose) have been detected in the ehm by using fluorescently labelled lectins FITC-ConA and FITC-RCA. Protein has been cytochemically detected in the ehm of *E. pisi* using eight fluorescent reagents specific for protein or for constituent groups of proteins (e.g. amino groups: Chard & Gay, 1984). Ca^{2+} was detected in the ehm of *E. pisi* by staining with the Ca^{2+}-specific compound chlorotetracycline.

Removal of the polysaccharide associated with the ehm renders the ehm much more permeable to water and susceptible to osmotic or detergent lysis (Gil & Gay, 1977). The associated polysaccharide would therefore appear to give the ehm its characteristic strength and resistance to various chemical and mechanical stresses. Ca^{2+} has also been implicated in the stability of the ehm (Chard & Gay, 1984).

Cytochemistry of the plasma membranes of infected pea epidermal cells has shown that they have high ATPase and β-glycerophosphate activities (Spencer-Phillips & Gay, 1981), which may be important in transmembrane transport and scavenging of solutes from the apoplast. However, the ehm of *E. pisi* lacks these activities and the abrupt transition is found to occur at the point of invagination where the plant plasma membrane joins the A-neckband. It has not yet been demonstrated whether the deficiency of ATPase activity is due to enzyme inactivation or to the absence of the enzyme protein in the ehm. Particles 10 nm in diameter, present in the normal plant plasma membranes, are absent when the ehm is studied by freeze fracture (Bushnell & Gay, 1978). It has been suggested that these may be a transport complex involving ATPase (Spencer-Phillips & Gay, 1981) or they could be involved in wall microfibril formation (Bushnell & Gay, 1978), which would agree with the reported absence of fibrillar material in the matrix (Bushnell & Gay, 1978; Chard & Gay, 1984). Their absence from the ehm would therefore indicate local suppression of host function(s) by the fungus.

Lateral diffusion of membrane components between the two membrane domains of the ehm and the uninvaginated plant plasma membrane is probably prevented by neckbands which may serve as 'domain delimiters' (Manners & Gay, 1983). However, experiments on the interface of the rust fungus *Puccinia poarum* with *Tussilago farfara* (Woods & Gay, 1987) appear to indicate that a neckband is not necessary for this type of domain separation and maintenance. Their observations show a more gradual transition over about 1.5 μm. It was suggested that this transition could be achieved by a decrease in membrane fluidity of the whole or part of the invaginated membrane, such as has been reported in senescing or damaged biological membranes (see e.g. Houslay & Stanley, 1982).

Functions of haustoria

In a biotrophic relationship the living plant provides various nutrients for the fungal parasite, thus sustaining heavy losses in yield. According to Manners & Gay (1982) about 20% of the total products of photosynthesis in an infected pea leaf enter the mycelium of the pathogen. Thus, host metabolism must be altered by the pathogen, though details on this are limited. The transport of host sugars to the fungus is mediated by haustoria with sucrose and glycerol reported to be either the main translocates or metabolic intermediates of this process (Manners & Gay, 1982, 1983). Recent reports have suggested that glucose is readily transported to the haustorium (Mendgen & Nass, 1988). Mannitol appears to be the major primary sink metabolite of the fungus and it is probable that it is in this form that organic material is translocated to spores and sites of growth for metabolism and subsequent storage or utilisation. A hypothesis for solute transport through the plasma membranes of epidermal cells infected by haustoria of *E. pisi* was first proposed by Spencer-Phillips & Gay (1981) and has since been applied to a range of pathogens and plants (for a review, see Gay & Manners, 1987; Manners, 1989). The epidermal cells in which the haustoria are confined contain only a few undeveloped chloroplasts, so that photoassimilates must first enter the epidermis before entering the haustoria. The neckbands seal the haustoria from the apoplast and therefore direct translocates through the epidermal cytoplasm. In pea, plasmodesmata only rarely connect the mesophyll with the epidermis (Bushnell & Gay, 1978) and it is proposed that the high ATPase activity of the plant plasma membrane of epidermal cells scavenges photosynthate from the apoplast. It is further argued that the absence of ATPase activity in the ehm results in solutes being unloaded at the face adjacent to the haustorium due to diminished control of solute retention (i.e no ATPase activity). Concentration gradients maintained by the fungus (e.g. conversion to mannitol and other fungal sugars for use by the mycelia) should continually draw solutes into the matrix. The phosphatase activity found on the hpm has the correct orientation to transport solutes actively from the matrix into the haustorium itself.

Objectives and approach

From the above it is clear that the haustorium is a complex structure; it is a physical and functional expression of a mutual recognition event whereby effectively the pathogen is recognised as 'self' by the host, precluding the induction of plant resistance mechanisms that would otherwise eliminate it. The ehm and its associated extrahaustorial matrix presumably play a critical role in the maintenance of this biotrophic relationship. The ehm is structurally and functionally different from the host plant plasma membrane so there must be differentiation in this region during its development and its formation must involve some cooperation between the fungus and the plant.

One approach to obtain more information on the molecular composition and architecture of the main structural components of the HC and to obtain more direct evidence for molecular differentiation within the HC is to prepare monoclonal antibodies (MAbs). MAbs can be used to detect hitherto uncharacterised antigens in complex mixtures and they have been used extensively in animals and plants to identify molecules specific to particular cell types and subcellular structures (Key & Weiler, 1988; Knox *et al.*, 1989; Jones *et al.*, 1988).

The key questions to be addressed are:

1. What is the molecular make-up of the hpm, haustorial wall, extrahaustorial matrix and ehm? Are there proteins/glycoproteins specifically associated with the haustorium? How do these structures relate to fungal/plant structures?

2. How does the haustorium develop? Are components inserted early or late in haustorial development? Are the components of the ehm of fungal origin, plant origin or a mixture of the two? Are most of the normal plant plasma membrane components excluded from the ehm?

3. What are the functions of the molecules that make up the different compartments of the haustorium?

4. Are there differences between haustoria formed during susceptible and resistant interactions? Are these differences apparent early or late in the infection process?

5. What changes occur in the pea epidermal cells during the infection process in order to generate a functional haustorium? What changes occur in epidermal cells during resistant interactions?

6. What types of recognition/signalling event occur during the formation and development of haustoria in susceptible and resistant interactions?

Isolated haustorial complexes

The approach taken in this work is possible because HCs can be isolated from infected leaves (Gil & Gay, 1977). Isolated HCs have been used to study the structural and physiological aspects of the host–haustorium interface (Gil & Gay, 1977; Manners & Gay, 1977). The ehm and extrahaustorial matrix are retained due to sealing neckbands and resistance of the ehm to both physical and biochemical treatments, and the haustorial cytoplasm is retained by the septal plug. It is worth noting that, in pathogens without this type of neckband system, the ehm, or its equivalent (e.g. the plasma membrane invaginating around rust haustoria, or infection vesicles of *Colletotrichum lindemuthianum*: Woods & Gay, 1987; O'Connell *et al.*, 1985) is difficult to isolate and matrical material is also often found incomplete. In addition the haustorial lobes of *E. pisi* wrap around the haustorial body and this renders these HCs far less fragile than HCs of the closely related *E. graminis*, with its finger-like, outwardly projecting haustorial lobes.

The method used for the isolation of HCs is based on that described by Gil & Gay (1977) and their appearance is as described (Fig. 2 a and b), all the major structures surviving the isolation procedure. The major contaminants are chloroplasts and the ratio of HCs to chloroplasts is approximately 2:1. However, the ratio of the surface area of one HC to the surface area of one chloroplast is approximately 9.4:1, so that the ratio of the total surface area of HCs in the preparation to the total surface area of chloroplasts is approximately 16.4:1 (Mackie, 1991). HCs can be routinely isolated in large enough numbers (approximately 2×10^5 HC per gram infected pea leaf material) for immunisation and screening procedures (Mackie *et al.*, 1991).

An important part of the MAb screening procedures requires that internal labelling with relevant probes can be observed and this was tested using fluorochrome-labelled lectins. About half of isolated HCs showed internal fluorescence when labelled with FITC-ConA or FITC-WGA (Fig. 3a). Accessibility was increased to >90% by drying down the HCs onto microscope slides, though often only localised areas were accessible. Labelling of sections of infected pea leaves with gold-conjugated lectins at the EM level, showed that they bound to the walls of the haustorial body and lobes. The ehm was also weakly labelled after direct labelling with FITC-ConA but did not fluoresce with FITC-WGA, even if disrupted (Mackie, 1991). This is in contrast to the findings of Chard & Gay (1984), who found that WGA bound to a small proportion of haustoria.

The isolated HCs were highly immunogenic and antibodies to internal components were obtained even though entire HCs were used for immunisations, presumably due to a proportion of the HCs being damaged either upon isolation or after immunisation. Previous work has suggested that both the ehm and the hpm are broken in approximately 40% of the HCs. MAbs have also been raised to plasma membrane enriched fractions isolated from uninfected pea leaves (A. M. Roberts *et al.*, unpublished results). Antibodies which label internal or peripheral structures of the HCs have been obtained and these are described below.

Molecular differentiation in haustoria

Internal components

Two MAbs which bind to internal components of the haustorium (see Fig. 3b) were selected after immunisation with isolated HCs (Mackie *et al.*, 1991). The key finding here was that one of the MAbs, UB8, recognised an antigen which is specifically located in the hpm within the haustorium. EM–immunogold labelling showed that UB8 binds specifi-

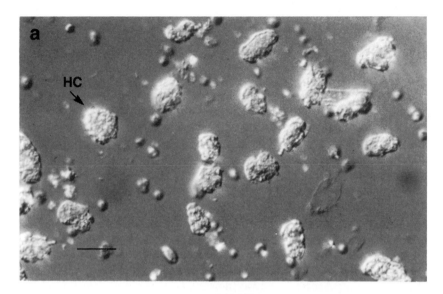

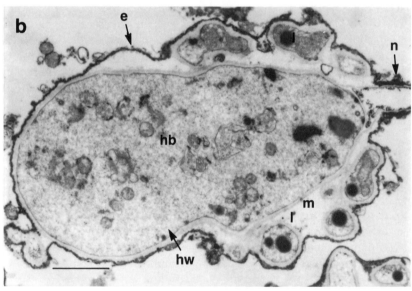

Fig. 2. Isolated haustorial complexes. (a) Light micrograph (differential interference contrast) of a preparation of haustorial complexes (HC) isolated from pea leaves infected with *Erysiphe pisi*. The bar represents 20 μm. (b) Transmission electron micrograph of an isolated HC showing the haustorial body (hb) and lobes (l), bounded by the haustorial plasma membrane and haustorial wall (hw). These structures are surrounded by the extrahaustorial matrix (m) and the extrahaustorial membrane (e). Part of the haustorial neck region can also be seen (n). The bar represents 3 μm.

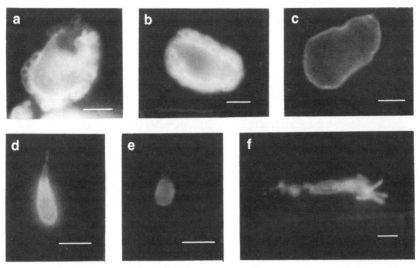

Fig. 3. Fluorescence images of haustoria treated with various antibody or lectin probes. (a) FITC-WGA labelling of an isolated HC showing fluorescence of haustorial walls of both the body and the lobes, but not the ehm. (b) to (e) Epidermal strips of pea infected with *Erysiphe pisi* labelled with MAbs followed by FITC-RAMIG. (b) MAb UB7 labelling of internal structures of a mature HC. (c) MAb UB11 labelling of the ehm of a mature HC. (d) MAb UB7 and (e) MAb UB11 labelling of HCs early in development. (f) Epidermal strip of wheat infected with *E. graminis* showing HC labelled with MAb UB7. The bars represent ((a) to (e)) 5 μm and (f) 10 μm.

cally to the haustorium, and the lack of cross-reaction with the surface hyphae was confirmed in Western blotting and enzyme-linked immunosorbent assays (ELISA). In contrast, immunogold labelling with a second MAb, UB7, showed binding primarily to fungal walls and plasma membranes of both the haustorium and surface hyphae; there was also some cytoplasmic labelling that could not be localised to any particular organelle. These results were in agreement with ELISA data obtained for binding of UB7 to crude fractions of hyphal wall, membranes and cytosol.

The EM–immunogold labelling patterns of antibody binding confirmed the initial observations made by indirect immunofluorescence (IIF) on isolated HCs, which showed that the MAbs were binding to internal structures. The evidence showed that the MAb UB8 recognises an antigen in the hpm that is specific to haustoria (Mackie *et al.*, 1991).

Western blotting showed that MAb UB8 recognises a 62 kDa antigen in HCs and, since the molecular mass is the same in reducing and non-reducing conditions, the antigen is probably a single polypeptide. The binding of UB8 to its antigen is sensitive to protease and insensitive to periodate, suggesting that the antibody is recognising a protein epitope. MAb UB7 recognises a 62 kDa antigen present in extracts of both haustoria and mycelia, showing that it is therefore a common antigen. An absence of binding of UB7 to blotted proteins separated on sodium dode-cyl sulphate/polyacrylamide gel electrophoresis (SDS/PAGE) in non-reducing conditions suggests that the antigen recognised is probably polymeric. Periodate sensitivity of antibody binding shows that UB7 recognises a carbohydrate epitope (Mackie *et al.*, 1991).

Endo-F treatment of antigens prior to their separation by SDS/PAGE showed that both UB7 and UB8 recognise glycoproteins with *N*-linked carbohydrate side-chains. Differences in the molecular size of the degradation products of UB7 and UB8 after endo-F treatment, together with differences in Western blots in non-reducing conditions, show that the 62-kDa antigens recognised by these antibodies are indeed distinct (Mackie *et al.*, 1991).

These results have shown that there is molecular differentiation within the HC, since UB8 recognises a glycoprotein unique to the hpm. The location at which UB8 begins to label the fungal membrane in the haustorium is at the neck region of the HC between the A- and B-neckbands and this coincides with the point at which ATPase activity originates in the fungal membrane of the haustorium (Spencer-Phillips & Gay, 1981). These results with UB8, and the previous data showing that this region is important for solute functions of the HC and is also the clear demarcation for the ehm from the plant plasma membrane (Gil & Gay, 1977; Bushnell & Gay, 1978), suggest that the membranes between the A- and B-neckbands form a key region in the differentiation of a functional infection structure.

Peripheral components

Two MAbs which label peripheral components of the HC have also been prepared (see Fig. 3c). MAb UB9 was raised against enriched pea plasma membrane fractions and this antibody binds to the pea membrane and to a proportion of HCs in infected leaves (A. M. Roberts *et al.*, unpublished results). EM–immunogold labelling shows that UB9 binds to the ehm, and IIF of isolated HCs (7 days after infection) and of epidermal strips derived from infected pea leaves shows that the antibody identifies a subpopulation of haustoria, labelling up to approximately 20% of HCs. UB9 recognises a carbohydrate epitope on a large glycoprotein

(>250 kDa) and the results show that at least one plant plasma membrane component is excluded from a proportion of ehms. This correlates with the results on the absence of particles and ATPase activity in this membrane (see p. 198, above).

MAb UB11 was raised to isolated HCs and binds to the periphery of all isolated HCs as shown by IIF. EM of pre-labelled HCs suggests that it binds to the outer face of the ehm (A. J. Mackie *et al.*, unpublished results). No binding of this MAb to mycelia, uninfected or infected plant tissue has been observed using ELISA and IIF techniques. The results therefore show that there is a specific antigen located in the ehm which must be synthesised during the formation of the HC. UB11 recognises a large glycoprotein (>200 kDa) and its binding characteristics show that it probably recognises a carbohydrate epitope.

Development of haustoria

Development in susceptible plants

The development of haustoria in pea leaves can be observed using epidermal strips labelled with MAbs and fluorescent probes. This system is particularly amenable to study using confocal laser scanning microscopy (A. J. Mackie *et al.*, unpublished results). The MAbs UB7, UB8 and UB11 all label haustoria at very early stages of development (Fig. 3d,e). Thus the specific glycoproteins recognised by UB8 (hpm) and UB11 (ehm) are inserted into their respective membranes shortly after infection, suggesting that new gene expression is switched on early. In contrast, UB9 labelling is clearly absent from the ehm of all haustoria at early stages of development. This antibody labels a subset of haustoria (approximately 20%) late in development, 5 to 7 days after infection. Thus at early stages of development the antigen recognised by UB9 is present on the plant membrane but is absent from the ehm.

Development in resistant interactions

Epidermal strips from several susceptible and resistant varieties of pea have been labelled with MAbs (A. J. Mackie *et al.*, unpublished results). Haustoria which have a healthy microscopical appearance in both resistant and susceptible plants, label with all the MAbs, though UB9 again recognises a subpopulation. Some haustoria are within autofluorescent cells in both resistant and susceptible cultivars and haustoria within these cells do not label with the MAbs. In resistant cultivars the labelled HCs tend to be found in increased numbers in the regions of the leaves where surface mycelial growth was observed and therefore it seems likely that these labelled HCs were functional at least at some stage. These haustoria therefore make up a 'compatible' population of haustorial–host interactions. The use of the term 'compatible' in this context considers each haustorial–epidermal cell interaction individually, such that a susceptible or resistant cultivar can give rise to both compatible and incompatible interactions. Infected susceptible cultivars show a predominance of compatible reactions and profuse fungal growth, with the majority of haustoria functioning normally, whereas the resistant cultivars tested, while showing a predominance of incompatible interactions, also supported a small proportion of compatible interactions.

Specificity of binding and other systems

Plants have a variety of interactions with microbes which include necrotrophic parasitic associations, biotrophic associations with fungal parasites, mycorrhizas and the *Rhizobium*–legume symbiosis (Smith & Smith, 1990). So far, it is only in legume root nodule formation that new gene products coded for by the plant and the bacterial partner have been found to be essential for a functional interaction (Gloudemans *et al.*, 1987; Verma *et al.*, 1986). There is some evidence for molecular differentiation in other systems. For example, changes in the polysaccharide/glycoprotein composition of haustorial walls during their maturation in the wheat stem rust fungus *Puccinia graminis* f. sp. *tritica* have been implicated using lectin probes (Chong *et al.*, 1986) and a MAb which specifically recognises the walls of the substomatal vesicles and infection hyphae of the rust fungus *Uromyces appendiculatus* has been described (Mendgen *et al.*, 1988). The presence of ectomycorrhizal specific polypeptides in the *Pisolithus/Eucalyptus* symbiosis have also been reported with molecular masses in the range 10–24 kDa (Hilbert & Martin, 1988). The results with UB8 and UB11 show that in the biotrophic interaction between peas and *E. pisi* glycoproteins restricted to haustoria are present and therefore show that specific gene expression is also required for a functional interaction in this system.

Binding of the MAbs UB7, 8, 9 and 11 have been tested in several other pathosystems and the evidence suggests that the MAbs which recognise specific haustorial glycoproteins (UB8, UB11) do not cross-react. However, UB7 labels wheat mildew (Fig. 3f), antirrhinum mildew (Mackie, 1991) and the mycelium and infection structures produced by *Colletotrichum lindemuthianum* in bean (N. Pain, R. J. O'Connell & J. R. Green, unpublished results). MAb UB9 binds to the haustoria formed in pea downy mildew (A. Beale & P. T. N. Spencer-Phillips, personal communication).

Conclusions

Cell biology of haustorial development

The results obtained with the set of MAbs described above have shown that there are several specific glycoproteins present in powdery mildew haustoria and these are expressed early in development. Some components normally present in the plant plasma membrane are excluded from the ehm, while a specific glycoprotein and at least one plant membrane glycoprotein become constituents of the ehm at different times during its development. The formation of the ehm is clearly a complex process.

The formation of a functional HC necessitates some cooperation between the plant and fungal partners, even though it is the fungus which benefits from the interaction. The host cytoplasm of infected cells usually shows evidence of high metabolic activity, e.g. there is an increase in the number of ribosomes and there is often an extensive endoplasmic reticulum developed adjacent to haustoria. Golgi vesicles are also profuse in this area and may fuse and augment the ehm while concomitantly contributing material to the ehm (Manners & Gay, 1983; Bushnell & Gay, 1978). During the formation of the ehm, macromolecules normally present in the plant membrane are not present (UB9 antigen and particles observed by freeze fracture), so there must be a drastic change in host targeting if the ehm is indeed of host origin. There is clearly a polarisation of activity of the epidermal cell during haustorial development and this aspect of HC formation should be amenable to study using infected epidermal strips and various probes.

Since the ehm increases in size with haustorial development, new membrane material must be produced and incorporated either with new antigens already formed or with novel antigen modifications occurring after incorporation. It may therefore seem more likely that the plant host provides this membrane material, possibly under fungal direction, in a way similar to that in which bacteroids can influence the plant to change the protein composition of the peribacteroid membrane in the *Rhizobium*–legume symbiosis (Werner *et al.*, 1988). It is also conceivable that this new membrane material could originate from the invading fungus with an incorporation mechanism into the ehm similar to that of the biogenesis of Gram-negative bacterial outer membranes (Bayer *et al.*, 1979; Davis & Tai, 1980). In the latter the inner membrane forms adhesion sites with the outer membrane, allowing membrane flow between the two. Close contact observed between the haustorial lobes and the ehm (Gil & Gay, 1977) could facilitate this and remove the need for a transport mechanism across the hpm, haustorial wall and extrahaustorial membrane. This is clearly an area that needs to be investigated.

It seems likely that the haustorial cytoplasm, hpm and haustorial wall are of fungal origin. Since the glycoprotein specific to the hpm, recognised by UB8, seems to arise on the fungal membrane in the region between the two neckbands it is probable that the antigen is synthesised by the fungus. The extrahaustorial matrix remains somewhat of a mystery, both in terms of its composition and origin. Whether it has any relationship to the plant cell wall or is made up of fungal components, or is a mixture of the two, will have consequences for models of haustorial development.

Functions of antigens identified by MAbs: recognition and signalling

The functions of the antigens recognised by the MAbs described above are not known, but the specific glycoproteins inserted into the hpm and ehm early in haustorial development must be crucial to haustorial function. Two possibilities are that they could be involved in nutrient transfer from the plant to the fungus, or may be part of cell signalling pathways. One way of elucidating the functions of antigens recognised by MAbs would be to set up functional assays and to perturb these with relevant antibodies (see e.g. Estrada-Garcia *et al.*, 1990). For example it may be possible to include MAbs in nutrient uptake experiments with isolated HCs or to use them to perturb development of HCs in epidermal strips *in vitro*. Another approach would be to determine the amino acid or coding gene sequences of the antigens and to look for homologies with known proteins or sequences: this is a clear goal in future work.

A variety of recognition events and signalling mechanisms must be operating between the plant and fungal partners during the development of the HC. It is possible to envisage signals involved in setting up the biotrophic interaction and a functional HC, and signals involved in generating resistant responses in the plant (for reviews, see Callow, 1984, 1987). The former must arise early in the infection process, since specific products (e.g. the glycoproteins recognised by MAbs UB8 and UB11) become components of the HC at very early stages of development. The latter may occur early or later in the infection process. Host responses may be activated after the formation of the HC, triggered by surface or extracellular components derived from the haustorium. It is interesting in this context that a host plasma membrane glycoprotein (identified by UB9) is incorporated into the ehm late in development in a subpopulation of HCs. Since haustoria are often found within autofluorescent cells late in the infection process, this subset of HCs may be an intermediate stage on the way to non-functionality within an 'incompatible' response (see pp. 205–60, above).

The use of MAb technology to study the host–parasite interface in pea powdery mildew has yielded important results concerning the differentiation and development of the HC. In addition to providing information on these aspects of this biotrophic interaction, it may be possible to use specific MAbs to prevent the disease from developing, since there have been reports of transformed plants expressing immunoglobulins (Hiatt *et al.*, 1989).

Acknowledgements

This work was supported by an AFRC grant and a studentship from the SERC.

References

Bayer, M. E., Thurow, H. & Bayer, M. H. (1979). Penetration of the polysaccharide capsule of *E. coli* by bacteriophage K29. *Virology* **94**, 95–118.

Bracker, C. E. (1968). Ultrastructure of the haustorial apparatus of *Erysiphe graminis* and its relationship to the epidermal cell of barley. *Phytopathology* **58**, 12–30.

Bushnell, W. R. (1971). The haustorium of *Erysiphe graminis*. An experimental study by light microscopy. In *Morphological and Biochemical Events in Plant–Parasite Interaction*, ed. S. Akai & S. Ouci, pp. 229–54. Tokyo: The Phytopathological Society of Japan.

Bushnell, W. R. & Gay, J. L. (1978). Accumulation of solutes in relation to the structure and function of haustoria in powdery mildews. In *The Powdery Mildews*, ed. D. M. Spencer, pp. 183–235. London: Academic Press.

Callow, J. A. (1984). Cellular and molecular recognition between higher plants and fungi. *Encyclopedia of Plant Physiology* **17**, 212–37.

Callow, J. A. (1987). Models for host–pathogen interaction. In *Genetics and Plant Pathogenesis*, ed. P. R. Day & G. J. Jellis, pp. 283–95. Oxford: Blackwell.

Chard, J. M. & Gay, J. L. (1984). Characterisation of the parasitic interface between *Erysiphe pisi* and *Pisum sativum* using fluorescent probes. *Physiological Plant Pathology* **25**, 259–76.

Chong, J., Harder, D. E. & Rohinger, R. (1986). Cytochemical studies on *Puccinia graminis* f.sp. *tritici* in a compatible wheat host. II. Haustorium mother cell walls at the host penetration site, haustorial walls, and the extrahaustorial matrix. *Canadian Journal of Botany* **64**, 2561–75.

Davis, B. D. & Tai, P. C. (1980). The mechanism of protein secretion across membranes. *Nature* **283**, 433–8.

Dickey, J. L. & Levy, M. (1979). Development of powdery mildew (*Erysiphe polygoni*) on susceptible and resistant races of *Oenothera biennis*. *American Journal of Botany* **66**, 1114–17.

Estrada-Garcia, T., Ray, T. C., Green, J. R., Callow, J. A. & Kennedy, J. F. (1990). Encystment of *Pythium aphanidermatum* zoospores is induced by root mucilage polysaccharides, pectin and a monoclonal antibody to a surface antigen. *Journal of Experimental Botany* **227**, 693–9.

Gay, J. L. & Manners, J. M. (1987). Permeability of the host–haustorium interface in powdery mildews. *Physiological and Molecular Plant Pathology* **30**, 389–99.

Gay, J. L., Salzberg, A. & Woods, A. M. (1987). Dynamic experimental evidence for the plasma membrane ATPase domain hypothesis of haustorial transport and for ionic coupling of the haustorium of *Erysiphe graminis* to the host cell (*Hordeum vulgare*). *New Phytologist* **107**, 541–8.

Gil, F. & Gay, J. L. (1977). Ultrastructural and physiological properties of the host interfacial components of haustoria of *Erysiphe pisi in vitro* and *in vivo*. *Physiological Plant Pathology* **10**, 1–12.

Gloudemans, T., de Vries, S., Bussink, H. J., Malik, N. S. A., Franssen, H. J., Louwerse, J. & Bisseling, T. (1987). Nodulin gene expression during soybean (*Glycine max*) nodule development. *Plant Molecular Biology* **8**, 395–405.

Hiatt, A., Cafferkey, R. & Bowdish, K. (1989). Production of antibodies in transgenic plants. *Nature* **343**, 76–8.

Hilbert, J. L. & Martin, F. (1988). Regulation of gene expression in ectomycorrhizas. I. Protein changes and the presence of ectomycorrhiza specific polypeptides in the *Psiolithus–Eucalyptus* symbiosis. *New Phytologist* **110**, 339–46.

Houslay, M. D. & Stanley, K. K. (1982). *Dynamics of Biological Membranes. Influence on Synthesis, Structure and Function*. Chichester, New York: J. Wiley & Sons.

Jones, J. L., Callow, J. A. & Green, J. R. (1988). Monoclonal antibodies to sperm surface antigens of the brown alga *Fucus serratus* exhibit region-, gamete-, species- and genus-preferential binding. *Planta* **176**, 298–306.

Key, G. & Weiler, E. W. (1988). Monoclonal antibodies recognise common and differentiation specific antigens on the plasma membrane of guard cells of *Vicia faba* L. *Planta* **176**, 472–81.

Knox, J. P., Day, S. & Roberts, K. (1989). A set of cell surface glycoproteins forms an early marker of cell position, but not cell type, in the root apical meristem of *Daucus carota* L. *Development* **106**, 47–56.

Mackie, A. J. (1991). Molecular analysis of the host–parasite interface in powdery mildew of pea using monoclonal antibodies. Ph.D. thesis, University of Birmingham.

Mackie, A. J., Roberts, A. M., Callow, J. A. & Green, J. R. (1991). Molecular differentiation in pea powdery-mildew haustoria. Identification of a 62 kDa *N*-linked glycoprotein unique to the haustorial plasma membrane. *Planta* **183**, 399–408.

Manners, J. M. (1989). The host–haustorium interface in powdery mildews. *Australian Journal of Plant Physiology* **16**, 45–52.

Manners, J. M. & Gay, J. L. (1977). The morphology of haustorial complexes isolated from apple, barley, beet and vine infected with powdery mildews. *Physiological Plant Pathology* **11**, 261–6.

Manners, J. M. & Gay, J. L. (1980). Autoradiography of haustoria of *Erysiphe pisi*. *Journal of General Microbiology* **116**, 529–33.

Manners, J. M. & Gay, J. L. (1982). Transport, translocation and metabolism of ^{14}C-photosynthates at the host-parasite interface of *Pisum sativum* and *Erysiphe pisi*. *New Phytologist* **91**, 221–44.

Manners, J. M. & Gay, J. L. (1983). The host–parasite interface and nutrient transfer in biotrophic parasitism. In *Biochemical Plant Pathology*, ed. J. A. Callow, pp. 163–95. Chichester: J. Wiley & Sons Ltd.

Mendgen, K. & Nass, P. (1988). The activity of powdery mildew haustoria after feeding the host cells with different sugars, as measured with a potentiometric cyanine dye. *Planta* **174**, 283–8.

Mendgen, K., Schneider, A., Sterk, M. & Fink, W. (1988). The differentiation of infection structures as a result of recognition events between some biotrophic parasites and their hosts. *Journal of Phytopathology* **123**, 259–72.

O'Connell, R. J., Bailey, J. A. & Richmond, D. V. (1985). Cytology and physiology of infection of *Phaseolus vulgaris* by *Colletotrichum lindemuthianum*. *Physiological Plant Pathology* **27**, 75–98.

Singh, H. B. & Singh, U. P. (1983). Development of *Erysiphe pisi* on susceptible and resistant cultivars of pea. *Transactions of the British Mycological Society* **81**, 275–8.

Singh, U. P., Singh, H. B. & Chauhan, V. B. (1982). Control of powdery mildew of pea with two new fungicides. *Phytopathologische Zeitschrift* **105**, 345–50.

Smith, S. E. & Smith, F. A. (1990). Structure and function of the interfaces in biotrophic symbioses as they relate to nutrient transport (Tansley Review No. 20). *New Phytologist* **114**, 1–38.

Spencer-Phillips, P. T. N. & Gay, J. L. (1981). Domains of ATPase in plasma membranes and transport through infected plant cells. *New Phytologist* **89**, 393–400.

Stumpf, M. A. & Gay, J. L. (1989). The haustorial interface in a resistant interaction of *Erysiphe pisi* with *Pisum sativum*. *Physiological and Molecular Plant Pathology* **35**, 519–33.

Verma, D. P. S., Fortin, M. G., Stanley, J., Mauro, V. P., Purohit, S. & Morrison, N. (1986). Nodulins and nodulin genes of *Glycine max.* *Plant Molecular Biology* **7**, 51–61.

Werner, D., Morschel, E., Garbers, C., Bassarab, S. & Mellor, R. B. (1988). Particle density and protein composition of the peribacteroid membrane from soybean root nodules is affected by mutation in the microsymbiont *Bradyrhizobium japonicum.* *Planta* **174**, 263–70.

Woods, A. M. & Gay, J. L. (1987). The interface between haustoria of *Puccinia poarum* (monokaryon) and *Tussilago farfara.* *Physiological and Molecular Plant Pathology* **30**, 167–85.

J. P. RIDE

Recognition signals and initiation of host responses controlling basic incompatibility between fungi and plants

Introduction

There is a basic incompatibility between higher plants and fungi at the species level, an incompatibility which is the norm in interactions between these two types of organism. Plants in general are resistant to most fungal species, even those that are pathogenic on other plants. This type of incompatibility is synonymous with non-host resistance. It is also sometimes termed 'heterologous incompatibility' to distinguish it from the specific incompatibility observed in homologous interactions between plants and pathogenic fungal species (Gabriel & Rolfe, 1990). In this chapter I summarise what we know about the recognition and response events which appear to be involved in this basic rejection of fungi by higher plants. The study of 'higher' levels of specificity (e.g. the specific incompatibility between races of some biotrophic pathogens and certain cultivars of their hosts) has perhaps attracted more attention, but it can be argued that a better understanding of basic incompatibility would enable more rapid progress to be made in understanding the mechanisms underlying successful infections (basic compatibility) and from thence the higher levels of specific resistance involved in homologous incompatibility.

Defence mechanisms controlling incompatibility

What types of mechanism operate at the level of basic incompatibility to inhibit fungal growth? Although we now seem to know much about the nature of the individual mechanisms that can operate during these interactions, there is certainly much that we do not understand. Plants have a battery of potential defence mechanisms against non-pathogens that dif-

Society for Experimental Biology Seminar Series 48: *Perspectives in Plant Cell Recognition*, ed. J. A. Callow & J. R. Green. © Cambridge University Press 1992, pp. 213–37.

fer from species to species, although themes are evident. However, we often know little about the co-ordination of these defences in a given interaction, i.e. their relative importance and how much they cooperate in a synergistic way. The situation is also complicated by the fact that the timing and mechanisms of resistance in a given plant species may vary depending on the invading species of fungus. Furthermore, different individuals of a given microbial population may be stopped at different sites and possibly by different mechanisms even within the same individual plant (Heath, 1977, 1981). Incompatibility is thus a complex phenomenon and it could be a mistake to overgeneralise.

On examining the types of resistance mechanism that have been identified, it is apparent that defence against non-pathogens may in part depend on passive, constitutive properties of the potential host. This appears to be more true for heterologous incompatibility (non-host resistance) than homologous incompatibility (e.g. varietal resistance). Defence may also depend on 'negative' features of the plant, i.e. the absence of characteristics that are important for successful differentiation and development of the potential pathogen, rather than on positive, inhibitory attributes. This is more often true for biotrophic fungi, which undergo complex structural differentiations in response to the presence of their hosts, than for necrotrophic fungi. For example, Wynn (1976) demonstrated that *Uromyces phaseoli*, the bean rust fungus, differentiates less successfully on the surfaces of leaves of non-host species, such as wheat and oats, than on host species. The number of haustoria produced, for instance, diminishes as the plant species becomes more distantly related to the natural hosts. The behaviour of the parasite on plastic replicas of the leaf surfaces almost exactly mimics its behaviour on the living leaf, indicating that the fungus responds primarily to the physical topography of the leaf surface. The 'mistakes' of the parasite on oats and wheat can therefore be attributed in part to the absence of suitable topographical triggers of differentiation on these plants, i.e. a negative, passive type of defence. This type of physical recognition is considered in more detail elsewhere (see Read *et al.*, this volume), but it is worthy of note that the 'mistakes' found in incompatible combinations rarely stop all of the members of the microbial population; other defence mechanisms must also be contributing to the resistance of the plant.

In fact, there is a substantial body of evidence to indicate that basic incompatibility depends heavily on responses induced in plant cells by fungi. Resistance can frequently be broken by pretreatment of plants with general inhibitors of metabolism or protein synthesis, e.g. heat shock, transcription and translation inhibitors (Heath, 1979; Tani *et al.*, 1976; Vance & Sherwood, 1976), indicating that an active response on the part of the plant is essential. Numerous active defence mechanisms have now been identified and they can be conveniently grouped into three main types: phytoalexins, antifungal proteins and cell wall alterations.

Phytoalexins are low molecular weight, broad-spectrum antimicrobial compounds which are known to be induced in many plants by fungal challenge and other stressful treatments. There have been many reviews on their nature, biosynthesis and the changes in gene expression underlying their induction (e.g. Dixon, 1986; Ebel, 1986). They are undoubtedly crucial to the disease resistance of many plant species. The demonstration by Snyder & Nicholson (1990) that they may be highly localised at the sites of attempted penetration by fungi must enhance our view of their significance. Their importance should not be overemphasised, however: not all plant species produce large quantities of phytoalexins when challenged and many other factors may play a role in resistance.

While phytoalexins have been deemed to be important for some time now, the potential importance of induced antifungal proteins has been appreciated only relatively recently. Many of these proteins were first noted as constitutive components of seeds, and the discovery of their induction, or the induction of similar proteins, in other tissues by microbial challenge followed later. In the case of the pathogenesis-related (PR) proteins, the induced proteins were identified by gel electrophoresis as being induced by viral infection long before any activity could be attributed to them. The list of defence proteins induced by microbial infection now includes the chitinases and $\beta(1–3)$-glucanases, which degrade fungal cell wall polymers, thionins, lectins and proteinase-inhibitors (for a review, see Bowles, 1990); the list will probably grow as new antifungal proteins are still being discovered in seeds (Roberts & Selitrennikoff, 1990). Lectins have had a rather chequered history as defence proteins. Mirelman *et al.* (1975) suggested that wheat germ agglutinin (WGA), a chitin-binding lectin, was antifungal. This suggestion was countered by Schlumbaum *et al.* (1986), who claimed that the activity was due to the presence of contaminating chitinase, both proteins being present at high levels in wheat germ, which was the source of the lectin. Interestingly, two lectins have recently been discovered, hevein from rubber trees and *Urtica dioica* agglutinin (UDA) from the rhizomes of stinging nettle, which show similarity to WGA and are strongly antifungal (Broekaert *et al.*, 1989; Van Parijs *et al.*, 1991). They are chitin-binding but have no chitinase activity associated with them. Although of lower molecular mass, they have a strong homology with WGA. They are known to be wound-induced (Broekaert *et al.*, 1990) and it will be interesting to see if they are pathogen-induced and whether homologous proteins exist in other plants.

Structural modifications of plant cell walls are known to occur widely in response to fungal challenge. The range includes callose deposition, lignification, suberisation, accumulation of silicon and calcium, deposition and cross-linking of wall phenolics, and the synthesis of hydroxyproline- or glycine-rich glycoproteins (for reviews, see Aist, 1983; Ride, 1983, 1986). Other types of structural response include tylosis, which involves the expansion of xylem parenchyma cells into the lumen of infection vessels, and renewed cambial activity to produce a barricade of cells, a response more commonly seen in tree species. Hypersensitive cell death could also be included in this category, death of host cells alone possibly having deleterious effects on biotrophic pathogens. Testing the effectiveness of any of these structural changes is extremely difficult. They could exert inhibitory effects in many different ways (Ride, 1983, 1986), many of which are not amenable to being tested *in vitro*.

Plants generally employ a combination of the above mechanisms. The relative importance of these mechanisms is often not well understood and is complicated by the fact that different combinations may operate in response to different fungi and that synergistic effects may occur. In some cases synergy between different active defences has been proven, e.g. in the action of chitinases and glucanases (Mauch *et al.*, 1988), and between these enzymes and ribosome-inactivating proteins (Leah *et al.*, 1991); in some cases synergy is strongly suspected, e.g. in the action of tyloses and phytoalexins (Bell & Mace, 1981). However, there are still many combinations that have not been examined, e.g. is the toxicity of phytoalexins enhanced by simultaneous treatment with chitinase/glucanase mixtures? There are also some 'defence-related responses' at whose effects on fungi we can only currently guess, e.g. hydroxyproline- and glycine-rich glycoproteins. There is therefore still much to be discovered about the ways in which plants inhibit fungal growth.

What induces these active defences?

Since the presence of fungi induces defensive reactions in plants, it seems sensible to assume that there are specific fungal molecules, the so-called elicitors, which are responsible for this induction. Most effort here has been concentrated on the phytoalexin response and a number of bioassays have been developed to detect such molecules, most of which are based on simple spectrophotometric tests. Less work has been expended on structural responses, but a number of assays have been developed based on histochemical reactions. Most of these are not quantitative, although it is possible to extract quantitative data from histochemical tests under suitable conditions (Barber & Ride, 1988). A different approach to elicitor bioassays has been suggested recently by Doerner *et al.* (1990). They fused a chalcone synthase promoter from bean with the *Escherichia coli* β-glucuronidase gene and then transformed tobacco with this construct. Chalcone synthase is a key enzyme in the production of the isoflavinoid phytoalexins characteristic of the legumes. Challenging such plants and assaying for glucuronidase activity, either by histochemical tests or assays of extracts, was proposed as a way to detect both elicitors and molecules which sensitise plants to elicitors. This is interesting but presumably a panel of such plants with different defence-activated promoters would be needed in order to dissect each system entirely.

Attempts to isolate signal molecules using bioassays has been somewhat complicated by the fact that many defence reactions in plants are stress-inducible. For instance, phytoalexin accumulation can be effected by all sorts of chemical, physical and biological treatments apart from fungal invasion (for a review, see Dixon, 1986), and many of the antifungal proteins are known to be induced by virus infection and chemical stresses (Bol *et al.*, 1990). It can thus be difficult to distinguish molecules which are involved in the specific recognition of fungal attributes by plants from those which are simply toxic. In fact it could be argued that recognition of 'stress' alone would be sufficient for a plant to detect attempted infection by fungi, particularly necrotrophic fungi. Again, however, it is dangerous to generalise too much. Studies on the induction of defensive lignification in wheat indicate a very high degree of specificity for filamentous fungi (Pearce & Ride, 1980). In this system, most chemicals that induce phytoalexins in other plants are inactive. It has also become apparent from other systems that even though defensive responses can be induced by a wide range of stresses, there is certainly an element of specificity for fungi, since a range of fungal products or structural components can elicit the response, and some of these treatments are not obviously toxic or stressful to plant tissues.

Two main categories of fungal elicitor have been identified: cell wall derived carbohydrates or glycoproteins, and secreted enzymes.

Elicitors derived from fungal cell walls

Of the molecules derived from cell walls, perhaps chitin, chitosan and the β-glucans are the most interesting. Because they are essential structural components of the walls of most fungi, they presumably are a source of recognition signals to plants which cannot easily be lost by, for example, mutation. Because of their insolubility, at least when complexed in the wall, it is necessary to envisage that plant enzymes cleave soluble oligosaccharides from these polymers and that these are the signal molecules which are recognised by the plant cells. Active oligosaccharides have, however, for practical reasons generally been identified first following partial acid hydrolysis. Relatively few defined oligosaccharide elicitors are known. A heptamer, released by acid hydrolysis from the β-glucan of *Phytophthora megasperma* f.sp *gycinea*, was isolated by Sharp *et al.* (1984) and is illustrated in Fig. 1. It was the smallest active fragment in the hydrolysate, inducing phytoalexin accumulation in soybean at nanomolar concentrations. Several closely related oligosaccharides were inactive. A wide range of similar synthetic oligosaccharides have recently been tested for elicitor activity (Cheong *et al.*, 1991), and it is clear that the trisaccharide at the non-reducing end of the oligosaccharide is essential for elicitor activity, as is the unbranched β(1–6)-linked glucose in the centre; however, alterations at the reducing terminus have little effect on activity. Soybean therefore is able to recognise a highly specific fragment from fungal cell walls, although it is clear that this fragment is not recognised by all plants (Barber & Ride, 1988).

Chitosan is the partially, or fully, deacetylated variant of chitin, a β(1–

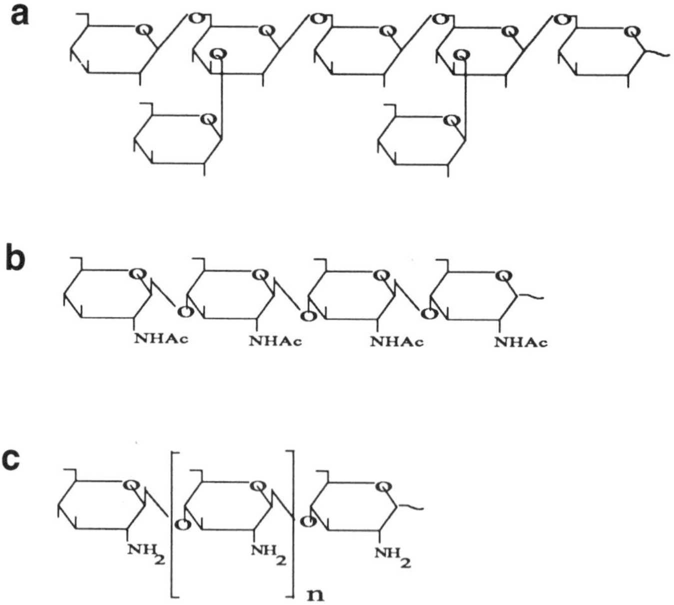

Fig. 1. Defined elicitor-active oligosaccharides from fungi: (a) β(1–3)-, β(1–6)-linked glucose heptamer from *Phytophthora megasperma*; (b) β(1–4)-linked *N*-acetylglucosamine tetramer from chitin; (c) β(1–4)-linked glucosamine oligomers from chitosan.

4) polymer of *N*-acetylglucosamine. Chitosan oligomers are active in a number of plant species, inducing the synthesis of phytoalexins (Hadwiger & Beckman, 1980; Keen *et al.*, 1983), callose (Kohle *et al.*, 1985; Conrath *et al.*, 1989), hydrolytic enzymes (Mauch *et al.*, 1984; Bernasconi *et al.*, 1986) and proteinase-inhibitor synthesis (Walker-Simmons & Ryan, 1984). The oligosaccharides (Fig. 1) have usually been generated by HCl hydrolysis or nitrous acid cleavage (which yields 2,5-anhydromannose end units) of chitosan. In general there does not seem to be a critical chain length, activity simply increasing with degree of polymerisation, and Kauss (1987) has concluded that chitosan and its oligosaccharides act primarily as polycations which interact with the negative phospholipids in the plant plasma membrane.

The partial acetylation of chitosan greatly reduced the ability of the molecule to elicit callose formation in *Catharanthus roseus* (Kauss *et al.*, 1989). This effect does not apply to all systems. The completely de-acetylated form of chitosan was found to be inactive in eliciting lignification in wheat (Barber & Ride, 1988), while the fully acetylated polymer, chitin, and oligosaccharides derived from it were active in this system (Barber *et al.*, 1989; Pearce & Ride, 1982). In this case a minimum chain length appeared to be important, with the tetrasaccharide and higher oligomers being active. This type of β(1–4)-linked oligosaccharide of *N*-acetylglucosamine is also active in inducing chitinase in melon plants (Roby *et al.*, 1987). Since chitin is frequently exposed at the hyphal tip, it seems likely that this type of signal oligosaccharide would be released by constitutive, extracellular plant chitinases upon attempted fungal invasion (Ride *et al.*, 1989). Certainly the four major constitutive chitinases in wheat leaves are all *endo* in action, and all release oligosaccharides from chitin which are of sufficient size to be elicitor-active (Ride & Barber, 1990). Similarly, plant-derived β(1–3)glucanases have been shown to release phytoalexin elicitors from fungal cell walls in other systems (Keen & Yoshikawa, 1983). The *endo*-enzymes which release elicitors, and additional glycosidases (Barber & Ride, 1989), may also be responsible for the turnover of the elicitors *in vivo* (Ride *et al.*, 1989); such degradation may be important in preventing unnecessary spread of the molecules within plant tissues, and hence in confining the energetically expensive defensive reactions to the immediate site of invasion.

Although cell wall-derived oligosaccharides are known to elicit defence reactions in plants, it is not clear how significant they are during natural infections. In most cases it is not known whether they are actually released *in planta*, or whether they are present in physiologically significant amounts. Because of concern over the relatively large amounts of chitin tetramer (when administered as a single dose) required to elicit lignification in wheat, we have recently measured the levels of all chitin oligosaccharides in wheat leaves treated with chitin or the tetrasaccharide (M. S. Barber & J. P. Ride, unpublished results). Following treatment with chitin, the profile of oligosaccharides *in planta* was found to be very similar to that observed *in vitro* when chitin was digested with purified chitinases, i.e. elicitor-active oligosaccharides were being released in the leaves. However, the concentrations of these signal molecules was several fold less in chitin-treated leaves than in leaves treated with a dose of tetramer of equivalent biological activity. While there are several possible explanations for these results, the general conclusion is that the straightforward $\beta(1-4)$-linked *N*-acetylglucosamine oligomers are of no real physiological significance. Preliminary results have indicated that a small number of deacetylated residues greatly enhances the activity of the oligosaccharides and that these modified forms may be of greater significance. Deacetylation of chitin is known to occur in fungal hyphae (Davis & Bartnicki-Garcia, 1984).

Secreted elicitors and plant-derived elicitors

Apart from the cell wall-derived products, pathogenic fungi also secrete some elicitors, most notable of which are enzymes which degrade plant cell walls. Several pectic enzymes have been shown to have elicitor activity, which in general seems to be mediated by the release of so-called endogenous elicitors from the plant cell wall (for reviews, see West *et al.*, 1985; Ryan, 1987). These are oligosaccharides of $\alpha(1-4)$-linked galacturonic acid, with a chain length of between 10 and 13 being required for optimum activity. Interestingly, oligosaccharides released from the cell walls of wheat appear to suppress defence responses rather than elicit them (Moerschbacher *et al.*, 1990), indicating again that different species of plants may have quite different recognition systems. Another type of fungal enzyme which has been shown to elicit defence responses is xylanase (Lotan & Fluhr, 1990; Dean & Anderson, 1991), although in this case the elicitor activity seems to depend directly on the enzyme's structure rather than its ability to release plant cell wall fragments.

In addition to the oligogalacturonides, certain other plant products have been shown to elicit or enhance resistance reactions – most notably the stress hormones ethylene and abscisic acid (Boller, 1988; Roberts & Kolattukudy, 1989), and stress-induced salicylate (Metraux *et al.*, 1990). The latter is interesting because of its association with the phenomenon of systemic acquired resistance, where challenge of one part of a plant may induce a higher degree of resistance in other quite distant parts. This type of response indicates that secondary signals, such as salicylate, can have quite profound effects at some distance from the site of the primary challenge.

Synergy between elicitors

In general the physiological significance of any of the known fungal elicitors has not been proven. A major problem is that measurement of *in vivo* levels is very difficult for most of the characterised elicitors, but the situation is also complicated by the fact that some of the elicitors may have synergistic effects resulting in a greater activity than might be anticipated. Synergy has certainly been noted for some of the purified elicitors. Davis *et al.* (1986) observed that in soybean the phytoalexin-inducing activity of a fungal hexa-β-glucosyl glucitol derived from *P. megasperma* cell walls was greatly enhanced by the presence of endopolygalacturonate lyase. Synergy was also observed between the fungal oligosaccharide and the purified decagalacturonide released from polygalacturonic acid by the lyase, indicating that the plant was much more sensitive to a mixture of endogenous and exogenous elicitors than to either alone. Similar results have been observed in parsley (Davis & Hahlbrock, 1987) and, with cruder preparations, in bean (Tepper & Anderson, 1990).

Synergy has also been observed between various stressful treatments and fungal elicitors. In soybean a pronounced synergy between acetate ions and fungal oligosaccharides has been recorded (Davis *et al.*, 1986), which was presumed to be acting via endogenous galacturonides, although this may not necessarily be the case. In wheat, prior wounding of leaves seems to greatly enhance the activity of several elicitors of lignification (Barber *et al.*, 1989). In general, it seems that plants have mechanisms for recognising damage, which results in an increase in their sensitivity to the products of potential pathogens.

There could be many explanations for synergy between treatments, but some clues can perhaps be extracted from studies on the changes in gene expression which accompany most of the induced defences in plants. Much of the work has been concerned with phytoalexin induction, and most has been conducted with tissue-cultured plant cells treated with relatively crude elicitor preparations from fungi, typically heat-released elicitors derived from fungal cell walls or mycelium. In parsley it has been shown that the changes observed in the cell culture system mimic very closely those oberved in the interaction between the whole plant and the live fungus (*P. megasperma*; Schmelzer *et al.*, 1989). The changes in mRNA and protein synthesis induced by elicitors can be very extensive and very rapid. For instance Bollmann & Hahlbrock (1990) reported that in elicitor-treated parsley they could detect 25 new proteins on two-dimensional gels after methionine labelling; they also noted that the synthesis of 10 proteins was suppressed. In general, enhanced protein synthesis in these systems is related to rapid and transient increases in transcripts from the appropriate genes. For some defence genes, activation of transcription has been detected as early as just a few minutes after elicitor treatment (see e.g. Lawton & Lamb, 1987; Hedrick *et al.*, 1988), although the activation of all genes is not this swift (Showalter *et al.*, 1985; Lawton & Lamb, 1987).

These response genes often exist as small families and there is increasing evidence that individual genes are regulated in different ways by different stresses, even within the same gene family. For example the chitinase gene family in peanut consists of four genes which are regulated in different ways by fungal elicitors and other stresses such as ultraviolet light (Herget *et al.*, 1990). Similar observations have been made for genes for phenylalanine ammonia-lyase (Cramer *et al.*, 1989; Jorrin & Dixon, 1990) and hydroxyproline-rich glycoproteins (Corbin *et al.*, 1987). Even elicitor subfractions from a given fungus may induce different responses in terms of phytoalexin accumulation and translatable mRNA populations (Hamdan & Dixon, 1987). One explanation for the synergy observed between treatments might therefore be a complementation effect: since different genes are regulated by different stimuli, the effect of a combination of treatments might be much greater than the sum of the effects of the individual treatments.

Interestingly, not only does the identity of the individual genes expressed vary with different treatments, but so also do the timing and spatial organisation. For instance Schmelzer *et al.* (1989) used antisense RNA probes for 4-coumarate:CoA ligase to locate transcripts in sections of parsley tissue by hybridisation *in situ*. Challenging leaves with an incompatible species of *Phytophthora* resulted in a highly localised induction of mRNA at the site of inoculation, while physically wounding a leaf resulted in a much larger zone of induction. Such results suggest the operation of several recognition and signal transduction pathways in the activation of plant defence genes (Lamb *et al.*, 1989).

It is clear therefore that despite the fact that many treatments may induce apparently the same gross result, e.g. phytoalexin synthesis or increase in chitinase activity, the mechanisms underlying these effects may actually be quite different.

So how do the fungal elicitors induce these rapid and specific changes in plant cells?

Are there specific receptors in plant cells for fungal elicitors?

This is the point at which hypotheses tend to become more abundant than experimental evidence, and there is a natural tendency to rely on mammalian systems for a source of ideas. In general animal cells are sensitive to external signal molecules, such as hormones, because of the interaction of the signal ligand with a specific protein receptor. Four general classes of receptor have been identified (Haga *et al.*, 1990). The first is typified by the receptors for certain polypeptide hormones, such as epidermal growth factor and insulin. These receptors consist of a monomeric (or heterooligomeric) protein with a single transmembrane segment per subunit, a ligand binding site on the external membrane surface, and an associated catalytic activity (usually protein tyrosine kinase) on the internal side. The second group include receptors for neurotransmitter ligands, e.g. γ-amino butyric acid. These are all oligomeric proteins, with each subunit having multiple membrane-spanning regions; a ligand binding site is located on the external face, and binding activates an ion channel with a defined specificity. A large number of receptors belong to the third group: these are all proteins with multiple (7) membrane-spanning regions and a ligand binding site on the external face; coupling to G-proteins occurs on the internal face eventually leading to activation of protein kinases via a number of routes. All three types mentioned so far are located in the plasma membrane and hence distinct from the fourth class, the steroid hormone receptors; these are internal proteins, nucleus-associated, and the receptor–ligand complex has direct DNA-binding properties.

Since the vast majority of the receptors seen in animal systems are plasma membrane proteins, and cell surface receptors are a common theme in cell–cell recognition events (Chapman *et al.*, 1988), it seems logical to propose that specific receptors for fungal elicitors are present at the surface of plant cells. Some indirect evidence has been obtained from observing the cellular location of fluorescein-labelled elicitor preparations after their application to soybean cell suspension cultures (Horn *et al.*, 1989). The label accumulated relatively rapidly at the cell surface, later concentrating in the vacuole. No such binding or internalisation was seen with fluorescein-labelled bovine serum albumin or insulin, which were inactive as elicitors. This was interpreted as indicating an interaction of the elicitor molecules with cell surface receptors, followed by receptor-mediated endocytosis as a way of destroying the elicitor after signal transduction.

More direct evidence for membrane receptors has come from work on the oligoglucoside elicitors of soybean phytoalexins derived from the cell walls of *P. megasperma* f.sp. *glycinea*. Most of this work has been conducted with a fraction obtained by gel filtration of an acid hydrolysate of the cell wall β-glucan; this fraction demonstrates specific, displaceable binding to a soybean microsomal preparation (Schmidt & Ebel, 1987; Cosio *et al.*, 1988). Solubilisation of the binding from a microsomal preparation has been achieved (Cosio *et al.*, 1990*a*). Binding of any size of oligosaccharide appears to be due to the hepta-β-glucoside structure that Sharp *et al.* (1984) had originally identified as being the smallest elicitor-active fragment in acid hydrolysates (Cosio *et al.*, 1990*b*). Recently this has been confirmed and extended by the work of Cheong & Hahn (1991), who compared the elicitor activities of a range of synthetic oligoglucosides with their binding affinities for a soybean microsomal preparation. As with previous work they took advantage of the fact that extending the reducing terminus of the oligomer has little effect on activity, and used an iodinated tyrosine residue at this position to generate a radiolabelled ligand. Inhibition of binding by a range of synthetic analogues was compared with their phytoalexin-eliciting capacities. The affinity of the different oligosaccharides for the soybean membrane preparation showed a remarkable correlation with elicitor activity, with the concentrations required for the two activities being very close. These results strongly suggest a physiologically significant receptor site in soybean membranes. The availability of more highly defined elicitors and analogues such as these would greatly assist the identification of receptors for other elicitors.

Preliminary evidence for a plasma membrane receptor has also recently been reported in wheat (Kogel *et al.*, 1991). A glycoprotein elicitor of lignification isolated from the germ-tube walls of *Puccinia graminis* f.sp. *tritici* has been shown to bind specifically to plasma membrane enriched preparations from wheat and barley leaves. The antigenicity of this elicitor has allowed an immunological approach to the measurement of binding, in addition to the more conventional radiolabelling approach.

There are several other approaches to the identification and purification of receptors which have been applied in other systems but have yet to appear in studies on elicitor receptors in plants. Affinity chromatography using covalently immobilised ligands or analogues has been widely applicable in other systems; the availability of heptaglucoside-based derivatives should mean the application of this technique to the *Phytophthora/* soybean system will be reported in the near future. Since some receptors in animal systems are tyrosine-phosphorylated on interaction with their ligand, the possibility of using antiphosphotyrosine antibody-based affinity chromatography in plant systems should perhaps also be considered.

Affinity or photo-affinity labelling have also not yet been reported in these systems. These techniques involve the synthesis of modified ligands which are capable of binding covalently to the other molecules with which they interact, the tagged molecules including it is hoped the natural receptor(s).

The use of anti-idiotype monoclonal antibodies in identifying the receptors for elicitors is also a possibility (Strosberg, 1989). These are antibodies which have been raised against antibodies to the ligand, i.e. anti-antiligand antibodies. A subpopulation of these antibodies may bind to the receptor; a problem here with plant systems may be how to recognise these antibodies. Identification is usually by functional inhibition tests, an approach which is made more difficult in plants by the presence of the cell wall, except in situations where protoplasts demonstrate a response to the ligand.

It is possible that cloning receptor genes will be feasible in the near future. Some potential approaches, such as cloning based on common sequences (Haga *et al.*, 1990), are only useful for identifying other members of a receptor family after the first has been cloned and sequenced. Other approaches, such as cloning based on screening mRNAs transcribed from cDNAs for a functional effect when micro-injected into cells, could be applied to plant cells but perhaps only with some difficulty. However, it is conceivable that inducing mutation to loss of sensitivity, e.g. in *Arabidopsis* by irradiation, or by transposon mutagenesis in other plants, could provide a route to receptor genes – although such mutants could prove difficult to culture except under axenic conditions. The simplest initial approach may be to purify the receptor, raise antibodies and screen an expression library.

Whichever approach is adopted, the task of identifying receptors for elicitors would undoubtedly be eased by the availability of more chemically defined elicitors of proven physiological significance. It must also be remembered that for some of the elicitors there may not necessarily be a defined receptor. As mentioned previously, Kauss (1987) has suggested that, for chitosan oligomers at least, there could be a relatively non-specific interaction of the positively charged elicitor with the negative phospholipid head groups on the plant plasmalemma. This is thought to result in changes in the movement of certain ions across the membrane, notably calcium ions moving inwards resulting in immediate changes in the activity of key enzymes such as the plasmalemma-bound glucan synthase involved in callose deposition. However, the involvement of calcium in chitosan-induced defences in pea has been questioned (Kendra & Hadwiger, 1987).

Signal–response coupling in incompatibility

This leads finally to what we know about the way in which the reception of fungal signal molecules at the cell surface is coupled within the cell to the expression of the response, which frequently involves a major change in gene expression. Again hypotheses are more plentiful than experimental results and it has proved tempting to look to animal cell systems for ideas.

For those receptors in animal cells with a single transmembrane segment per subunit, ligand binding results in enhancement of a catalytic activity on the inner membrane face. The activity is often a protein kinase specific for tyrosine residues, and the receptors are usually capable of autophosphorylation. Phosphorylation of proteins in animal cells is a major mechanism for regulating activity, although it is not yet known which are the key substrates for these receptors (Haga *et al.*, 1990). Since no receptors for elicitors have yet been characterised, it is impossible to know whether this type of coupling occurs in the responses of plants to fungi.

Different mechanisms operate in animal cells in those cases where the receptor has multiple membrane spanning regions and interacts with G proteins. Binding of ligand generally results in dissociation of the G-protein complex with the released α-subunit binding GTP. This complex can then activate other transducers via several pathways. Activation of adenylate cyclase may occur, resulting in enhanced levels of intracellular cyclic AMP (cAMP). cAMP-dependent protein kinase is then activated resulting in protein phosphorylation and an increase in another activity, e.g. an enzyme. Again there is no conclusive evidence that this type of mechanism operates in plant cells in response to fungi, and the role of cyclic nucleotides in general in stimulus–response coupling in plants has been dismissed (Boss, 1989; Morre, 1989). Neither adenylate cyclase (Yunghans & Morre, 1977) nor cAMP-dependent kinases appear to be present in plants (Brown & Newton, 1981), and levels of cAMP in elicitor-treated and untreated soybean cells have been reported to be similar and extremely low (Hahn & Grisebach, 1983). However, exogenously supplied cAMP or dibutyryl cAMP can elicit phytoalexin accumulation in some systems (Oguni *et al.*, 1976; Kurosaki *et al.*, 1987), and elicitor-induced changes in endogenous cAMP levels have been reported (Kurosaki *et al.*, 1987). These facts coupled with the recent cloning of plant genes encoding proteins resembling cAMP responsive proteins from animal cells (Katagiri *et al.*, 1989; Lawton *et al.*, 1989) indicate that it is too early to dismiss cyclic nucleotides altogether from response-coupling in plant cells.

Activation of G-proteins in animal cells may alternatively result in enhanced phospholipase C activity and the release of diacylglycerol (DAG) and inositol-1,4,5-trisphosphate (InsP_3) from phosphatidylinositol-4,5-bisphosphate (PtdInsP_2). Each of these metabolites is a potential second messenger, initiating a cascade of metabolic events (for a review, see Boss, 1989). For example InsP_3 mobilises calcium from intracellular stores, hence activating calcium- and calcium/calmodulin-dependent enzymes such as protein kinases; DAG increases the affinity of protein kinase C for calcium and thus activates the enzyme. There is some evidence that the components of these systems are present in plants; for instance, the plasma membranes of plant cells contain phosphoinositides and the associated enzymes involved in their biosynthesis and metabolism. However, it is clear that phosphoinositide cycling in plants may differ in several important ways from that in animal systems (Boss, 1989), and there is certainly as yet no conclusive evidence that these second messengers operate in the response of plants to fungal elicitors. Strasser *et al.* (1986) found no significant changes in phosphoinositides or inositol phosphates when elicitors were added to parsley or soybean cell cultures. Chen & Boss (1990) also found no changes in inositol-1,4-bisphosphate or InsP_3 in elicitor-treated carrot cells, although a possibly significant increase in PtdInsP_2 was observed.

There is more substantive evidence for a role for calcium as a second messenger in plant cells (Marme, 1989). I have already mentioned the proposed role for calcium in callose deposition (Kauss, 1987), a defence response which is unusual in that it is not necessary to invoke a change in gene expression. However, indirect evidence also exists that calcium is involved in elicitation of phytoalexin synthesis. Thus, experiments with Ca^{2+} channel blockers such as verapamil, ionophores such as A23187, and direct modifications of extracellular Ca^{2+} levels, generally support the view of a role for Ca^{2+} somewhere in phytoalexin elicitation (Stab & Ebel, 1987; Kurosaki *et al.*, 1987), although there are exceptions (Kendra & Hadwiger, 1987). The application of Ca^{2+}-sensitive fluorescence and patch clamp techniques to these systems will, I hope, yield more direct information on calcium changes within these cells.

Some receptors in animal cells are ligand-gated ion channels with a specificity for certain ions e.g. Na^+ or Cl^-. The involvement of this type of receptor in the response of plants to fungi has been proposed by Gabriel *et al.* (1988), with various receptors with different ligand specificities being invoked to explain the various levels of specificity in host–pathogen interactions. The hypothesis is attractively simple but there is as yet no direct evidence to support it.

It is apparent that in many of the response-coupling systems in animal cells there is a role for protein phosphorylation, either at tyrosine residues (in cases where the receptor itself possesses protein kinase activity) or serine/threonine residues; sometimes the receptor itself is rapidly phosphorylated. There have been reports that protein phosphorylation can occur in membrane preparations treated with elicitors (Farmer *et al.*, 1989). However, more convincing evidence for a role of phosphorylation comes from the recent report by Dietrich *et al.* (1990) on responses of living parsley cells. They showed that treatment of cultured cells with fungal elicitor triggers rapid, transient and sequential phosphorylation of a number of polypeptides. For instance, a 45 kDa polypeptide was phosphorylated as early as 1–2 min after elicitor treatment. The effect was rapidly reversed by removal of elicitor from the medium. Pronase treatment destroyed both elicitor activity and phosphorylation-inducing ability of fungal preparation. Interestingly, the response did not occur in response to other phytoalexin-inducing stresses, e.g. ultraviolet light or $HgCl_2$, again supporting the view that different stresses may operate via different signal transduction pathways. Phosphorylation apparently depended on the presence of calcium, which suggests that elicitation involves a calcium-dependent protein kinase.

Additional support for the involvement of protein phosphorylation in responses to elicitors comes from the use of inhibitors. Grosskopf *et al.* (1990) demonstrated that the microbial metabolite 'K-252a', a known protein kinase inhibitor, blocks the response of tomato cells to fungal elicitor, as well as inhibiting a microsomal protein kinase *in vitro*. Protein kinase inhibitors also suppress production of the phytoalexin pisatin (Shiraishi *et al.*, 1990). It should be interesting to identify and characterise the protein kinase implicated in these interactions, although it should perhaps be remembered that elicitor-induced dephosphorylation has also been observed in some systems (Grab *et al.*, 1989), and that this may also be involved in signal transduction.

While protein phosphorylation seems to be the earliest known event in plant cells following elicitor treatment or microbial challenge, there are several other rapid changes which have been observed which might play a role in response coupling. Rapid changes in membrane potential have been observed in several systems (Tomiyama *et al.*, 1983; Pelissier *et al.*, 1986; Mayer & Ziegler, 1988). Changes in vacuolar and cytoplasmic pH have been observed in bean suspension cultures within 10 min of elicitor treatment (Ojalvo *et al.*, 1987), such changes possibly accounting for some of the metabolic alterations observed in elicitor-treated cells (Kneusel *et al.*, 1989). There also have been suggestions that hydrogen peroxide (Lindner *et al.*, 1988; Apostol *et al.*, 1989), superoxide (Doke & Chai, 1985; Lindner *et al.*, 1988), hydroxyl radicals (Epperlein *et al.*,

1986) and lipid peroxidation (Rogers *et al.*, 1988; Adam *et al.*, 1989) may be involved in plant cell responses. Other, novel, signal transduction mechanisms (Anderson, 1989; Scherer, 1989) could perhaps also be involved in responses to elicitors.

Conclusions

Incompatibility between plants and fungi clearly involves major changes in the activity of affected plant cells. All of the phenomena described above may have a role to play somewhere in the incompatibility response; the difficulty is in distinguishing cause and effect. Also, since it is clear that different types of treatment may induce defence responses in plants by quite different routes, it would be a mistake to expect the same recognition and coupling systems to be active in all fungus–plant interactions. Unfortunately, most of the fungal elicitor preparations used so far have not been fully characterised. For those that have been defined, we still have little idea of their spectrum of activity, i.e. how many different defence responses they can elicit, in which plant species, and how many fungi contain them? The availability of a greater number of well-characterised elicitors and synthetic analogues would greatly advance our attempts to dissect these complex signalling systems. It must also not be forgotten that, particularly for necrotrophic fungi, the stress induced by attempted fungal infection is likely to release constitutive and endogenous elicitors which may act synergistically with fungal products as well as having elicitor activity in their own right. Thus, incompatibility in these systems involves an intricate and continuing interaction of various fungal-specific and non-specific signals, evoking response networks by mechanisms whose complexity we are only just beginning to understand.

References

Adam, A., Farkas, T., Somlyai, G., Hevesi, M. & Kiraly, Z. (1989). Consequence of O_2^- generation during a bacterially induced hypersensitive reaction in tobacco: deterioration of membrane lipids. *Physiological and Molecular Plant Pathology* **34**, 13–26.

Aist, J. R. (1983). Structural responses as resistance mechanisms. In *The Dynamics of Host Defence*, ed. J. A. Bailey & B. J. Deverall, pp. 33–70. Sydney: Academic Press.

Anderson, J. M. (1989). Membrane-derived fatty acids as precursors to second messengers. In *Second Messengers in Plant Growth and Development*, ed. W. F. Boss & D. J. Moree, pp. 181–212. New York: Alan R. Liss, Inc.

Apostol, I., Heinstein, P. F. & Low, P. S. (1989). Rapid stimulation of an oxidative burst during elicitation of cultured plant cells. Role in defense and signal transduction. *Plant Physiology* **90**, 109–16.

Barber, M. S., Bertram, R. E. & Ride, J. P. (1989). Chitin oligosaccharides elicit lignification in wounded wheat leaves. *Physiological and Molecular Plant Pathology* **34**, 3–12.

Barber, M. S. & Ride, J. P. (1988). A quantitative assay for induced lignification in wounded wheat and its use to survey potential elicitors of the response. *Physiological and Molecular Plant Pathology* **32**, 185–97.

Barber, M. S. & Ride, J. P. (1989). Purification and properties of a wheat leaf *N*-acetyl-β-ᴅ-hexosaminidase. *Plant Science* **60**, 163–72.

Bell, A. A. & Mace, M. E. (1981). Biochemistry and physiology of resistance. In *Fungal Wilt Diseases of Plants*, ed. M. E. Mace., A. A. Bell & C. H. Beckman, pp. 431–86. London: Academic Press.

Bernasconi, P., Jolles, P. & Pilet, P. E. (1986). Increase of lysozyme and chitinase in *Rubus calli* caused by infection and some polymers. *Plant Science* **44**, 79–83.

Bol, J. F., Linthurst, H. J. M. & Cornelissen, B. J. C. (1990). Plant pathogenesis-related proteins induced by virus infection. *Annual Review of Phytopathology* **28**, 113–38.

Boller, T. (1988). Ethylene and the regulation of antifungal hydrolases in plants. *Oxford Surveys of Plant Molecular and Cell Biology* **5**, 145–74.

Bollmann, J. & Hahlbrock, K. (1990). Timing of changes in protein synthesis pattern in elicitor-treated cell suspension cultures of parsley (*Petroselinum crispum*). *Zeitschrift für Naturforschung: C-A Journal of Biosciences* **45**, 1011–20.

Boss, W. F. (1989). Phosphoinositide metabolism: its relation to signal transduction in plants. In *Second Messengers in Plant Growth and Development*, ed. W. F. Boss & D. J. Morre, pp. 29–56. New York: Alan R Liss Inc.

Bowles, D. J. (1990). Defence-related proteins in higher-plants. *Annual Review of Biochemistry* **59**, 873–907.

Broekaert, W., Lee, H. I., Kush, A., Chua, N. H. & Raikhel, N. (1990). Wound-induced accumulation of messenger RNA containing a hevein sequence in laticifers of rubber tree (*Hevea brasiliensis*). *Proceedings of the National Academy of Sciences, USA* **87**, 7633–7.

Broekaert, W. F., Van Parijs, J., Leyns, F., Joos, H. & Peumans, W. J. (1989). A chitin-binding lectin from stinging nettle rhizomes with antifungal properties. *Science* **245**, 1100–2.

Brown, E. G. & Newton, R. P. (1981). Cyclic AMP and higher plants. *Phytochemistry* **20**, 2453–63.

Chapman, G. P., Ainsworth, C. C. & Chatham, C. J. (1988). *Eukaryote Cell Recognition. Concepts and Model Systems*. Cambridge: Cambridge University Press.

Chen, Q. Y. & Boss, W. F. (1990). Short-term treatment with cell wall degrading enzymes increases the activity of the inositol phospholipid kinases and the vanadate-sensitive ATPase of carrot cells. *Plant Physiology* **94**, 1820–9.

Cheong, J. J., Birberg, W., Fugedi, P., Pilotti, A., Garegg, P. J., Hong, N., Ogawa, T. & Hahn, M. G. (1991). Structure–activity relationships of oligo-β-glucoside elicitors of phytoalexin accumulation in soybean. *Plant Cell* **3**, 127–36.

Cheong, J. J. & Hahn, M. G. (1991). A specific, high-affinity binding site for the hepta-β-glucoside elicitor exists in soybean membranes. *Plant Cell* **3**, 137–47.

Conrath, U., Domard, A. & Kauss, H. (1989). Chitosan-elicited synthesis of callose and of coumarin derivatives in parsley cell-suspension cultures. *Plant Cell Reports* **8**, 152–5.

Corbin, D. R., Sauer, N. & Lamb, C. J. (1987). Differential regulation of a hydroxyproline-rich glycoprotein gene family in wounded and infected plants. *Molecular and Cellular Biology* **7**, 4337–44.

Cosio, E. G., Frey, T. & Ebel, J. (1990a). Solubilization of soybean membrane binding sites for fungal beta-glucans that elicit phytoalexin accumulation. *FEBS Letters* **264**, 235–8.

Cosio, E. G., Frey, T., Verduyn, R., Van Boom, J. & Ebel, J. (1990b) High-affinity binding of a synthetic heptaglucoside and fungal glucan phytoalexin elicitors to soybean membranes. *FEBS Letters* **271**, 223–6.

Cosio, E. G., Popperl, H., Schmidt, W. E. & Ebel, J. (1988). High-affinity binding of fungal beta-glucan fragments to soybean (*Glycine max* L.) microsomal fractions and protoplasts. *European Journal of Biochemistry* **175**, 309–15.

Cramer, C. L., Dron, M., Dixon, R. A., Dildine, S. L., Edwards, K., Lamb, C. J., Liang, X. W., Schuch, W. & Bolwell, G. P. (1989). Phenylalanine ammonia-lyase gene organization and structure. *Plant Molecular Biology* **12**, 367–83.

Davis, K. R., Darvill, A. G. & Albersheim, P. (1986). Host-pathogen interactions. XXXI. Several biotic and abiotic elicitors act synergistically in the induction of phytoalexin accumulation in soybean. *Plant Molecular Biology* **6**, 23–32.

Davis, K. R. & Hahlbrock, K. (1987). Induction of defence responses in cultured parsley cells by plant cell wall fragments. *Plant Physiology* **85**, 1286–90.

Davis, L. L. & Bartnicki-Garcia, S. (1984). Chitosan synthesis by the tandem action of chitin synthetase and chitin deacetylase from *Mucor rouxii. Biochemistry* **23**, 1065–73.

Dean, J. F. D. & Anderson, J. D. (1991). Ethylene biosynthesis-inducing xylanase. 2. Purification and physical characterization of the enzyme produced by *Trichoderma viride. Plant Physiology* **95**, 316–23.

Dietrich, A., Mayer, J. E. & Hahlbrock, K. (1990). Fungal elicitor triggers rapid, transient, and specific protein-phosphorylation in parsley cell-suspension cultures. *Journal of Biological Chemistry* **265**, 6360–8.

Dixon, R. A. (1986). The phytoalexin response: elicitation, signalling and control of host gene expression. *Biological Reviews* **61**, 239–91.

Doerner, P. W., Stermer, B., Schmid, J., Dixon, R. A. & Lamb, C. J. (1990). Plant defense gene promoter-reporter gene fusions in transgenic plants: tools for identification of novel inducers. *Bio/Technology* **8**, 845–8.

Doke, N. & Chai, H. B. (1985). Activation of superoxide generation and enhancement of resistance against compatible races of *Phytophthora infestans* in potato plants treated with digitonin. *Physiological Plant Pathology* **27**, 323–34.

Ebel, J. (1986). Phytoalexin synthesis: the biochemical analysis of the induction process. *Annual Review of Phytopathology* **24**, 235–64.

Epperlein, M. M., Noronha-Dutra, A. A. & Strange, R. N. (1986). Involvement of the hydroxyl radical in the abiotic elicitation of phytoalexins in legumes. *Physiological and Molecular Plant Pathology* **28**, 67–77.

Farmer, E. E., Pearce, G. & Ryan, C. A. (1989). *In vitro* phosphorylation of plant plasma membrane proteins in response to the proteinase inhibitor inducing factor. *Proceedings of the National Academy of Sciences, USA* **86**, 1539–42.

Gabriel, D. W., Loschke, D. C. & Rolfe, B. G. (1988). Gene-for-gene recognition: the ion channel defense model. In *Molecular Genetics of Plant Parasite Interactions*, ed. R. Palacios & D. P. S. Verma, pp. 3–14. St Paul, MN: APS Press.

Gabriel, D. W. & Rolfe, B. G. (1990). Working models of specific recognition in plant–microbe interactions. *Annual Review of Phytopathology* **28**, 365–91.

Grab, D., Ebel, J. & Feger, M. (1989). An endogenous factor from soybean (*Glycine max* L.) cell-cultures activates phosphorylation of a protein which is dephosphorylated *in vivo* in elicitor-challenged cells. *Planta* **179**, 340–8.

Grosskopf, D. G., Felix, G. & Boller, T. (1990). K-252a inhibits the response of tomato cells to fungal elicitors *in vivo* and their microsomal protein kinase *in vitro*. *FEBS Letters* **275**, 177–80.

Hadwiger, L. A. & Beckman, J. M. (1980). Chitosan as a component of pea–*Fusarium solani* interactions. *Plant Physiology* **66**, 205–11.

Haga, T., Haga, K. & Hulme, E. C. (1990). Solubilization, purification, and molecular characterization of receptors: principles and strategy. In *Cellular and Molecular Biology of Plant Stress*, ed. J. L. Key & T. Kosuge, pp. 215–24. New York: Alan R. Liss, Inc.

Hahn, M. G. & Grisebach, H. (1983). Cyclic AMP is not involved as a secondary messenger in the response of soybean to infection by *Phytophthora megasperma* f.sp. *glycinea*. *Zeitschrift für Naturforschung* **38c**, 578–82.

Hamdan, M. A. M. S. & Dixon, R. A. (1987). Differential pattern of protein synthesis in bean cells exposed to elicitor fractions from *Colletotrichum lindemuthianum. Physiological and Molecular Plant Pathology* **31**, 105–21.

Heath, M. C. (1977). A comparative study of non-host interactions with rust fungi. *Physiological Plant Pathology* **10**, 73–88.

Heath, M. C. (1979). Effects of heat shock, actinomycin D, cycloheximide and blasticidin S on non-host interactions with fungi. *Physiological Plant Pathology* **15**, 211–18.

Heath, M. C. (1981). Resistance of plants to rust infection *Phytopathology* **71**, 971–4.

Hedrick, S. A., Bell, J. N., Boller, T. & Lamb, C. J. (1988). Chitinase cDNA cloning and mRNA induction by fungal elicitor, wounding, and infection. *Plant Physiology* **86**, 182–6.

Herget, T., Schell, J. & Schreier, P. H. (1990). Elicitor-specific induction of one member of the chitinase gene family in *Arachis hypogaea. Molecular and General Genetics* **224**, 469–76.

Horn, M. A., Heinstein, P. F. & Low, P. S. (1989). Receptor-mediated endocytosis in plant cells. *Plant Cell* **1**, 1003–9.

Jorrin, J. & Dixon, R. A. (1990). Stress responses in alfalfa (*Medicago sativa* L.). 2. Purification, characterization, and induction of phenylalanine ammonia-lyase isoforms from elicitor-treated cell-suspension cultures. *Plant Physiology* **92**, 447–55.

Katagiri, F., Lam, E. & Chua, N. H. (1989). Two tobacco DNA-binding proteins with homology to the nuclear factor CREB. *Nature* **340**, 727–9.

Kauss, H. (1987). Some aspects of calcium-dependent regulation in plant metabolism. *Annual Review of Plant Physiology* **38**, 47–72.

Kauss, H., Jeblick, W. & Domard, A. (1989). The degree of polymerization and *N*-acetylation of chitosan determine its ability to elicit callose formation in suspension cells and protoplasts of *Catharanthus roseus. Planta* **178**, 385–92.

Keen, N. T. & Yoshikawa, M. (1983). β-1,3-Endoglucanase from soybean releases elicitor active carbohydrates from fungus cell walls. *Plant Physiology* **71**, 460–5.

Keen, N. T., Yoshikawa, M. & Wang, M. C. (1983). Phytoalexin elicitor activity of carbohydrates from *Phytophthora megasperma* f.sp. *glycinea* and other sources. *Plant Physiology* **71**, 466–71.

Kendra, D. F. & Hadwiger, L. A. (1987). Calcium and calmodulin may not regulate the disease resistance and pisatin formation responses of *Pisum sativum* to chitosan or *Fusarium solani. Physiological and Molecular Plant Pathology* **31**, 337–48.

Kneusel, R. E., Nicolay, K. & Matern, U. (1989). Formation of *trans*-caffeoyl-CoA from *trans*-4-coumaroyl-CoA by Zn^{2+}-dependent enzymes in cultured plant-cells and its activation by an elicitor-induced pH shift. *Archives of Biochemistry and Biophysics* **269**, 455–62.

Kogel, G., Beissmann, B., Reisener, H. J. & Kogel, K. (1991). Specific binding of a hypersensitive lignification elicitor from *Puccinia graminis* f.sp. *tritici* to the plasma membrane from wheat (*Triticum aestivum* L.). *Planta* **183**, 164–9.

Kohle, H., Jeblick, W., Poten, F., Blaschek, W. & Kauss, H. (1985). Chitosan-elicited callose synthesis in soybean cells as a Ca^{2+}-dependent process. *Plant Physiology* **77**, 544–51.

Kurosaki, F., Tsurusawa, Y. & Nishi, A. (1987). The elicitation of phytoalexins by Ca^{2+} and cyclic AMP in carrot cells. *Phytochemistry* **26**, 1919–23.

Lamb, C. J., Lawton, M. A., Dron, M. & Dixon, R. A. (1989). Signals and transduction mechanisms for activation of plant defenses against microbial attack. *Cell* **56**, 215–24.

Lawton, M. A. & Lamb, C. J. (1987). Transcriptional activation of plant defense genes by fungal elicitor, wounding and infection. *Molecular and Cellular Biology* **7**, 334–41.

Lawton, M. A., Yamamoto, R. T., Hanks, S. K. & Lamb, C. J. (1989). Molecular cloning of plant transcripts encoding protein kinase homologs. *Proceedings of the National Academy of Sciences, USA* **86**, 3140–4.

Leah, R., Tommerup, H., Svendsen, I. & Mundy, J. (1991). Biochemical and molecular characterization of three barley seed proteins with antifungal properties. *Journal of Biological Chemistry* **266**, 1564–73.

Lindner, W. A., Hoffmann, C. & Grisebach, H. (1988). Rapid elicitor-induced chemiluminescence in soybean cell suspension cultures. *Phytochemistry* **27**, 2501–3.

Lotan, T. & Fluhr, R. (1990). Xylanase, a novel elicitor of pathogenesis-related proteins in tobacco, uses a nonethylene pathway for induction. *Plant Physiology* **93**, 811–17.

Marme, D. (1989). The role of calcium and calmodulin in signal transduction. In *Second Messengers in Plant Growth and Development*, ed. W. F. Boss & D. J. Morre, pp. 57–80. New York: Alan R. Liss, Inc.

Mauch, F., Hadwiger, L. E. & Boller, T. (1984). Ethylene: symptom, not signal for the induction of chitinase and β-1,3-glucanase in pea pods by pathogens and elicitors. *Plant Physiology* **76**, 607–11.

Mauch, F., Mauch-Mani, B. & Boller, T. (1988). Antifungal hydrolases in pea tissue. II. Inhibition of fungal growth by combinations of chitinase and beta-1,3-glucanase. *Plant Physiology* **88**, 936–42.

Mayer, M. G. & Ziegler, E. (1988). An elicitor from *Phytophthora megasperma* f.sp. *glycinea* influences the membrane potential of soybean cotyledonary cells. *Physiological and Molecular Plant Pathology* **33**, 397–407.

Metraux, J. P., Signer, H., Ryals, J., Ward, E., Wyss Benz, M., Gaudin, J., Raschdorf, K., Schmid, E., Blum, W. & Inverardi, B. (1990). Increase in salicylic acid at the onset of systemic acquired resistance in cucumber. *Science* **250**, 1004–6.

Mirelman, D., Galun, E., Sharon, N. & Lotan, R. (1975). Inhibition of fungal growth by wheat germ agglutinin. *Nature* **256**, 414–16.

Moerschbacher, B. M., Schrenk, F., Graessner, B., Noll, U. & Reisener, H. J. (1990). A wheat cell wall fragment suppresses elicitor-induced resistance responses and disturbs fungal development. *Journal of Plant Physiology* **136**, 761–4.

Morre, D. J. (1989). Stimulus–response coupling in auxin regulation of plant cell elongation. In *Second Messengers in Plant Growth and Development*, ed. W. F. Boss & D. J. Morre, pp. 29–56. New York: Alan R. Liss Inc.

Oguni, I., Suzuki, K. & Uritani, I. (1976). Terpenoid induction in sweet potato roots by cyclic-3,5-adenosine monophosphate. *Agricultural and Biological Chemistry* **40**, 1251–2.

Ojalvo, I., Rokem, J. S., Navon, G. & Goldberg, I. (1987). ^{31}P NMR study of elicitor treated *Phaseolus vulgaris* cell suspension cultures. *Plant Physiology* **85**, 716–19.

Pearce, R. B. & Ride, J. P. (1980). Specificity of induction of the lignification response in wounded wheat leaves. *Physiological Plant Pathology* **16**, 197–204.

Pearce, R. B. & Ride, J. P. (1982). Chitin and related compounds as elicitors of the lignification response in wounded wheat leaves. *Physiological Plant Pathology* **20**, 119–23.

Pelissier, B., Thibaud, J. B., Grignon, C. & Esquerre-Tugaye, M. T. (1986). Cell surfaces in plant–microorganism interactions. VII. Elicitor preparations from two fungal pathogens depolarize plant membranes. *Plant Science* **46**, 103–9.

Ride, J. P. (1983). Cell walls and other structural barriers in defence. In *Biochemical Plant Pathology*, ed. J. A. Callow, pp. 215–36. Chichester: John Wiley & Sons Ltd.

Ride, J. P. (1986). Induced structural defences in plants. In *Natural Antimicrobial Systems* Part 1. *Antimicrobial Systems in Plants and Animals*, ed. G. W. Gould, M. E. Rhodes-Roberts, A. K. Charnley, R. M. Cooper & R. G. Board, pp. 159–75. Bath: Bath University Press.

Ride, J. P. & Barber, M. S. (1990). Purification and characterisation of multiple forms of endochitinase from wheat leaves. *Plant Science* **71**, 185–97.

Ride, J. P., Barber, M. S. & Bertram, R. E. (1989). Infection-induced lignification in wheat. In *Plant Cell Wall Polymers. Biogenesis and Biodegradation*, ed. N. G. Lewis & M. G. Paice, pp. 361–9. Washington, DC: ACS.

Roberts, E. & Kolattukudy, P. E. (1989). Molecular cloning, nucleotide sequence and abscisic acid induction of a suberization-associated highly anionic peroxidase. *Molecular and General Genetics* **217**, 217–23.

Roberts, W. K. & Selitrennikoff, C. P. (1990). Zeamatin, an antifungal protein from maize with membrane-permeabilizing activity. *Journal of General Microbiology* **136**, 1771–8.

Roby, D., Gadelle, A. & Toppan, A. (1987). Chitin oligosaccharides as elicitors of chitinase activity in melon plants. *Biochemical and Biophysical Research Communications* **143**, 885–92.

Rogers, K. R., Anderson, A. J. & Albert, F. (1988). Lipid-peroxidation is a consequence of elicitor activity. *Plant Physiology* **86**, 547–53.

Ryan, C. A. (1987). Oligosaccharide signalling in plants. *Annual Review of Cell Biology* **3**, 295–317.

Scherer, G. F. E. (1989). Ether phospholipid platelet-activating factor (PAF) and a proton-transport-activating phospholipid (PAP): potential new signal transduction constituents for plants. In *Second Messengers in Plant Growth and Development*, ed. W. F. Boss & D. J. Morre, pp. 167–79. New York: Alan R. Liss, Inc.

Schlumbaum, A., Mauch, F., Vogeli, U. & Boller, T. (1986). Plant chitinases are potent inhibitors of fungal growth. *Nature* **324**, 365–7.

Schmelzer, E., Kruger-Lebus, S. & Hahlbrock, K. (1989). Temporal and spatial patterns of gene expression around sites of attempted fungal infection in parsley leaves. *Plant Cell* **1**, 993–1001.

Schmidt, W. E. & Ebel, J. (1987). Specific binding of a fungal phytoalexin elicitor to membrane fractions from soybean *Glycine max*. *Proceedings of the National Academy of Sciences, USA* **84**, 4117–21.

Sharp, J. K., McNeil, M. & Albersheim, P. (1984). The primary structures of one elicitor-active and seven elicitor-inactive hexa (β-D-glucopyranosyl)-D-glucitols isolated from the mycelial walls of *Phytophthora megasperma* f.sp. *glycinea*. *Journal of Biological Chemistry* **259**, 11321–36.

Shiraishi, T., Hori, N., Yamada, T. & Ohu, H. (1990). Suppression of pisatin accumulation by an inhibitor of protein kinase. *Annals of the Phytopathological Society of Japan* **56**, 261–4.

Showalter, A. M., Bell, J. N., Cramer, C. L., Bailey, J. A., Varner, J. E. & Lamb, C. J. (1985). Accumulation of hydroxyproline-rich glycoprotein mRNAs in response to fungal elicitor and infection. *Proceedings of the National Academy of Sciences, USA* **82**, 6551–5.

Snyder, B. A. & Nicholson, R. L. (1990). Synthesis of phytoalexins in sorghum as a site-specific response to fungal ingress. *Science* **248**, 1637–9.

Stab, M. R. & Ebel, J. (1987). Effects of Ca^{2+} on phytoalexin induction by fungal elicitor in soybean cells. *Archives of Biochemistry and Biophysics* **257**, 416–23.

Strasser, H., Hoffman, C., Grisebach, H. & Matern, U. (1986). Are polyphosphoinositides involved in signal transduction of elicitor-induced phytoalexin synthesis in cultured plant cells? *Zeitschrift für Naturforschung C* **41**, 717–24.

Strosberg, A. D. (1989). Interaction of antiidiotypic antibodies with membrane-receptors – practical considerations. *Methods in Enzymology* **178**, 179–91.

Tani, T., Yamamoto, H., Kadota, G. & Naito, N. (1976). Development of rust fungi in oat leaves treated with blasticidin S, a protein synthesis inhibitor. *Technical Bulletin of Faculty of Agriculture Kagawa University* **27**, 95–103.

Tepper, C. S. & Anderson, A. J. (1990). Interactions between pectic fragments and extracellular components from the fungal pathogen *Colletotrichum lindemuthianum*. *Physiological and Molecular Plant Pathology* **36**, 147–58.

Tomiyama, K., Okamoto, H. & Katou, K. (1983). Effect of infection by *Phytophthora infestans* on the membrane potential of plant cells. *Physiological Plant Pathology* **22**, 233–43.

Van Parijs, J., Broekaert, W. F., Goldstein, I. J. & Peumans, W. J. (1991). Hevein – an antifungal protein from rubber-tree (*Hevea brasiliensis*) latex. *Planta* **183**, 258–64.

Vance, C. P. & Sherwood, R. T. (1976). Cycloheximide treatments implicate papilla formation in resistance of reed canarygrass to fungi. *Phytopathology* **66**, 498–502.

Walker-Simmons, M. & Ryan, C. A. (1984). Proteinase inhibitor synthesis in tomato leaves. Induction by chitosan oligomers and chemically modified chitosan and chitin. *Plant Pathology* **76**, 787–90.

West, C. A., Moesta, P., Jin, D. F., Lois, A. F. & Wickham, K. A. (1985). The role of pectic fragments of the plant cell wall in the response to biological stresses. In *Cellular and Molecular Biology of Plant Stress*, ed. J. L. Key & T. Kosuge, pp. 335–49. New York: Alan R. Liss, Inc.

Wynn, W. K. (1976). Appressorium formation over stomates by the bean rust fungus: response to a surface contact stimulus. *Phytopathology* **66**, 136–46.

Yunghans, W. N. & Morre, D. J. (1977). Adenylate cyclase activity not found in soybean hypocotyl and onion meristem. *Plant Physiology* **60**, 144–9.

P. BONFANTE-FASOLO, R. PERETTO
AND S. PEROTTO

Cell surface interactions in endomycorrhizal symbiosis

Mycorrhizae are the most widespread type of association established between plants and soil microorganisms. The roots of 90% of land plants associate with many soil fungi to form complex systems, whose structure and function depend on the specific combination of the eukaryotic partners. During the formation of mycorrhizae, numerous events of specificity and recognition occur at different levels: species, plant organ, root tissue and cell type.

Specificity and recognition in mycorrhizae

Among the different 260 000 plant species, 200 000–240 000 have been estimated to have the potential to form mycorrhizal associations (Law & Lewis, 1983). On the other hand, many fungal representatives of the Zygomycotina, Ascomycotina, Basidiomycotina and Deuteromycotina depend on the mycorrhizal association for the completion of their life cycle (Harley, 1989). The single word 'mycorrhiza' encompasses a complexity of forms of interaction, since at least six main types of mycorrhiza can be recognised (Table 1). Extensive observations indicate that some specificity exists in the interaction, as a range of compatible and incompatible hosts can be defined. According to Smith & Douglas (1987) specificity of a symbiosis refers 'to the degree of taxonomic difference between acceptable partners and may vary from very low to high or very high'. From Table 1 it is clear that mycorrhizae have a low degree of specificity compared to other associations, as a single fungus may associate with plants of more than one class and also a single host plant may associate with different fungal endophytes. The processes leading to the selection of the correct partners, or to discrimination against the wrong ones, are collectively known as *recognition* (Smith & Douglas, 1987). Many experiments have demonstrated that recognition takes place in

Society for Experimental Biology Seminar Series 48: *Perspectives in Plant Cell Recognition*, ed. J. A. Callow & J. R. Green. © Cambridge University Press 1992, pp. 239–55.

Table 1. *The most important types of mycorrhiza and their related symbionts*

Type of mycorrhiza	Plant symbiont	Fungal symbiont
Vesicular–arbuscular mycorrhizae	Representatives of Bryophytes, Pteridophytes Gymnosperms and many Angiosperms	Glomales
Ectomycorrhizae	Trees and shrubs especially of temperate regions	25 families Basidiomycotina, 7 families of Ascomycotina, 1 genus of Zygomycotina (*Endogone*)
Ericoid	Ericales with fine hair roots	Discomycete *Hymenoschyphus ericae*, Hyphomycetes, Onygenales (*Oidiodendron*), dark sterile mycelia
Arbutoid	Ericales such as *Arbutus* and *Arctostaphylos*	Ascomycotina and Basidiomycotina related to ectomycorrhizal fungi
Monotropoid	Achlorophyllous members of Ericales	Basidiomycotina related to ectomycorrhizal fungi
Orchid mycorrhizae	All members of Orchidaceae	8 genera of Basidiomycotina referable to *Rhizoctonia*

Modified from Smith & Douglas (1987).

mycorrhizae and the association is unsuccessful when the partners are inappropriate. For example, fungal species which form ericoid mycorrhizae on Ericales can act as weak pathogens on non-host plants, such as *Trifolium*, commonly involved in another type of endomycorrhiza (Bonfante-Fasolo *et al.*, 1984). The same fungi cause a darkening of the short root tips in *Arbutus*, which usually hosts ectendomycorrhizae (Giovannetti & Lioi, 1990).

Mycorrhizae are naturally formed between the roots of plants and soil fungi. The existence of organ specificity has been confirmed by studies *in vitro*. Organs different from the roots are free from the symbionts, and isolated cells or undifferentiated calli are not infected. On the contrary, isolated roots, including roots transformed by *Agrobacterium rhizogenes*, represent good hosts for fungi with vesicular–arbuscular mycorrhizae (VAMs) (Becard & Pichet, 1989). This suggests that expression of root specific genes could be required for the establishment of mycorrhizae. The root is composed of different tissues and the fungus invades some tissues (i.e. cortical ones) but not others (i.e. meristems and central cylinder), indicating a high degree of tissue specificity. The inaccessibility of the central cylinder to fungal penetration may be due to the presence of a physico-chemical barrier such as the root endodermis and to the low hydrolysing capacities of the mycorrhizal fungal enzymes (Bonfante-Fasolo & Perotto, 1991). However, some ericoid fungi have been shown to overcome this cellular barrier in non-host plants, suggesting the existence of an active control on the fungus by the host plant (Bonfante-Fasolo *et al.*, 1984). The lack of colonisation of the root meristem is far more difficult to explain on the basis of enzymic activity from the fungus and it is in fact a common phenomenon of symbiosis in roots (Bonfante-Fasolo & Perotto, 1991).

During the establishment of a symbiosis, the two partners come in close contact and identify themselves as compatible or incompatible with each other. This stage is referred to as *cell-to-cell recognition*. It involves specific physical and chemical signals which result in tolerance or suppression of the host defence response (Smith & Douglas, 1987). A variety of cell-to-cell physical interactions can be identified in mycorrhizae, including simple intercellular contacts and more complex intracellular contacts.

In this chapter we analyse some aspects of plant–fungal recognition at the cellular level in endomycorrhizae and in particular in VAMs.

The role of cell surfaces in cell-to-cell recognition in endomycorrhizae

In VAMs an important step of the infection process is represented by the intracellular colonisation by the fungus, a zygomycete belonging to the *Glomales* (Morton & Benny, 1990). The hyphae penetrate root cortical cells and develop into different specialised structures. They form coils in the outer root layers, intercellular hyphae and lastly arbuscules, which are highly branched intracellular structures (for a review of the infection process, see Bonfante-Fasolo, 1984). An investigation based largely on electron microscopy, cytochemistry and immunological techniques has suggested that the development of the intracellular fungal structures requires an adjustment of the cell surfaces of both partners. All cell-to-cell interactions are in fact mediated through walls that must enable passage of nutritional and signalling molecules to the partners' plasma membranes. This situation gives rise to the establishment of interface structures, where molecules typical of the apoplast are deeply involved (Bonfante-Fasolo & Scannerini, 1991; Smith & Smith, 1990). In VAMs, two different types of interface are recognised, depending on whether the fungus develops among the host cells or penetrates them. In the first case, the walls of both the partners are directly in contact, while in the second case the specialised intracellular hyphae are separated from the host cytoplasm by the invaginated host membrane and by an interfacial material (Figs. 1, 2 and 3).

We focus on this second type of interface, since its establishment is accompanied by defined adjustments of the partners' cell surfaces.

The plant–fungal interface as an example of specialised cell surfaces

The fungal wall

From a structural point of view the interface consists of the fungal plasma membrane and wall, of the interfacial material and of the host membrane (Figs. 1 and 3). Detailed analysis of fungal morphology showed important

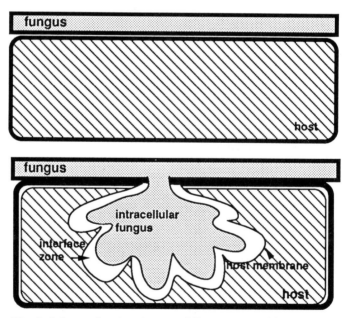

Fig. 1. Scheme showing the two different types of interface between a VAM fungus and its host. (Top) Interface of the intercellular phase. (Bottom) Interface of the intracellular phase.

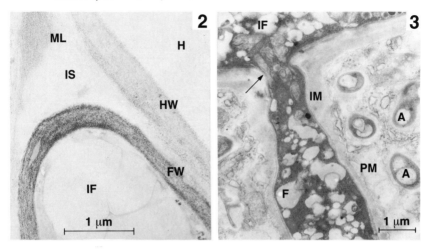

Figs. 2 and 3. Ultrastructural aspects of the two different types of interface between a VAM fungus, *Glomus versiforme*, and its host plant
Allium porrum.
Fig. 2. The intercellular fungus (IF) with a fibrillar wall (FW) passes
through the intercellular space (IS), where only remnants of the middle
lamella (ML) are visible. H, host cortical cell; HW, host wall.
Fig. 3. The intercellular fungus (IF) produces a penetrating hypha (F),
causing the invagination of the host plasma membrane (PM) at the point
of penetration (arrow). An interface space is created where interfacial
material (IM) is present. The penetrating hypha branches to produce
smaller hyphae (the arbuscule branches, A). Each branch is surrounded
by the host membrane.

modifications of cell walls in VAM fungi during their infection process.
There is a striking reduction in thickness (from 0.5 µm to 20–30 nm)
associated with a change of texture from fibrillar in the intercellular
hyphae into amorphous in the arbuscules. Chitin represents the skeletal
molecule in the wall of VAM fungi, and *N*-acetylglucosamine residues
can be identified at each step of fungal morphogenesis by using specific
lectins bound to fluorescent or electron dense markers (Fig. 4). These
experiments suggest that chitin or oligomers of chitin are regularly synthesised during all stages of fungal development (Bonfante-Fasolo *et al.*,
1990*a*). Concanavalin A (Con-A) binding sites are easily identified at the
surface of all the infection structures, indicating the presence of mannose
and glucose residues. However, the Con-A binding sites, which occur at
the arbuscule surface, are fully removed by treatments with oxidising

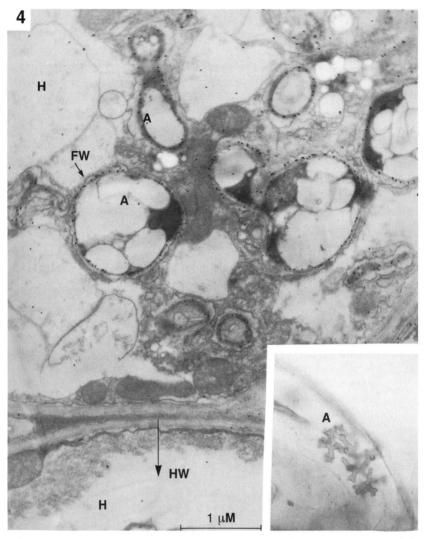

Fig. 4. Ultrastructural aspects of the interface in a young arbuscule (A) (in the inset it is shown at the light microscope). The fungal wall (FW) is regularly labelled by wheat germ agglutinin which is bound to gold granules. No labelling occurs over the host wall (HW). H, host cell.

agents. We suggest that the thin and amorphous wall of arbuscules contains chitin in non-crystalline form and glucans which are not yet bound to chitin.

The interfacial material

The observation that wall material is deposited by the plant around the hyphae during their intracellular growth dates back many years (Scannerini & Bonfante-Fasolo, 1979). Affinity probes, such as enzymes, lectins or antibodies conjugated to colloidal gold, allowed us to define better some of the molecules occurring in this compartment. They have revealed the presence of molecules which are common to the primary wall, even if they do not assemble into a structure which resembles the architecture of the peripheral cell wall. This means that the cytochemical continuity between the peripheral host wall and the interfacial material around the living arbuscular branches does not correspond to a morphological continuity (Bonfante-Fasolo *et al.*, 1990*b*).

β(1-4)-Glucans were localised around the intracellular fungus by using an enzyme, cellobiohydrolase (CBH1), whose substrate is cellobiose (Chanzy *et al.*, 1983). When used as a gold probe over the thin sections of resin embedded VAM fungal roots, it binds specifically to the host, as labelling is detected over the host wall, but not over the fungal wall (Bonfante-Fasolo *et al.*, 1990*b*). This allows us to identify the glucans found in the interface as molecules of host origin. In contrast, in ericoid mycorrhizae, CBH1-gold labels not only the host cell walls – as expected – but also the fungal wall. This shows that fungal glucans are co-localised with chitin over the fungal wall. The cross-reactivity between plant and fungal glucans makes it impossible to define with this probe the origin of the material occurring in the region that separates ericoid partners.

Non-esterified and methylesterified polygalacturonans were identified in the interfacial compartment in pea, leek and *Ginkgo* roots by using two monoclonal antibodies, JIM7 and JIM5, respectively (Knox *et al.*, 1990). Labelling was present over the host cell wall and followed a precise distribution pattern, which appeared to be species- and tissue-specific (Bonfante-Fasolo & Perotto, 1991). In mycorrhizal cells, pectic molecules were easily detected over the interface and were particularly abundant among the collapsed fungal branches (Bonfante-Fasolo *et al.*, 1990*b*; Figs. 5–8). Cell wall proteins were also localised on the interfacial material: in particular, an antibody raised against an extensin-like hydroxyproline-rich glycoprotein ($HRGP_{2b}$) from melon revealed related molecules over the walls and in the interface space. The labelling was particularly evident over and near the host membranes (Bonfante-Fasolo *et al.*, 1991). It has been suggested that extensin, which can be immobilised in the wall as an insoluble component, could control the wall rigidity (Lamport, 1989). In addition to this, there is a soluble extensin pool, which can be isolated by a simple saline extraction (Lamport, 1989).

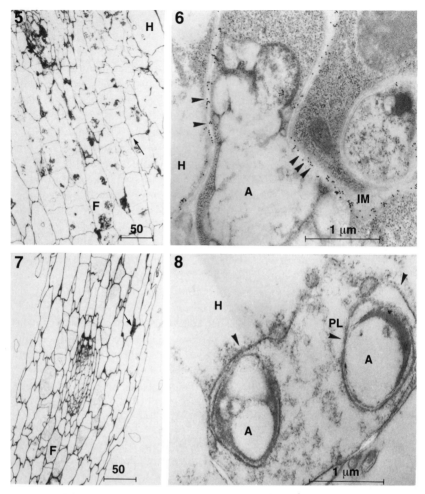

Figs. 5–8. Features of the interfacial material in pea infected by *Glomus versiforme*, after immunolabelling with monoclonal antibodies recognising non-esterified (JIM5) and methylesterified (JIM7) pectins.

Fig. 5. After treatment with JIM5 and the silver intensification, the labelling is found over the host cell walls (arrow) and around the fungal arbuscules (F). H, host cell.

Fig. 6. After immunogold treatment the labelling is evident in the interfacial material (IM) around the arbuscular branch (A). H, host cell.

Fig. 7. After treatment with JIM7 and the silver intensification, the labelling is found over the host cell walls (arrow), but is not intense around the fungus (F).

Fig. 8. After immunogold treatment with JIM7 no labelling is evident in the interfacial material (IM) around the arbuscular branch (A). H, host cell; PL, perifungal membrane.

By performing dot blots, we observed a cross-reaction with the soluble protein fraction of pea and leek roots, thus revealing the existence of a soluble fraction. We suggest that the molecules which are revealed by the antibody against the $HRGP_{2b}$ fraction and occur in the interface space may represent a soluble fraction, after its secretion and before its insertion into the wall material.

These results allow us to define the host cell that contains a symbiotic fungus as a cell which possesses: (1) a peripheral membrane with a regular cell wall; and (2) an inner membrane which invaginates around the fungus but keeps its capability to control the secretion of new cell wall molecules at its external side, like the membrane surrounding a protoplast (Fig. 9). The characterisation of the interfacial compartment as an apoplastic compartment opens the question of new targeting for vesicles carrying cell wall components. The problem of the formation and transport of cell wall polysaccharides by endomembrane flow has been only recently approached in some simple experimental systems, such as *in vitro* cultures of cells and elongating cells in mung bean hypocotyl (Brummel *et al.*, 1990; Vian & Roland, 1991).

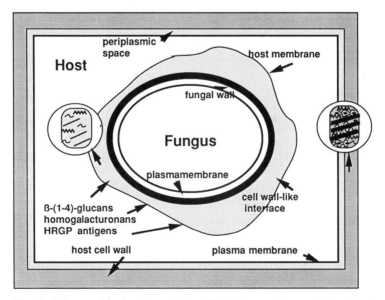

Fig. 9. Scheme showing the relationships between the peripheral host wall and the interfacial material, on the basis of *in situ* labelling experiments. The scheme suggests that the cell wall molecules are assembled in the cell wall, while they are not fully assembled in the interfacial material.

In conclusion the interface can be seen as a structural expression of the symbiotic status: the partners work alongside each other, since they are separated by a space which does not represent a drawback for nutrient exchanges, but prevents the direct cell-to-cell contact which could cause the plant defence reaction.

The periarbuscular membrane

The interfacial compartment is limited by a membrane of host origin which surrounds the fungus. The morphological continuity between the host plasma membrane and the invaginated membrane suggests their relatedness, even if important differences exist concerning their functional activities (Smith & Smith, 1990; Gianinazzi-Pearson *et al.*, 1991). The presence of an invaginated membrane around a penetrating organism is a common pattern in the heterologous interactions in the plant kingdom. According to Smith (1979), the sequestration of endosymbionts by the host membrane offers some form of control. The invaginated membranes have been named in different ways in the diverse associations; perihaustorial membrane, which is described in pathogenic associations formed by biotrophic fungi (Harder, 1989), peribacteroid membrane which surrounds *Rhizobium* bacteroids (Brewin, 1990) and periarbuscular membrane which envelops the VAM fungal arbuscule. All of these are actively involved in cell surface interactions during the establishment of the intracellular phase of the symbiosis and in metabolite exchange with the endosymbiont. Many differences exist between the perihaustorial membrane and the periarbuscular membrane (for reviews, see Smith & Smith, 1990; Bonfante-Fasolo & Scannerini, 1992). On the contrary, some similarities can be traced inside the mutualistic associations, which provide an interesting way to investigate and compare the molecular basis of the plant response during symbiosis.

Monoclonal antibodies (MAbs) can be used to investigate complex structures such as membranes, exploiting their ability to dissect these structures into single molecular components. MAbs provide useful probes to be used in different systems to detect specific molecules. They also allow the subsequent characterisation and identification of the specific components recognised.

A set of MAbs was raised against molecular components of the peribacteroid membrane in pea nodules (Brewin *et al.*, 1985; Bradley *et al.*, 1988; Peretto *et al.*, 1991). Some of the epitopes recognised by these MAbs were partially characterised and shown to be developmentally regulated during the formation of pea root nodules, probably being involved with different stages of the interaction with *Rhizobium* (Peretto *et al.*, 1991). By using these MAbs as probes to study VAM fungi in pea roots it was possible to demonstrate that most components which were present on the peribacteroid membranes were also found to be expressed on the periarbuscular membrane (S. Perotto, unpublished results). These results suggest a strong relatedness between these two plant-derived membranes and confirm recent data of Wyss *et al.* (1990) which detect similar products in VAMs and nodules by using antibodies to nodule-specific proteins. They also extend previous reports by Gianinazzi-Pearson *et al.* (1990). One MAb in particular, MAC 266, which recognises some membrane and soluble glycoproteins present in the peribacteroid compartment, also reacts strongly with components expressed around the arbuscules (Figs. 10 and 11). The same MAb can detect on a Western blot a series of bands present in the mycorrhizal plant but absent in the uninfected control (data not shown).

Mechanism of regulation of cell-to-cell recognition

The whole set of the events described, from cell surface modification to formation of an interface and synthesis of specific molecules, is the

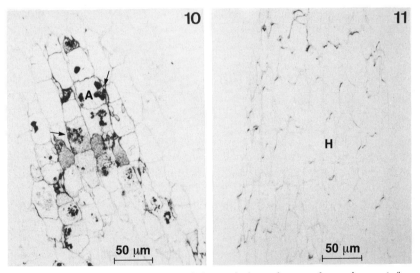

Figs. 10 and 11. Features of the periarbuscular membrane in pea infected by *Glomus versiforme*, after immunolabelling with a monoclonal antibody MAC 266 and silver intensification.
Fig. 10. A strong reaction occurs in the arbuscule (A) infected cells; (arrow). In the uninfected cells, the membrane does not react.
Fig. 11. In the uninfected pea, there is only a very weak reaction. H, host cell.

expression of cell-to-cell recognition. In a functioning mycorrhiza, it leads to a balance between the partners. One question is: what mechanisms regulate this balance? By definition, a mutualistic symbiosis is characterised by a dominance of non-aggressive strategies. Having identified some of these strategies, it is challenging to investigate these molecular mechanisms. Many reports have shown that VAM fungi do not elicit the expression of plant pathogenesis related proteins, such as chitinase and glucanases (Gianinazzi & Gianinazzi-Pearson, 1990; Bonfante-Fasolo & Spanu, 1991). Our tentative explanation is that a higher level of elicitation is prevented by a contained production of cell wall loosening enzymes by the fungus, or possibly by some features of the fungal cell surface that act as negative feed-back on the plant response.

The growth of some fungi (for example, *Trichoderma*) can be inhibited *in vitro* by lytic enzymes, which are normally produced by the plant as a defence response to pathogens (Schlumbaum *et al.*, 1986). We investigated the cellular basis of the growth inhibition caused by enzymes such as chitinase and glucanase. It was found that these cell wall loosening enzymes have a dramatic effect on the fungal tips. Their action induces a thinning of the fungal cell wall, a detachment of the plasma membrane and, lastly, wall breakdown: all these events lead to an apical swelling. This pattern is evident in fungi such as *Trichoderma*, characterised by quick growth. On the contrary, some mycorrhizal fungi which are characterised by a reduced growth rate, are not sensitive to the action of these cell wall loosening enzymes (Arlorio *et al.*, 1992). In addition to the different growth rate, the different susceptibility to plant enzymes may be due to the variability in the fungal cell wall composition (Bartnicki-Garcia, 1970). This may create a different degree of substrate accessibility to the lytic enzymes *in vivo*. Mycorrhizal fungi could therefore be protected by the action of lytic enzymes, for example due to strong covalent linkages occurring between glucans and chitin inside their walls.

A further 'non aggressive' strategy seems to be adopted by mutualistic fungi in the production of cell wall loosening enzymes. The role of cell wall degrading enzymes produced by several pathogenic fungi in penetration and elicitation of plant responses has been established (Cooper, 1989). Strong correlation was found between the ability of the fungus to produce wall degrading enzymes *in vitro* and its ability to penetrate the host walls and the effects of purified enzymes on plants in generating elicitors. However, very little is known about cell wall loosening enzymes produced by mycorrhizal fungi. They must be able to penetrate the wall of root cells, but without creating severe disturbance to their host, as their survival depends on the host cell viability. Ultrastructural observations confirm that there are only signs of a limited loosening of the middle lamella during cortical cell penetration (Bonfante-Fasolo & Perotto, 1991; Fig. 2). The penetration of epidermal cells probably requires alternative strategies. The inability to grow VAM fungi in pure culture makes it difficult to determine their enzymic activity *in vitro*. Biochemical data are so far restricted to measurements of pectinolytic activity in spore extracts and indirect observations on the ability of *Glomus mosseae* to penetrate roots in the presence of various pectic substances (Garcia-Romera *et al.*, 1990). Preliminary results show that polygalacturonase (PG) activity detected in the soluble fraction of extracts from leek mycorrhizae is comparable to that observed in uninfected samples. The presence of PG activity in the roots makes detection of any fungal PG very difficult. However, immunological studies carried out by using a polyclonal antibody against a purified PG from *Fusarium moniliforme* (De Lorenzo *et al.*, 1987) show a reaction over the extraradical hyphae, while no activity was found over the mycelium obtained from the germinating spore. These results are not yet conclusive for true PG activity from the fungus, but show that this activity, if present, is very low.

Conclusions

Mycorrhizal associations represent complex systems and, similarly to many other symbiotic systems, they represent a biological novelty (Margulis, 1990) in comparison with their separated partners. According to the first definition by De Bary (1879), symbiosis is defined as 'unlike organisms which live together', covering therefore a variety of biological interactions, from pathogenic to mutualistic. Mycorrhizae fall in the mutualistic type of symbiosis, which means that, at an organismic level, they are the expression of a cooperative strategy of both partners.

As already established for many animal systems, a cooperative strategy can be highly successful from an evolutionary point of view (Margulis, 1990): this could also be true for mycorrhizal associations. VAM systems are surely the oldest form of symbiosis, found in several fossil remnants (Pirozinsky, 1981) and are currently the most widespread in the world. Thanks to their mutualistic status, the fungi complete their life cycle and the plants improve their vegetative phase. The cooperative strategy in the formation of a mycorrhizal symbiosis requires a cascade of multistep events, which can be described at the cellular level as cell-to-cell recognition events. Non-aggressive strategies, mostly on the part of the fungus, seem to be essential for the establishment and function of mycorrhizae, even if the understanding of the molecular mechanisms involved awaits the development of further knowledge on the regulation of plant and fungal gene expression during symbiosis.

Acknowledgement

This research was supported by the National Research Council of Italy, Special Project RAISA, subproject no. 2; paper N. 194.

References

Arlorio, M., Ludwig, A., Boller, T., Mischiati, P. & Bonfante-Fasolo, P. (1992). Effects of chitinase and β-1,3-glucanase from pea on the growth of saprophytic, pathogenic and mycorrhizal fungi. *Giorn. Bot. Ital.* (in press).

Bartnicki-Garcia, S. (1970). Cell wall composition and other biochemical markers in fungal phylogeny. In *Phytochemical Phylogeny*, ed. B. Harborne, pp. 81–103. London: Academic Press.

Becard, G. & Pichet, Y. (1989). New aspects on the acquisition of biotrophic status by a vesicular–arbuscular mycorrhizal fungus *Gigaspora margarita*. *New Phytologist* **112**, 77–83.

Bonfante-Fasolo, P. (1984). Anatomy and morphology. In *V.A. Mycorrhizas*, ed. C. L. Powell & D. J. Bagyaraj, pp. 5–33. Boca Raton, FL: CRC Press.

Bonfante-Fasolo, P., Faccio, A., Perotto, S. & Schubert, A. (1990*a*). Correlation between chitin distribution and cell wall morphology in the mycorrhizal fungus *Glomus versiforme*. *Mycological Research* **94**, 157–65.

Bonfante-Fasolo, P., Gianinazzi-Pearson, V. & Martinengo, L. (1984). Ultrastructural aspects of endomycorrhiza in the Ericaceae. IV. Comparison of infection by *Pezizella ericae* in host and non-host plants. *New Phytologist* **98**, 329–33.

Bonfante-Fasolo, P. & Perotto, S. (1991). Plants and endomycorrhizal fungi: the cellular and molecular basis of their interaction. In *Molecular Signals in Plant–Microbe Communication*, ed. D. P. Verma. Boca Raton, FL: CRC Press, pp. 445–70.

Bonfante-Fasolo, P. & Scannerini, S. (1991). The cellular basis of plant–fungus interchanges in mycorrhizal associations. In *Functioning in Mycorrhizae*, ed. M. Allen. London: Academic Press, in press.

Bonfante-Fasolo, P. & Spanu, P. (1991). Pathogenic and endomycorrhizal associations. *Methods in Microbiology* **24**, 141–68.

Bonfante-Fasolo, P., Tamagnone, L., Peretto, R., Esquerré-Tugayé, M. T., Mazau, D., Mosiniak, M. & Vian, B. (1991). Immunocytochemical location of hydroxyproline rich glycoproteins in the interface between a mycorrhizal fungus, and its host plants. *Protoplasma*, **165**, 127–38.

Bonfante-Fasolo, P., Vian, B., Perotto, S., Faccio, A. & Knox, J. P. (1990*b*). Cellulose and pectin localization in roots of mycorrhizal *Allium porrum*: labelling continuity between host cell wall and interfacial material. *Planta* **180**, 537–47.

Bradley, D. J., Wood, E. A., Larkins, A. P., Galfrè, G., Bucker, G. W. & Brewin, N. J. (1988). Isolation of monoclonal antibodies reacting with peribacteroid membranes and other components of pea root nodules containing *Rhizobium leguminosarum*. *Planta* **173**, 149–60.

Brewin, N. J. (1990). The role of the plant plasma membrane in symbiosis. In *The Plant Plasma Membrane*, ed. C. Larsson & I. M. Moller, pp. 351–75. Berlin, Heidelberg, New York: Springer-Verlag.

Brewin, N. J., Robertson, J. G., Wood, E. A., Wells, B., Larkins, A. P., Galfrè, G. & Butcher, G. W. (1985). Monoclonal antibodies to antigens in the peribacterois membrane from *Rhizobium*-induced root nodules of pea- cross-react with plasmamembrane and Golgi bodies. *EMBO Journal* **4**, 605–11.

Brummel, D. A., Camirand, A. & MacLachlan, G. A. (1990). Differential distribution of xyloglucan glycosyl transferases in pea Golgi dictyosomes and secretory vesicles. *Journal of Cell Science* **96**, 705–10.

Chanzy, H., Henrissat, B., Voung, R. & Schulein, M. (1983). The action of 1,4-β-D-glucan cellobiohydrolase on *Valonia* cellulose microcrystals. *FEBS Letters* **153**, 113–18.

Cooper, R. M. (1989). Host cell wall loosening and separation by plant pathogens. In *Cell Separation in Plants*, NATO ASI series, vol. H35, ed. D. J. Osborne & M. B. Jackson, pp. 165–78. Berlin, Heidelberg, New York: Springer-Verlag.

De Bary, H. A. (1879). *Die Erscheinung der Symbiose*. Strasbourg: R. J. Trubner.

De Lorenzo, G., Salvi, G., Degrà, L., D'Ovidio, R. & Cervone, F. (1987). Induction of extracellular polygalacturonase and its mRNA in the phytopathogenic fungus *Fusarium moniliforme*. *Journal of General Microbiology* **133**, 3365–73.

Garcia-Romera, I., Garcia-Garrido, J. M., Martinez-Molina, E. & Ocampo, J. A. (1990). Possible influence of hydrolytic enzymes on vesicular–arbuscular mycorrhizal infection of alfalfa. *Soil Biology and Biochemistry* **22**, 149–52.

Gianinazzi, S. & Gianinazzi-Pearson, V. (1990). Cellular interactions in vesicular–arbuscular (VA) endomycorrhizae. The host's point of view. In *Endocytobiology IV*, ed. P. Nardon, V. Gianinazzi-Pearson, A. M. Grenier, L. Margulis & D. C. Smith, pp. 83–90. Paris: INRA Press.

Gianinazzi-Pearson, V., Gianinazzi, S. & Brewin, N. J. (1990). Immunocytochemical localisation of antigenic sites in the perisymbiotic membrane of vesicular–arbuscular endomycorrhiza using monoclonal antibodies reacting against the peribacteroid membrane of nodules. In *Endocytobiology IV*, ed. P. Nardon, V. Gianinazzi-

Pearson, A. M. Grenier, L. Margulis & D. C. Smith, pp. 127–31. Paris: INRA Press.

Gianinazzi-Pearson, V., Smith, S. E., Gianinazzi, S. & Smith, F. A. (1991). Enzymatic studies on the metabolism of vesicular–arbuscular mycorrhizas. V. Is H⁺-ATPase a component of ATP-hydrolysing enzyme activities in plant-fungus interfaces? *New Phytologist* **117**, 61–74.

Giovannetti, M. & Lioi, L. (1990). The mycorrhizal status of *Arbutus unedo* in relation to compatible and incompatible fungi. *Canadian Journal of Botany* **68**, 1239–44.

Harder, D. E. (1989). Rust fungal haustoria – past, present, future. *Canadian Journal of Plant Pathology* **11**, 91–9.

Harley, J. L. (1989). The significance of mycorrhiza. *Mycological Research* **92**, 92–129.

Knox, J. P., Linstead, P. J., King, J., Cooper, C. & Roberts, K. (1990). Pectin esterification is spatially regulated both within cell walls and between developing tissues of root apices. *Planta* **181**, 512–21.

Lamport, D. T. A. (1989). Extensin peroxidase ties the knots in the extensin network. In *Cell Separation in Plants*, NATO ASI series, vol. H35, ed. D. J. Osborne & M. B. Jackson, pp. 101–13. Berlin, Heidelberg, New York: Springer-Verlag.

Law, R. & Lewis, D. H. (1983). Biotic environments and the maintenance of sex-some evidence from mutualistic symbioses. *Biological Journal of the Linnean Society* **20**, 249–76.

Margulis, L. (1990). Words as battle cries – symbiogenesis and the new field of endocytobiology. *BioScience* **40**, 673–7.

Mazau, D., Rumeau, D. & Esquerré-Tugayé, M. T. (1988). Two different families of hydroxyproline-rich glycoproteins in melon callus. *Plant Physiology* **86**, 540–6.

Morton, J. B. & Benny, G. L. (1990). Revised classification of arbuscular mycorrhizal fungi (Zygomycetes): a new order, Glomales, two new suborders, Glomineae and Gigasporineae, and two new families, Acaulosporinaceae and Gigasporaceae, with an emendation of Glomaceae. *Mycotaxon* **37**, 471–91.

Peretto, S., VandenBosch, K. A., Butcher, G. W. & Brewin, N. J. (1991). Molecular composition and development of the plant glycocalyx associated with the peribacteroid membrane of pea root nodules. *Development* **112**, 763–73.

Pirozinsky, K. A. (1981). Interactions between plants and fungi through the ages. *Canadian Journal of Botany* **59**, 1824–7.

Scannerini, S. & Bonfante-Fasolo, P. (1979). Ultrastructural cytochemical demonstration of polysaccharides and proteins within host–arbuscule interfacial matrix in an endomycorrhiza. *New Phytologist* **83**, 739–44.

Schlumbaum, A., Mauch, F., Vogeli, U. & Boller, T. (1986). Plant chitinases are potent inhibitors of fungal growth. *Nature* **324**, 365–7.

Smith, D. C. (1979). From extracellular to intracellular: the establishment of a symbiosis. In *The Cell as a Habitat*, ed. M. H. Richmond & D. C. Smith, pp. 1–16. Cambridge: Cambridge University Press.

Smith, D. C. & Douglas, A. E. (1987). *The Biology of Symbiosis.* London: Edward Arnold.

Smith, S. E. & Smith, F. A. (1990). Structure and function of the interfaces in biotrophic symbioses as they relate to nutrient transport. *New Phytologist* **114**, 1–38.

Vian, B. & Roland, J. C. (1991). Affinodetection of the sites of formation and of the further distribution of polygalacturonans and native cellulose in growing plant cells. *Biology of the Cell* **71**, 43–53.

Wyss, P., Mellor, R. B. & Wiemken, A. (1990). Vesicular–arbuscular mycorrhizas of wild type soybean and non-nodulating mutants with *Glomus mosseae* contain symbiosis specific polypeptides (mycorrhizins), immunologically cross-reactive with nodulins. *Planta* **182**, 22–6.

J. A. DOWNIE, M.-A. BARNY, T. M. CUBO,
A. DAVIES, A. ECONOMOU, J. L. FIRMIN,
A. W. B. JOHNSTON, C. MARIE,
A. MAVRIDOU, R. RIVILLA, A.-K. SCHEU,
J. M. SUTTON AND K. E. WILSON

Host recognition in the *Rhizobium leguminosarum*–pea symbiosis

Abstract

Recognition between leguminous plants and the specific rhizobial strains that nodulate them is mediated via a regulon of nodulation (*nod*) genes present in the bacteria. These *nod* genes are induced by flavonoids secreted from legume roots. Many of the *nod* gene products are involved in the synthesis of host-specific signals that are recognised by appropriate legume hosts. Recently (Lerouge *et al.*, 1990), the signal molecule made by one strain of *Rhizobium meliloti* was identified as an acylated and sulphated, tetraglucosamine glycolipid and there is strong evidence that *Rhizobium leguminosarum* makes related but structurally distinct signals.

On the basis of these observations it is now possible to make sense of several similarities that have been recognised between *nod* gene products and enzymes of known function. Thus, for example, it appears that the *nodM* gene product is involved in the formation of glucosamine precursors of the signal molecule, whilst other gene products are likely to be involved in specific substitutions that confer host specificity to the signal molecule.

In addition to those *nod* gene products that are involved in the synthesis of the glycolipid, it is evident that there are other genes which may carry out a different role. Of particular interest is the *nodO* gene which encodes a secreted Ca^{2+}-binding protein that has the potential to interact directly with plant cells. In the absence of the *nodFEL* genes, *nodO* is necessary for nodulation, indicating that the NodO protein can compensate for the loss of *nodFEL* function during infection. Possible roles of several of the *nod* gene products are discussed.

Society for Experimental Biology Seminar Series 48: *Perspectives in Plant Cell Recognition*, ed. J. A. Callow & J. R. Green. © Cambridge University Press 1992, pp. 257–66.

Introduction

In *Rhizobium leguminosarum* biovar *viciae*, 13 nodulation (*nod*) genes have been identified as being involved in the nodulation of legumes such as peas and vetch. These genes (Fig. 1) are arranged in five operons, four of which are under the control of the *nodD* gene product which functions as a positively acting transcriptional regulatory protein. The NodD protein binds (Hong *et al.*, 1987) to the highly conserved promoter elements (about 50 nucleotides in length) upstream from *nodA*, *nodF*, *nodM* and *nodO*. NodD probably interacts directly with flavonoid molecules (Burn *et al.*, 1987; Spaink *et al.*, 1989*b*) secreted from legume roots, thereby stimulating transcription of the other *nod* operons. Not all of the induced *nod* genes shown in Fig. 1 are necessary for nodulation to occur (Downie & Johnston, 1988). Whereas mutation of the *nodABC* genes totally blocks nodulation, mutation of the *nodF* or *nodE* genes reduces and delays nodulation (Downie *et al.*, 1985; Spaink *et al.*, 1989*a*), mutation of *nodL* inhibits nodulation of peas, but not of some vetch plants (Surin & Downie, 1988) and mutations in the other *nod* genes only slightly affect nodulation (Downie *et al.*, 1985; Wijffelman *et al.*, 1985; Economou *et al.*, 1989; de Maagd *et al.*, 1989; Surin *et al.*, 1990).

Recognition by leguminous plants of the specific rhizobial strains that nodulate them is mediated primarily by low molecular weight signal molecules. It is now clear that several of the rhizobial *nod* gene products are involved in the biosynthesis of low molecular weight host-specific signals that are recognised by appropriate legume hosts. These molecules induce root hair deformation (the first observed plant response in the interaction) and initiation of a programme that results in the development of a

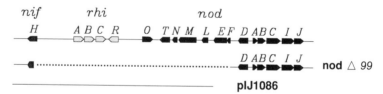

Fig. 1. Map of the nodulation region of *Rhizobium leguminosarum* bv. *viciae*. The nodulation (*nod*), rhizosphere expressed (*rhi*) and nitrogen fixation (*nif*) genes from the symbiotic plasmid pRL1JI are represented as arrows, their size and direction indicating the length and orientations of the open reading frames identified from their DNA sequences. The region represented corresponds to about 30 000 bases of DNA. The size of the deletion *nodΔ99* in strain A69 is represented by the dotted line and extent of the complementing cosmid pIJ1086 is also shown.

new plant-made organ – the nodule. Lerouge *et al.* (1990) have purified and characterised a signal molecule made by *R. meliloti* as being an acylated and sulphated derivative of tetra-*N*-acetylglucosamine. This molecule is unique amongst the plant signal molecules identified to date, since it induces a plant response at very low concentrations (10^{-11} M), it is specific for those legumes nodulated by *R. meliloti* and it initiates a complete developmental programme leading to nodule formation (Roche *et al.*, 1991).

The groundwork leading to the identification of this compound was laid by Bhuvaneswari & Solheim (1985), who found that a low molecular weight compound with root-hair-curling activity could be isolated from the liquid around the roots of *Rhizobium*-inoculated legumes. Using this root-hair-curling response as a bioassay, the signal molecule was purified from the growth medium of axenically grown *R. meliloti* strongly induced for the expression of the *nod* genes (Lerouge *et al.*, 1990).

Although the structure molecule made by *R. meliloti* is highly specific for only some legumes (such as *Medicago* spp.), it was already clear that different but closely related molecules must be made by other rhizobial strains, and recently Spaink *et al.* (1991) have shown that the molecule made by *R. leguminosarum* is closely related to, but different from, the glycolipid made by *R. meliloti*. It had been established that the *nodABC* genes are essential for root hair curling and that these genes are functionally conserved in all rhizobia studied (Long, 1989). It is likely that they are involved in the formation of a 'core structure' that is chemically modified by other 'host specific' *nod* genes.

However, although the signal molecule is a prerequisite for nodulation to proceed, it is also clear that there are other nodulation gene products which are unlikely to be directly involved in the synthesis of low molecular weight signal molecules. In this regard the *R. l.* bv. *viciae* appears to be somewhat unusual, since it secretes a nodulation protein (de Maagd *et al.*, 1989; Economou *et al.*, 1990) that plays an important role in nodulation (Downie & Surin, 1990). In addition, we have identified other genes that are not nodulation genes (since they are not under NodD control) but appear to influence nodulation. These genes are found in different isolates of *R. l.* bv. *viciae* but we have not detected similar genes in related biovars or other rhizobia. These (*rhi*) genes are expressed within the rhizosphere and may influence the efficiency of the nodulation process.

In this chapter we briefly overview the possible roles of several genes involved in the early stages of the *R. l.* bv. *viciae*–plant interaction.

Results

Expression of *nod* genes

It has been established that rhizobial *nod* genes are induced by flavonoids such as naringenin and hesperetin secreted from legume roots (for a review, see Long, 1989). It appears that these flavonoids interact with the *nodD* gene products (Burn *et al.*, 1987) which then activate transcription from the promoters preceding the various *nod* operons such as *nodAB-CIJ*, *nodFEL*, *nodMNT* and *nodO* in *R. l.* bv. *viciae*. However, it is not yet clear how the NodD protein stimulates *nod* gene transcription. To identify non-symbiotic plasmid genes that may be involved in *nod* gene expression, we have used a strain cured of its symbiotic plasmid to identify mutants altered in *nod* gene expression. Two different loci (not on the symbiotic plasmid) have been identified in which mutations reduce the level of *nod* gene expression. When the symbiotic plasmid pRL1JI was transferred to these strains, normal nodulation was initiated. However, in one of the mutants, later stages of the symbiosis appear to be affected, since the mutant has greatly reduced levels of nitrogen fixation. The cloned genes that complement this mutant are currently under investigation.

The *nodABCFELM* genes are involved in the formation of a signal molecule

The predicted protein sequence of the *nodE* gene product suggests that it is homologous to the condensing enzyme subunit of the fatty-acid synthase (Hopwood & Sherman, 1990); as such it probably functions in conjunction with the *nodF* gene product, an acyl-carrier-like protein (Shearman *et al.*, 1986). Furthermore, protein sequence homologies suggest that the *nodL* gene product is probably an acetyltransferase (Downie, 1989).

On the basis of these observations it appeared to us likely that the formation of their metabolic products could be followed using acetate labelling experiments. After [^{14}C]acetate or [^{14}C]glucose labelling of the wild-type strain used as a control, the growth medium supernatant was passed through a preparative C-18 column which was eluted with methanol and then fractionated on an analytical C-18 column using a methanol gradient. At about 75% methanol a radioactive peak was found to co-elute with those fractions that had the ability to induce strong root hair deformation on vetch. Both the root hair deformation activity and peak of radioactivity were absent from strains carrying mutations in *nodA*, *nodB* or *nodC* confirming that the radioactively labelled peak corresponds with the root hair curling factor made by the *nodABC* gene products. When strains carrying mutations in *nodE* or *nodL* were used in similar experiments, the elution profile was altered indicating that their gene products modify the root hair curling molecule. Therefore the *nodABCFEL* gene products are all involved in the formation of the root hair deformation signal molecule.

We have purified the root-hair-curling activity from the growth medium supernatant of *R. l.* bv. *viciae* and found that it is closely related to the glucosamine-containing glycolipid molecule made by *R. meliloti*. The major peak of activity was found by mass spectroscopy to have mass ion of 1089 (M+H^+) and by chemical analysis was found to contain glucosamine residues and an acyl group. Unlike the *R. meliloti*-made signal, that made by *R. l.* bv. *viciae* is not sulphated. Similar observations have been reported by Spaink *et al.* (1991), who also showed that the reducing end of the glucosamine-containing polysaccharide was *O*-acetylated and that the attachment of the acetyl group was mediated by the *nodL* gene product.

The *nodM* gene product appears to play a role in the formation of precursor molecules that are used in the formation of the signal molecule. NodM bears strong sequence similarity to the *Escherichia coli* glucosamine synthase encoded by the *glmS* gene and biochemical experiments have confirmed that NodM is involved in the formation of glucosamine 6-phosphate from fructose 6-phosphate and glutamine. We have isolated glucosamine auxotrophs of *R. leguminosarum* and found that the auxotrophy can be corrected by either the cloned *R. leguminosarum glmS* gene or the *nodM* gene. Therefore, it appears that for efficient nodulation, *R. leguminosarum* has a 'backup' system for the synthesis of glucosamine intermediates that are used as precursors in the synthesis of the glycolipid signal molecule.

Additive effects of various genes on nodulation

Individual mutations in the *nodF,E,L,M,N,T* or *O* genes do not block nodulation. As described above, the reason that *nodM* is not essential for nodulation is that there is a homologous gene on the *Rhizobium* chromosome. However, this does not appear to be the case with several other *nod* genes. In order to determine why mutations in other *nod* genes do not block nodulation, a mutant strain (A69) was made which lacks the *nodFELMNTO*, *rhiABCR* and *nifH* gene regions (Fig. 1) due to an approximately 20 000 base deletion (*nodΔ99*) of the symbiotic plasmid pRL1JI. When this mutant strain was inoculated into vetch plants no nodulation was observed (Downie & Surin, 1990) even after prolonged periods of growth. This result indicates that, although mutations in the individual *nod* genes in this region cause leaky phenotypes, these genes are collectively essential for nodulation. The mutant strain could be partially restored for nodulation by recombinant plasmids carrying the *nodFE* or *nodFEL* genes. Moreover, the mutant could also be partially complemented for nodulation by a cosmid clone (pIJ1086) carrying the *nodO* gene (but lacking the *nodFEL* genes) and this complementation was shown to require *nodO*, since a mutation in the *nodO* gene on pIJ1086 blocked nodulation.

To confirm these observations a double mutant strain was made carrying both the *nodO93*::Tn3*HoHo1* allele and the *nodE68*::Tn5 allele on pRL1JI. When this mutant was inoculated into vetch plants, the level of nodulation was significantly less than that observed with strains carrying single mutations in either *nodO* or *nodE*. Therefore, these genes must have a synergistic effect during the nodulation process. Nevertheless, a low level of nodulation was observed with the double mutant strain. Taking this result together with the observation that the deletion mutant (A69) is Nod⁻ we conclude that even in the absence of *nodO* and *nodE* other genes in the region deleted in strain A69 can allow a low level of nodulation of vetch.

Genes that could play a role in this residual nodulation include the *rhi* genes. The *rhi* gene region was identified previously as being expressed in the rhizosphere but not in the nodules of peas (Dibb *et al.*, 1984). Genetic experiments and DNA sequencing of the *rhi* region has identified four *rhi* genes (Fig. 1) and on the basis of DNA hybridisation experiments, and probing with antibody to the *rhiA* gene product, the *rhi* genes appear to be specific to *R. l.* bv. *viciae*. The *rhiABC* genes have been shown to be under the control of the *rhiR* gene product which is a positively acting regulator.

These *rhi* genes are present on pIJ1086, which complements strain A69(*nodΔ99*) for nodulation on vetch. A derivative of pIJ1086 carrying a Tn5 insertion in the *rhiA* gene was constructed and transferred to strain A69. In this case the nodulation observed was significantly lower than that observed with the strain carrying the unmutated derivative of pIJ1086. On the basis of these observations it appears that the *rhi* genes do play a role in the nodulation process. This fits with previous observations which indicate that the *rhi* genes are found only in bv. *viciae*. However, it is not clear how these genes contribute to the overall process of nodulation. It is possible for example that the *rhi* genes are involved in the formation of a second signal or alternatively that they help the growth of the bacteria, either in the rhizosphere or in the infection threads.

nod gene products that are exported or secreted

Although the major requirement for normal nodulation is the secretion of the low molecular weight glycolipid signal molecule, it is now evident that some *nod* gene products are externalised such that they have the potential to interact directly with plant cell walls or membranes.

The *nodO* gene product is one such protein (de Maagd *et al.*, 1989; Economou *et al.*, 1990) and its mechanism of secretion does not require an N-terminal transit peptide. Instead, it appears that NodO is secreted by a mechanism analogous to the secretion of haemolysin. Thus, whereas *Escherichia coli* cannot normally secrete the NodO protein, we have established that NodO can be recognised and secreted from *E. coli* via the haemolysin secretion complex encoded by the *hlyBD* genes. It is not yet clear how the secreted NodO protein interacts with plants but it is clear that this Ca^{2+}-binding protein (Economou *et al.*, 1990) does play an important role in the nodulation process (Downie & Surin, 1990).

The predicted protein sequence of the *nodT* gene product indicated that it was likely to have an N-terminal transit sequence that would direct it to a periplasmic or outer-membrane location in *Rhizobium* (Surin *et al.*, 1990). The *nodT* gene has been fused in frame to a derivative of the *phoA* gene which lacks an N-terminal transit sequence. The derived protein fusion is exported by *R. leguminosarum* and by *E. coli*, as judged from the high levels of alkaline phosphatase activity found in cells expressing the NodT–PhoA hybrid protein. Therefore it is possible that NodT also has the potential to interact with the plant cells during the symbiosis.

Similarly, using *nodC-phoA* gene fusions it has been established that a major domain of the NodC protein is located in the periplasmic space. It is very likely that the NodC protein plays a catalytic role in the formation of the glycolipid signal molecule, but it has also been suggested that NodC could act as a receptor that recognises plant-derived signals (Schmidt *et al.*, 1991; John *et al.*, 1988).

Discussion

It is now apparent that the early stages of the interaction between rhizobia and roots of leguminous plants are dependent upon a glycolipid molecule whose synthesis requires several of the *nod* gene products. It appears likely that the NodM protein provides glycosamine precursors that are incorporated into a glucosamine polymer in a process that probably involves the *nodABC* gene products. The *nodFE* gene products are likely to be involved in determining the type of acyl group that is present on the polymer whilst the *nodL* gene product is involved in *O*-acetylating the reducing end of the glycolipid (Spaink *et al.*, 1991).

The secreted signal molecule induces a plant response at very low concentrations (10^{-11} M) and can induce nodule organogenesis even in the absence of rhizobia (Roche *et al.*, 1991). However, it is clear that there are a number of *nod* gene products and *rhi* gene products that also play a role in efficient nodulation. Some of these proteins are secreted or exported such that they have the potential to interact directly with the legume roots. Thus, for example: NodO is a secreted protein (Economou *et al.*, 1990; de Maagd *et al.*, 1989); NodT is an exported protein; a major domain of NodC is external to the bacterial cytoplasmic membrane (John *et al.*, 1988); and the *nodJ* gene product is a transmembrane protein that has a domain external to the cytoplasmic membrane (Surin *et al.*, 1990). Some of these proteins could play a role in the secretion and stabilisation of the secreted glycolipid signal molecule, or alternatively could interact directly with legume roots. Future biochemical work will be required to determine the function of these gene products in recognition events that occur between *R. leguminosarum* and pea roots.

Acknowledgements

This work was supported by a grant from the AFRC. We are also indebted to support from the John Innes Foundation (to C.M. and A.M.), the Spanish Government (to R.R. and T.M.C.) and the Gatsby Charitable Trust (to J.M.S.).

References

Bhuvaneswari, T. V. & Solheim, B. (1985). Root hair deformation in the white clover/*Rhizobium trifolii* symbiosis. *Physiological Plant* **63**, 25–34.

Burn, J. E., Hong, G.-F. & Johnston, A. W. B. (1987). Four classes of mutations in the NodD gene of *Rhizobium leguminosarum* and *R. phaseoli*. *Genes and Development* **1**, 456–64.

de Maagd, R. A., Wijfes, A. H. M., Spaink, H. P., Ruiz-Sainz, J. E., Wijffelman, C. A., Okker, R. J. H. & Lugtenberg, B. J. J. (1989). *nodO*, a new *nod* gene of the *Rhizobium leguminosarum* biovar *viciae* Sym plasmid encodes a secreted protein. *Journal of Bacteriology* **172**, 4255–62.

Dibb, N. J., Downie, J. A. & Brewin, N. J. (1984). Identification of a rhizosphere protein encoded by the symbiotic plasmid of *Rhizobium leguminosarum*. *Journal of Bacteriology* **158**, 621–7.

Downie, J. A. (1989). The *nodL* gene from *Rhizobium leguminosarum* is homologous to the acetyl transferases encoded by *lacA* and *cysE*. *Molecular Microbiology* **3**, 1649–51.

Downie, J. A. & Johnston, A. W. B. (1988). Nodulation of legumes by rhizobium. *Plant, Cell and Environment* **11**, 403–12.

Downie, J. A., Knight, C. D. & Johnston, A. W. B. (1985). Identification of genes and gene products involved in nodulation of peas by *Rhizobium leguminosarum*. *Molecular and General Genetics* **198**, 255–62.

Downie, J. A. & Surin, B. P. (1990). Either of two *nod* gene loci can complement the nodulation defect of a *nod* deletion mutant of *Rhizobium leguminosarum* bv. *viciae*. *Molecular and General Genetics* **222**, 81–6.

Economou, A., Hamilton, W. D. O., Johnston, A. W. B. & Downie, J. A. (1990). The *Rhizobium* nodulation gene *nodO* encodes a Ca^{2+}-binding protein that is exported without N-terminal cleavage and is homologous to haemolysin and related proteins. *EMBO Journal* **9**, 349–54.

Economou, A., Hawkins, F. K. L., Downie, J. A. & Johnston, A. W. B. (1989). Transcription of *rhiA*, a gene on a *Rhizobium leguminosarum* bv. *viciae* Sym plasmid, requires *rhiR* and is repressed by flavonoids that induce *nod* genes. *Molecular Microbiology* **3**, 87–93.

Hong, G.-F., Burn, J. E. & Johnston, A. W. B. (1987). Evidence that DNA involved in the expression of nodulation *nod* genes in *Rhizobium* binds to the product of the regulation gene *nodD*. *Nucleic Acids Research* **15**, 9677–90.

Hopwood, D. A. & Sherman, D. H. (1990). Molecular genetics of polyketides and its comparison to fatty acid biosynthesis. *Annual Review of Genetics* **24**, 37–66.

John, M., Schmidt, J., Wieneke, U., Krussmann, H.-D. & Schell, J. (1988). Transmembrane orientation and receptor-like structure of the *Rhizobium meliloti* common nodulation protein NodC. *EMBO Journal* **7**, 583–8.

Lerouge, P., Roche, P., Faucher, C., Maillet, F., Truchet, G., Prome, J. C. & Denarie, J. (1990). Symbiotic host-specificity of *Rhizobium meliloti* is determined by a sulphated and acylated glucosamine oligosaccharide signal. *Nature* **344**, 781–4.

Long, S. R. (1989). *Rhizobium*-legume nodulation: life together in the underground. *Cell* **56**, 203–14.

Roche, P., Lerouge, P., Prome, J. C., Faucher, C., Vasse, J., Maillet, F., Camut, S., De Billy, Dénarié, J. & Truchet, G. (1991). NodRm-1, a sulphated lipo-oligosaccharide signal of *Rhizobium meliloti*, elicits hair deformation, cortical cell divisions and nodule organogenesis on alfalfa roots. In *Molecular Genetics of Plant–Microbe Interactions*, ed. H. Hennecke & D. P. S. Verma, pp. 119–26. Dordrecht: Kluwer Academic Publishers.

Schmidt, J., John, M., Wienecke, G., Stacey, G., Röhrig, H. & Schell, J. (1991). Studies on the function of *Rhizobium meliloti* nodulation genes. In *Molecular Genetics of Plant–Microbe Interactions*, ed. H. Hennecke & D. P. S. Verma, pp. 150–5. Dordrecht: Kluwer Academic Publishers.

Shearman, C. A., Rossen, L., Johnston, A. W. B. & Downie, J. A. (1986). The *Rhizobium* gene *nodF* encodes a protein similar to acyl carrier protein and is regulated by *nodD* plus a factor in pea root exudate. *EMBO Journal* **5**, 647–52.

Spaink, H. P., Geiger, O., Sheeley, D. M., van Brussel, A. A. N., York, W. S., Reinhold, V. N., Lugtenberg, B. J. J. & Kennedy, E. P. (1991). The biochemical function of the *Rhizobium leguminosarum* proteins involved in the production of host specific signal molecules. In *Advances in Molecular Genetics of Plant–Microbe Interactions*, ed. H. Hennecke & D. P. S. Verma, pp. 142–9. Dordrecht: Kluwer Academic Publishers.

Spaink, H. P., Okker, R. J. H., Wijffelman, C. A., Tak, T., Goosen-de Roo, L., Pees, E., Van Brussel, A. A. N. & Lugtenberg, B. J. J. (1989*a*). Symbiotic properties of rhizobia containing a flavonoid-independent hybrid *nodD* product. *Journal of Bacteriology* **171**, 4045–53.

Spaink, H. P., Weinman, J., Djordjevic, M. A., Wijffelman, C. A., Okker, R. J. H. & Lugtenberg, B. J. J. (1989*b*). Genetic analysis and cellular localization of the *Rhizobium* host specificity-determining NodE protein. *EMBO Journal* **8**, 2811–18.

Surin, B. P. & Downie, J. A. (1988). Characterization of the *Rhizobium leguminosarum* genes *nodLMN* involved in efficient host specific nodulation. *Molecular Microbiology* **2**, 173–83.

Surin, B. P., Watson, J. M., Hamilton, W. D. O., Economou, A. & Downie, J. A. (1990). Molecular characterization of the nodulation gene *nodT* from two biovars of *Rhizobium leguminosarum*. *Molecular Microbiology* **4**, 245–52.

Wijffelman, C. A., Pees, E., Van Brussel, A. A. N., Okker, R. J. H. & Lugtenberg, B. J. J. (1985). Genetic and functional analysis of the nodulation region of the *Rhizobium leguminosarum* Sym plasmid pRL1JI. *Archives of Microbiology* **143**, 225–32.

Note added in proof

After submission of this work, Spaink *et al.* (1991, Nature **354**, 125–30) described the structures of glycolipid signals made by *R. leguminosarum* bv. *viciae*. They are tetramers or pentamers of *N*-acetyl glucosamine in which the first glucosamine carries a novel, highly unsaturated, N-linked acyl group (C18:4). Studies with mutant strains showed that the *nodE* gene product is required for the formulation of signals carrying this novel lipid.

J. W. KIJNE, R. BAKHUIZEN, A. A. N. VAN
BRUSSEL, H. C. J. CANTER CREMERS,
C. L. DIAZ, B. S. DE PATER, G. SMIT, H. P.
SPAINK, S. SWART, C. A. WIJFFELMAN
AND B. J. J. LUGTENBERG

The *Rhizobium* trap: root hair curling in root–nodule symbiosis

Introduction

The root-nodule bacteria *Rhizobium*, *Bradyrhizobium* and *Azorhizobium* (collectively rhizobia) invade and nodulate the roots of their host plants via either wounds or root hairs. The choice is made by the host plant, e.g. the same rhizobial strain infects *Vigna* roots via root hairs and *Arachis* roots via wounds (Sen & Weaver, 1984), whereas another strain infects *Parasponia* via root epidermal cracks and *Macroptilium* via root hairs (Marvel *et al.*, 1985). Shortly before or during root invasion, rhizobia induce cell divisions in the root cortex, resulting in formation of a nodule primordium. Through infection threads (tip-growing tubular structures containing invading rhizobia) and/or between cortical cells the rhizobia migrate towards the growing primordium, are endocytosed by young nodule cells, and differentiate into dinitrogen-fixing bacteroids (see also Brewin *et al.*, this volume).

Rhizobial invasion of most agronomically important legumes such as pea (*Pisum sativum*), soybean (*Glycine max*) and bean (*Phaseolus vulgaris*) occurs through root hairs. Infection of a living plant cell is an unusual phenomenon in plant–bacteria interactions. Plants are open organisms. At many sites, the intercellular space of a plant is in direct contact with the environment, e.g. in stomata, hydathodes or wounds resulting from emergence of lateral roots. A plant is used to regular visits of (plant-associated) bacteria to its interior. Therefore, wound-infection by rhizobia is a normal phenomenon whereas root hair infection is special.

Interestingly, rhizobia must be trapped before expressing their ability

Society for Experimental Biology Seminar Series 48: *Perspectives in Plant Cell Recognition*, ed. J. A. Callow & J. R. Green. © Cambridge University Press 1992, pp. 267–84.

to penetrate a root hair. Sometimes the bacteria are (accidentally?) caught between two root hairs or between a root hair and, for example, a soil particle. Mostly, however, the rhizobia are trapped in the inner curve of tightly curling young root hairs. Such root hairs have already been described by Ward (1887). The curling process is induced by the bacteria. In other words, the bacteria let themselves lock in and then 'escape' into the plant root by formation of an infection thread. This infection mechanism is host plant-specific. As a rule, legume root hairs curl tightly around only their homologous symbiotic partners (Haack, 1964; Yao & Vincent, 1969).

Host plant specificity

Intimate associations between plants and bacteria are usually host plant-specific. For those associations characterised by entry through plant openings such as stomata, this specificity is expressed in molecular interactions enabling colonisation of the intercellular spaces of the plant by the bacteria (Klement *et al.*, 1964). The pea blight bacterium *Pseudomonas syringae* pathovar *pisi*, for example, can proliferate between parenchymous cells in pea leaves but not in bean leaves, whereas the contrary applies to *P.s.* pv. *phaseolicola*. Similar interactions may play a role in host-specific root invasion by wound-infecting rhizobia. However, for root hair-infecting rhizobia, host plant specificity is expressed at the trapping stage. *Rhizobium leguminosarum* biovar *viciae*, for example, induces tight root hair curling (phenotype Hac) and infection thread formation (phenotype Inf) in pea and vetch (*Vicia*) but induces only root hair deformations (phenotype Had) in alfalfa (*Medicago sativa*). In contrast, the alfalfa symbiont *Rhizobium meliloti* can be trapped in tightly curled alfalfa root hairs but not in those of pea. At present, it is unclear if host-specific interactions also enable proliferation of rhizobia in an infection thread and/or between root cortical cells.

Root hair growth

Root hairs are tip-growing protuberances of plant root epidermal cells, and are involved in uptake of water and nutrients. Root hair formation by an epidermal cell starts with preparation for cell division. There is a strong topological correlation between the site of root hair emergence and the cell division site in foregoing root epidermal cytokinesis (Sinnott & Bloch, 1939). Before root hair emergence, the nucleus of the epidermal cell enlarges and synthesises DNA (Tschermak-Woess & Hasitschka, 1953; Cutter & Feldman, 1970; Nagl, 1978), a microtubule array radiating from the nuclear envelope is established, and the nucleus migrates to the presumptive plane of cell division (Bakhuizen, 1989). Next, instead of a cell division site, a polar tip growth area is established with which the nucleus is connected by endoplasmic microtubules (Bakhuizen, 1989). During the root hair emergence which follows, the nucleus usually migrates in step with the extending hair tip (Nutman *et al.*, 1973; Dart, 1974), probably because nucleus and hair tip are connected by microtubules (Lloyd *et al.*, 1988). The tip growth area is characterised by accumulation of cytoplasm, polar fusion of secretory vesicles and cell wall synthesis (see e.g. Ridge, 1988). At the tip of the cell, a visco-elastic wall is formed that expands under internal pressure. During growth, newly assembled root hair wall segments are displaced in subapical direction and undergo gradual maturation and rigidification by a cross-linking process (for a model of apical wall growth, see Wessels, 1988). A similar growth mechanism has been described for fungal hyphae, pollen tubes and moss caulonema cells (for a review, see Sievers & Schnepf, 1981). According to Picton & Steer (1982, 1983), tip growth is controlled by the rate of vesicle fusion and by the state of relaxation of a microfilament network in the growth area, both factors being sensitive to changes in the local Ca^{2+} concentration. However, plant molecules determining initiation of a tip growth area are still to be characterised.

Very young root hairs are most susceptible to rhizobial infection. It is the dynamic microenvironment of a newly initiated tip growth area towards which many rhizobia are attracted by root metabolites (Gaworzewska & Carlile, 1982; Aguilar *et al.*, 1988; Armitage *et al.*, 1988).

A model for root hair curling

Computer simulation predicts that tight curling of a growing root hair results from induction and redirection of tip growth by attached rhizobia (Van Batenburg *et al.*, 1986). This model is based on the following elements: (1) ability of rhizobia to initiate and/or sustain tip growth, (2) attachment of one or a few rhizobia within the growth area of the root hair, (3) one-sided displacement of these rhizobia by root hair growth, and (4) a gradual shift in growth activity from the original root hair tip to the site of rhizobial attachment. When supplied with these elements and assuming a small rhizobia-induced growth area, the computer program produces a model of a perfect tight root hair curl enclosing the active rhizobia. Concentration of the plant cell tip growth machinery at the site of the trapped rhizobia may subsequently result in establishment of an infection focus and in initiation of infection thread growth (Fig. 1). Since, in addition to a tip growth area, infection thread growth is likely to

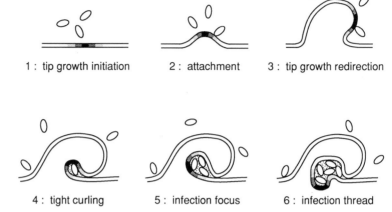

1 : tip growth initiation   2 : attachment   3 : tip growth redirection

4 : tight curling   5 : infection focus   6 : infection thread

Fig. 1. Successive steps in curling and infection of a legume root hair by *Rhizobium* bacteria.

require a growth pressure (just as turgor pressure is needed for root hair growth), entrapment of rhizobia may be essential for generation of this growth pressure.

The following observations are consistent with this model: (1) rhizobia produce factors able to enhance root hair formation and to induce branching and deformation of growing root hairs (Yao & Vincent, 1969, 1976; Bhuvaneswari & Solheim, 1985; Van Brussel *et al.*, 1986; Faucher *et al.*, 1989; Lerouge *et al.*, 1990; Spaink *et al.*, 1991); (2) such factors, if applied in (semi)purified form in absence of rhizobia, are unable to induce Hac; (3) presence of rhizobia at the root hair surface is required for induction of Hac (Yao & Vincent, 1969); (4) initiation of root hair infection results in inhibition of root hair growth, and, instead of moving along with the root hair tip, the nucleus now accompanies the tip of the growing infection thread (Lloyd *et al.*, 1988).

Which of these elements in tight root hair curling are involved in determination of host plant specificity?

Attachment

Rhizobia can attach to various surfaces including the cell walls of legume and non-legume root hairs. Attachment of rhizobia to the surface of a plant root hair is probably mediated by a rhizobial calcium-binding 14 kDa protein, rhicadhesin (Smit *et al.*, 1989). At present, evidence for this statement originates from the following observations: (1) purified rhicadhesin inhibits attachment of rhizobia to the surface of root hairs in a dose-dependent way; (2) with the use of mild shearing forces, rhicadhesin can be isolated from the surface of rhizobia, while leaving the bacteria viable; (3) such shearing strongly reduces the attachment ability of rhizobia; (4) growth of rhizobia under Ca^{2+}-limiting conditions results in release of rhicadhesin into the growth medium, in the absence of rhicadhesin from cell surface preparations, and in loss of attachment ability; and (5) pretreatment of root hairs with rhicadhesin restores attachment of rhizobia grown under Ca^{2+}-limiting conditions (Smit *et al.*, 1991). Rhicadhesin appears to be common among Rhizobiaceae, including *Agrobacterium*, and is most probably chromosomally encoded. Rhizobial mutants lacking rhicadhesin have not been identified yet. Furthermore, the identity of the plant receptor(s) of rhicadhesin remains to be determined.

Following binding to the root hair surface, rhizobia can form aggregates on root hair tips. For *R.l.* bv. *viciae*, this aggregation was found to be mediated by bacterial cellulose fibrils (Smit *et al.*, 1987). However, since fibril-negative mutants normally nodulate, fibril-mediated aggregation appears not to be essential for successful root infection.

Root hair attachment itself is not sufficient for induction of Hac by rhizobia. Growth of infective *R.l.* bv. *viciae* bacteria in the carbon-limited TY-medium (Beringer *et al.*, 1978), for example, yields cells perfectly able to show the rhicadhesin/cellulose-type of attachment on (host) plant root hairs but unable to induce curling and infection thread formation (Kijne *et al.*, 1988; Diaz, 1989). For induction of these processes, presence and expression of nodulation (*nod*) genes are required.

Nodulation genes and signal molecules

In fast-growing *Rhizobium* species, many essential *nod* genes are clustered on large Sym(biosis) plasmids, whereas in the slow-growing *Bradyrhizobium* species similar symbiotic genes are located on the chromosome (Long, 1989; and references therein). These *nod* genes encode both common and host plant-specific nodulation functions. Most *nod* genes are inducible, and their expression requires the presence of a functional *nodD* gene as well as a flavonoid inducer released by the host plant. The *nodABC* genes appear to be functionally interchangeable among rhizobia and therefore represent common *nod* genes. These genes constitute the minimal genetic requirement for induction of both root hair curling and mitotic activity in the plant root cortex. Other *nod* genes such as *nodFEL*, *nodMN* and *nodH* are not common and represent host range genes (*hsn* genes, for host specificity of nodulation). Mutations in *hsn* genes often result in abnormal root hair curling, disturbed root hair infection, and/or delayed nodulation (see e.g. Downie *et al.*, 1983). Tight curling of root hairs induced by rhizobia appears to result from the concerted action of common *nod* and *hsn nod* gene products (see also Downie *et al.*, this volume).

Recently, it was shown that both common *nod* and *hsn nod* genes are involved in the synthesis of lipooligosaccharide molecules, produced and released in detectable amounts by rhizobia upon *nod* gene induction by an appropriate flavonoid (Lerouge *et al.*, 1990; Roche *et al.*, 1991; Spaink *et al.*, 1991). The first identified lipooligosaccharide, NodRm-1, was shown to be a sulphated β-1,4-tetrasaccharide of D-glucosamine in which three amino groups are acetylated and one amino group is acylated with a bis-unsaturated C_{16} fatty acid chain (Lerouge *et al.*, 1990). This signal molecule, produced by *R. meliloti*, has a molecular weight of 1102 and induces root hair deformation and branching as well as nodule primordium initiation in the host plant alfalfa. According to Dénarié and colleagues (Faucher *et al.*, 1989; Roche *et al.*, 1991), the heterologous host plant vetch does not respond to this molecule unless the sulphate group is absent. Addition of this group to the oligosaccharide part of the compound is encoded by the *R. meliloti hsn* gene *nodH* (Roche *et al.*, 1991). These observations show that *R. meliloti* produces a signal molecule that is able specifically to influence growth of alfalfa root hairs and that contains a structure (i.e. a sulphate group) that prevents functional recognition by heterologous plants. Similar specific signal molecules have now been identified for *R.l.* bvs. *viciae* and *trifolii* (Spaink *et al.*, 1991; A. Aarts, personal communication).

According to the model of Van Batenburg *et al.* (1986), attachment of *Rhizobium*, together with production of NodRm-1-like lipooligosaccharides, can be sufficient for induction of Hac in root hairs of a host plant. However, NodD/flavonoid-mediated induction of *nod* genes may not be necessary for induction of curling. Spaink *et al.* (1989) found that *nodD* mutants of *R.l.* bv. *trifolii* carrying a heterologous *nodD* gene are able to induce Hac but not Inf in red clover (*Trifolium pratense*), although *nod* gene-inducing compounds in exudate of red clover roots do not interact functionally with heterologous *nodD* genes (Spaink *et al.*, 1987). This observation suggests that a basic level of NodD-mediated *nod* gene expression is sufficient for induction of Hac, and that flavonoid-induced activation of *nod* genes is necessary for infection thread initiation rather than for root hair curling. The level of lipooligosaccharide production by rhizobia at the host plant surface remains to be established.

Supercurling

As an exception, the *R.l.* bvs. *viciae* and *trifolii* induce extensive curling in root hairs of each other's host plants. However, this type of curling, called supercurling (phenotype Hac^{++}, according to Van Brussel *et al.*, 1988), is not the normal tight curling found for homologous symbionts. According to the model of Van Batenburg *et al.* (1986), supercurling results from the inability of the rhizobia to attract all tip growth activity to the site of attachment. The residual tip growth activity of the root hair is probably responsible for continuous extensive curling.

The difference in host plant specificity between *R.l.* bv. *viciae* and *R.l.* bv. *trifolii* is primarily determined by the *hsn* gene *nodE* (Spaink *et al.*, 1989). Like some other *nod* genes, this gene is involved in production of a specific bioactive lipooligosaccharide (Spaink *et al.*, 1991). Mutation of *nodE* in these biovars results in cells inducing supercurling instead of tight curling in root hairs of their homologous host plants vetch and white clover, respectively. These observations present additional evidence for an essential role of *nod*-gene related lipooligosaccharides in tight root hair curling and in generation of an infection focus.

Branching and curling factors

Before discovery of the lipooligosaccharide signal molecules, branching and curling factors produced by rhizobia had been reported by several authors (e.g. Yao & Vincent, 1969, 1976; Bhuvaneswari & Solheim, 1985). Interestingly, Yao & Vincent (1969, 1976) describe properties of a crude preparation of (a) rhizobial factor(s) considered to be responsible for branching of clover root hairs and most probably different from a *nod* gene-related lipooligosaccharide. This factor was prepared by suspending infective *R.l.* bv. *trifolii* cells, harvested from yeast-mannitol (YM) agar (lacking flavonoid inducers!), in distilled water for 1 h, followed by removal of the bacteria from the water-extract by centrifugation and filtration. Root hair branching activity was stable at 100 °C and partly dialysable. A dried residue of the active water extract contained a mixture of nucleic acid, protein and carbohydrate, including minor amounts of lipopolysaccharide. Recent results from our laboratory demonstrate that *nod* gene-related lipooligosaccharides are unlikely to be present in an extract prepared under these conditions. The same extract also showed moderate curling (hair deformation) activity. Remarkably, the branching/curling activity of an extract prepared from homologous strains appeared to be superior to that prepared from heterologous strains.

These observations suggest that infective rhizobia with unactivated *nod* genes can produce a water-soluble, heat-stable, root hair-branching factor, with host plant-specific activity. Since induction of root hair branching is equivalent to induction of tip growth, such a branching factor may contribute to induction of tight root hair curling and infection thread growth. The nature of the branching factor remains to be established.

Considering the present state of knowledge, the results of Yao & Vincent merit re-examination by studying factor production by various nodulation mutants of *Rhizobium* in the presence or absence of appropriate flavonoids.

Plant lectin

Production of symbiotically bioactive compounds by rhizobia grown on YM-agar correlates positively with the observation of Kijne *et al.* (1988) that YM-grown rhizobia are infective, in contrast to TY-grown rhizobia. YM-grown *R.l.* bv. *viciae* cells show accelerated accumulation at pea root hair tips. This accumulation is mediated by pea lectin (Psl) molecules (Diaz, 1989). Lectins are sugar-binding (glyco)proteins that are not enzymes or antibodies (Barondes, 1988). Interestingly, host plants of a particular rhizobial species or biovar produce lectin molecules with a high-affinity sugar binding-specificity different from that of lectins produced by other groups of host plants (Van Driessche, 1988). Psl is a 49 kDa glucose/mannose-binding protein, abundantly present in pea seeds and also produced and secreted by pea roots (Diaz *et al.*, 1990). Involvement of Sym plasmid-encoded (*nod*) genes in Psl-enhanced accumulation of *R.l.* bv. *viciae* cells could not be demonstrated (Kijne *et al.*, 1988; Diaz, 1989). Moreover, several *Rhizobium* species are able to bind Psl (Van der Schaal *et al.*, 1983), and Psl can precipitate extracellular polysaccharides (EPS) of various fast-growing rhizobia, including those of *R.l.* bv. *viciae* (Kamberger, 1979*a*,*b*). In contrast, binding of Psl to lipopolysaccharide (LPS) of *R.l.* bv. *viciae* has been reported to be biovar-specific (Wolpert & Albersheim, 1976; Kamberger, 1979*a*,*b*; Kato *et al.*, 1979). Since Sym plasmid-cured *R.l.* bv. *viciae* cells show Psl-enhanced accumulation but are Hac-negative, attachment together with production of lectin-binding EPS/LPS, is apparently not sufficient for induction of root hair curling.

On the other hand, clover lectin (trifoliin A: Dazzo *et al.*, 1978) binds specifically to encapsulated and infective *R.l.* bv. *trifolii* bacteria (for a review, see Dazzo *et al.*, 1984). Trifoliin A is a 53 kDa (glyco)protein that recognises 2-deoxy-D-glucose and 2-amino-2,6-dideoxy-D-glucose as haptenic monosaccharides (Hrabak *et al.*, 1981), but also binds to certain saccharides lacking these sugars (Hollingsworth *et al.*, 1988). Like Psl in pea, trifoliin A is synthesised in clover seeds and roots and is secreted into the rhizosphere (Dazzo & Hrabak, 1981; Sherwood *et al.*, 1984*a*). *R.l.* bv. *trifolii* produces several extracellular polysaccharide ligands for trifoliin A, including EPS-derivatives and LPS, each changing in composition with culture age (Dazzo *et al.*, 1979; Hrabak *et al.*, 1981; Abe *et al.*, 1984; Sherwood *et al.*, 1984*b*). In contrast to the situation for *R.l.* bv. *viciae*, trifoliin-binding properties of these ligands are dependent on the presence of *nod* genes (Zurkowski, 1980; Philip-Hollingsworth *et al.*, 1989*b*). However, as in *R.l.* bv. *viciae*, most studies on lectin-binding by *R.l.* bv. *trifolii* have been performed using cells grown in the absence of *nod* gene-inducing flavonoids.

These observations show that, under appropriate culture conditions, rhizobia may produce (biovar-specific) extracellular ligands able to interact with host plant-specific lectins at the root hair surface. Production of such ligands in the absence of *nod* gene inducers suggest the absence of detectable amounts of *nod* gene-related lipooligosaccharides among these ligands. Since EPS and EPS-derivatives are water-soluble and heat-stable, the relationship between lectin-binding ligands and the branching/curling factor(s) of Yao & Vincent (1969, 1976) is worth studying. Additional support for this is provided by the results of Abe *et al.* (1984), who found that lectin-binding oligosaccharide fragments of EPS of *R.l.* bv. *trifolii* stimulate infection thread formation in clover root hairs. Again, the bacteria used in these experiments were grown in the absence of *nod* gene-inducing flavonoids. The precise role of certain *nod* genes in synthesis of lectin-binding ligands remains to be established. Whether *nod* genes play a role in modification of the structure of EPS is currently a matter of debate (see e.g. Philip-Hollingsworth *et al.*, 1989*b*; Canter Cremers, 1990).

EPS-deficient mutants

Rhizobium leguminosarum produces a high molecular weight EPS that can be recovered from the growth medium or can be washed off from the cells (Robertsen *et al.*, 1981; Carlson, 1982; Philip-Hollingsworth *et al.*, 1989*a*; Canter Cremers, 1990). Many genes control EPS-biosynthesis, and EPS-deficient mutants (*exo* mutants) of the *R.l.* bvs. *viciae* and *trifolii* show different phenotypes, varying from normal nodulation (see e.g. Canter Cremers, 1990), abortive infection thread formation (see e.g. Borthakur *et al.*, 1986; Diebold & Noel, 1989), to poor root hair curling and avirulence (Canter Cremers, 1990). An interesting example of the latter class of mutants is *R.l.* bv. *viciae* RBL5515 *exo4* : : Tn5 pRL1JI (i.e. *exo4*; Canter Cremers, 1990). In this mutant, Tn5 is inserted in the middle of gene *pssA*, which is involved in EPS biosynthesis (Borthakur *et al.*, 1986, 1988). The biochemical step at which EPS synthesis is blocked by a mutation of *pssA* is not known. *exo4* produces about 1% of the amount of EPS produced by the parental strain, and this EPS lacks acetyl groups esterified to the glucose residue carrying the EPS side-chain (Canter Cremers, 1990). Expression of the Hac⁻ phenotype by this mutant on vetch was somewhat surprising, since *exo4* shows normal attachment to host plant root hairs and, upon *nod* gene induction, produces bioactive lipooligosaccharide signal molecules. Interestingly, the mutation has a biovar-specific effect, in that replacement of the *R.l.* bv. *viciae* Sym plasmid by the *R.l.* bv. *trifolii* Sym plasmid pSym5 results in a normal nodulation phenotype on white clover (Canter Cremers, 1990). In another *exo* mutant, *R.l.* bv. *viciae* RBL5515 *exo2* : : Tn5 pRL1JI (i.e. *exo2*), Tn5 is located in the region immediately upstream from the *pssA* gene which may contain the *pssA* promoter (Canter Cremers, 1990). Since *exo2* normally nodulates vetch plants, it has been suggested that *exo2* is still able to synthesise small amounts of active PssA protein, in contrast to *exo4*. Both mutants are indistinguishable with respect to EPS production, EPS structure, LPS profile, cell envelope protein pattern, and production of β-1,2-glucan and capsular polysaccharide. Apparently, the differences in structure and amount of EPS biosynthesis between *exo4* and its parental strain are not responsible for the difference in nodulation phenotype. It is tempting to suggest that, in addition to being involved in EPS production, the PssA protein is also involved in the synthesis of a biovar-specific root hair branching/curling factor different from the usual set of *nod* gene-related lipooligosaccharides.

In order to test whether *pss* mutants of *R.l.* bv. *viciae* could be rescued by the presence of an EPS-producing strain, Borthakur *et al.* (1988) performed co-inoculation experiments on pea plants using a mixture of *R.l.* bv. *viciae pss* : : Tn5 pRL1JI cells and a non-infective *R. leguminosarum* strain lacking a Sym plasmid. These experiments resulted in the formation of normal numbers of root nodules that were occupied exclusively by the strain lacking the Sym plasmid. This result indicates that the presence of *nod* genes (and thus the production of *nod* gene-related lipooligosaccharides) is essential only during tight curling and initiation of root hair infection. In contrast, production of the putative EPS-related branching/curling factor(s) may be essential during the entire infection process.

Roles of lipooligosaccharides and EPS-related factors

Genetic, biochemical and physiological studies have now provided strong evidence that production of *nod* gene-related lipooligosaccharides by rhizobia is necessary for induction of Hac, initiation of Inf and induction of mitotic activity in the host root cortex (see also Downie *et al.*, this volume). The influence of these signal molecules on the local Ca^{2+} concentration, vesicle transport and fusion, and the cell cytoskeleton has yet to be determined. However, lipooligosaccharide production, concurrent with attachment of the rhizobia to the root hair surface, appears not to be sufficient for successful root infection. Putative EPS-related factors play an additional (host-specific?) role, most notably in tip growth initiation and continuation. Purification and characterisation of these factors will reveal their relationship with rhizobial surface polysaccharides, such as the acidic high molecular weight EPS and LPS, and with *nod*-gene related lipooligosaccharides. Lectins are obvious tools for affinity-purification of these factors. Lectins may also interact with lipooligosaccharides, with either the sugar-binding sites or the conserved hydrophobic pockets being involved. The relationship between lectin (ligands) and plant cell tip growth is still unclear.

Referring to studies on *R. meliloti exo* mutants, Pühler *et al.* (1991) suggested that a signal molecule from the EPS biosynthetic pathway is necessary for the invading rhizobial cells to be recognised as symbionts. Inability to produce such signal(s) appears to be correlated with induction of a local plant defence reaction against the rhizobial mutants, which may inhibit infection thread formation and proliferation of rhizobia in the host plant tissue. Interestingly, a similar role for EPS and/or EPS-related compounds in prevention of plant defence has been proposed for compatible interactions between phytopathogenic bacteria and their host plant cells (see e.g. Geider *et al.*, 1991). Interactions between EPS from these bacteria and host plant tissue can have a high level of host plant specificity. Studying blight diseases of various plants, El-Banoby & Rudolph (1979) found that infiltration of leaves with a crude EPS-preparation from blight-inducing pseudomonads results in formation of persistent water-soaked spots only for compatible host–pathogen combinations. These spots appeared to result from the presence of a sponge-like fibrillar network in the intercellular spaces of the plant tissue (El-Banoby *et al.*, 1981). It was suggested that formation of this hydrophilic network enables host plant-specific proliferation of pathogenic pseudomonads. Production of matrix material during and following formation of an infection focus in curling root hairs of legume host plants may be equivalent to formation of water-soaked spots.

In conclusion, rhizobia differ from phytopathogenic bacteria in their ability to infect living plant cells, i.e. root hairs, by attachment and production of *nod* gene-related lipooligosaccharides. However, the interactions during and following enclosure in the root hair trap (and interactions during wound-infection by rhizobia) may represent compatible interactions such as those between phytopathogens and their host plant cells.

References

Abe, M., Sherwood, J. E., Hollingsworth, R. I. & Dazzo, F. B. (1984). Stimulation of clover root hair infection by lectin-binding oligosaccharides from the capsular and extracellular polysaccharides of *Rhizobium trifolii*. *Journal of Bacteriology* **160**, 517–20.

Aguilar, J. J. M., Ashby, A. M., Richards, A. J. M., Loake, G. J., Watson, M. D. & Shaw, C. H. (1988). Chemotaxis of *Rhizobium leguminosarum* biovar *phaseoli* towards flavonoid inducers of the symbiotic nodulation genes. *Journal of General Microbiology* **134**, 2741–6.

Armitage, J. P., Gallagher, A. & Johnston, A. W. B. (1988). Comparison of the chemotactic behaviour of *Rhizobium leguminosarum* with and without the nodulation plasmid. *Molecular Microbiology* **2**, 743–8.

Bakhuizen, R. (1989). The plant cytoskeleton in the *Rhizobium*–legume symbiosis. Ph.D. thesis. Leiden University.

Barondes, S. H. (1988). Bifunctional properties of lectins: lectins redefined. *Trends in Biochemical Sciences* **13**, 480–4.

Beringer, J. E., Beynon, J. L., Buchanan-Wollaston, A. V. & Johnston, A. W. B. (1978). Transfer of the drug-resistence transposon Tn*5* to *Rhizobium*. *Nature* **276**, 633–4.

Bhuvaneswari, T. V. & Solheim, B. (1985). Root hair deformations in white clover/*Rhizobium trifolii* symbiosis. *Physiologia Plantarum* **68**, 1144–9.

Borthakur, D., Barber, C. E., Lamb, J. W., Daniels, M. J., Downie, J. A. & Johnston, A. W. B. (1986). A mutation that blocks exopolysaccharide synthesis prevents nodulation of peas by *Rhizobium leguminosarum* but not of beans by *Rhizobium phaseoli* and is corrected by cloned DNA from *Rhizobium* or the phytopathogen *Xanthomonas*. *Molecular and General Genetics* **203**, 320–3.

Borthakur, D., Barker, R. F., Latchford, J. W., Rossen, L. & Johnston, A. W. B. (1988). Analysis of *pss* genes of *Rhizobium leguminosarum* required for exopolysaccharide synthesis and nodulation of peas: their primary structure and their interaction with *psi* and other nodulation genes. *Molecular and General Genetics* **213**, 155–62.

Canter Cremers, H. C. J. (1990). Role of exopolysaccharide in nodulation by *Rhizobium leguminosarum* bv viciae. Ph.D. thesis, Leiden University.

Carlson, R. W. (1982). Surface chemistry. In *Ecology of Nitrogen Fixation*, vol. 2 *Rhizobium*, ed. W. Broughton, pp. 199–234. Oxford: Oxford University Press.

Cutter, E. C. & Feldman, L. J. (1970). Trichoblasts in *Hydrocharis*. II. Nucleic acids, proteins and consideration of cell growth in relation to endoploidy. *American Journal of Botany* **57**, 202–11.

Dart, P. J. (1974). The infection process. In *The Biology of Nitrogen*

Fixation, ed. A. Quispel, pp. 381–429. Amsterdam, Oxford: North Holland Publ. Cy.

Dazzo, F. B. & Hrabak, E. M. (1981). Presence of trifoliin A, a *Rhizobium*-binding lectin, in clover root exudate. *Journal of Supramolecular Structure and Cell Biochemistry* **16**, 133–8.

Dazzo, F. B., Truchet, G. L., Sherwood, J. E., Hrabak, E. M., Abe, M. & Pankratz, S. H. (1984). Specific phases of root hair attachment in the *Rhizobium trifolii*–clover symbiosis. *Applied and Environmental Microbiology* **48**, 1140–50.

Dazzo, F. B., Urbano, M. R. & Brill, W. J. (1979). Transient appearance of lectin receptors on *Rhizobium trifolii*. *Current Microbiology* **2**, 15–20.

Dazzo, F. B., Yanke, W. E. & Brill, W. J. (1978). Trifoliin: a *Rhizobium* recognition protein from white clover. *Biochimica et Biophysica Acta* **539**, 276–86.

Diaz, C. L. (1989). Root lectin as a determinant of host-plant specificity in the *Rhizobium*–legume symbiosis. Ph.D. thesis, Leiden University.

Diaz, C. L., Hosselet, M., Logman, G. J. J., Van Driessche, E., Lugtenberg, B. J. J. & Kijne, J. W. (1990). Distribution of glucose/mannose-specific isolectins in pea (*Pisum sativum* L.) seedlings. *Planta* **181**, 451–61.

Diaz, C. L., Melchers, L. S., Hooykaas, P. J. J., Lugtenberg, B. J. J. & Kijne, J. W. (1989). Root lectin as a determinant of host-plant specificity in the *Rhizobium*–legume symbiosis. *Nature* **338**, 579–81.

Diebold, R. & Noel, K. D. (1989). *Rhizobium leguminosarum* exopolysaccharide mutants: biochemical and genetic analyses and symbiotic behavior on three hosts. *Journal of Bacteriology* **171**, 4821–30.

Downie, J. A., Hombrecher, G., Ma, Q. S., Knight, C. D., Wells, B. & Johnston, A. W. B. (1983). Cloned nodulation genes of *Rhizobium leguminosarum* determine host range specificity. *Molecular and General Genetics* **190**, 359–65.

El-Banoby, F. E. & Rudolph, K. (1979). Induction of water-soaking in plant leaves by extracellular polysaccharides from phytopathogenic pseudomonads and xanthomonads. *Physiological Plant Pathology* **15**, 341–9.

El-Banoby, F. E., Rudolph, K. & Mendgen, K. (1981). The fate of extracellular polysaccharide from *Pseudomonas phaseolicola* in leaves and leaf extracts from halo-blight susceptible and resistant bean plants (*Phaseolus vulgaris* L.). *Physiological Plant Pathology* **18**, 91–8.

Faucher, C., Camut, S., Dénarié, J. & Truchet, G. (1989). The *nodH* and *nodQ* host range genes of *Rhizobium meliloti* behave as avirulence genes in *R. leguminosarum* bv. *viciae* and determine changes in the production of plant-specific extracellular signals. *Molecular Plant–Microbe Interactions* **2**, 291–300.

Gaworzewska, E. T. & Carlile, M. J. (1982). Positive chemotaxis of *Rhizobium leguminosarum* and other bacteria towards root exudates from legumes and other plants. *Journal of General Microbiology* **128**, 1179–88.

Geider, K., Bellemann, P., Bernhard, F., Chang, J.-R., Geier, G., Metzger, M., Pahl, A., Schwartz, T. & Theiler, R. (1991). Exopolysaccharides in the interaction of the fire-blight pathogen *Erwinia amylovora* with its host cells. In *Advances in Molecular Genetics of Plant–Microbe Interactions*, vol. 1, ed. H. Hennecke & D. P. S. Verma, pp. 90–3. Dordrecht: Kluwer Academic Press.

Haack, A. (1964). Über den Einfluss der Knöllchenbakterien auf die Wurzelhaare von Leguminosen und Nichtleguminosen. *Zentralblatt für die Bakteriologie und Parasitenkunde*, Abt II **117**, 343–66.

Hollingsworth, R. I., Dazzo, F. B., Hallenga, K. & Musselman, B. (1988). The complete structure of the trifoliin A lectin-binding capsular polysaccharide of *Rhizobium trifolii* 843. *Carbohydrate Research* **172**, 97–112.

Hrabak, E. M., Urbano, M. R. & Dazzo, F. B. (1981). Growth-phase dependent immuno-determinants of *Rhizobium trifolii* lipopolysaccharide which bind trifoliin A, a white clover lectin. *Journal of Bacteriology* **148**, 697–711.

Kamberger, W. (1979*a*). An Ouchterlony double diffusion study on the interaction between legume lectins and rhizobial cell surface antigens. *Archiv für Microbiologie* **121**, 83–90.

Kamberger, W. (1979*b*). Role of cell surface polysaccharides in the *Rhizobium*–pea symbiosis. *FEMS Microbiology Letters* **6**, 361–5.

Kato, G., Maruyama, Y. & Nakamura, M. (1979). Role of lectins and lipopolysaccharides in the recognition process of specific legume–*Rhizobium* symbiosis. *Agricultural and Biological Chemistry* **43**, 1085–92.

Kijne, J. W., Smit, G., Diaz, C. L. & Lugtenberg, B. J. J. (1988). Lectin enhanced accumulation of manganese-limited *Rhizobium leguminosarum* cells on pea root hair tips. *Journal of Bacteriology* **170**, 2994–3000.

Klement, Z., Farkas, G. L. & Lovrekovich, L. (1964). Hypersensitive reaction induced by phytopathogenic bacteria in tobacco leaf. *Phytopathology* **54**, 474–7.

Lerouge, P., Roche, P., Faucher, C., Maillet, F., Truchet, G., Prome, J. C. & Dénarié, J. (1990). Symbiotic host-specificity of *Rhizobium meliloti* is determined by a sulphated and acylated glucosamine oligosaccharide signal. *Nature* **344**, 781–4.

Lloyd, C. W., Pearce, K. J., Rawlins, D. J., Ridge, R. W. & Shaw, P. J. (1988). Endoplasmic microtubules connect the advancing nucleus to the tip of legume root hairs, but F-actin is involved in basipetal migration. *Cell Motility and the Cytoskeleton* **8**, 27–36.

Long, S. R. (1989). Rhizobium–legume nodulation: life together in the underground. *Cell* **56**, 203–14.

Marvel, D. J., Kuldau, G., Hirsch, A., Richards, E., Torrey, J. C. & Ausubel, F. M. (1985). Conservation of nodulation genes between *Rhizobium meliloti* and a slow-growing *Rhizobium* strain that nodulates a non-legume host. *Proceedings of the National Academy of Sciences, USA* **82**, 3841–5.

Marvel, D. J., Torrey, J. G. & Ausubel, F. M. (1987). *Rhizobium* symbiotic genes required for nodulation of legume and non-legume hosts. *Proceedings of the National Academy of Sciences, USA* **84**, 1319–23.

Nagl, W. (1978). *Endoploidy and polyteny in differentiation and evolution.* Amsterdam, New York, Oxford: North Holland Publ. Cy.

Nutman, P. S., Doncaster, C. C. & Dart, P. J. (1973). *Infection of clover by root nodule bacteria.* Black and white 16 mm optical sound track film, available from the British Film Institute, London, UK.

Philip-Hollingsworth, S., Hollingsworth, R. I. & Dazzo, F. B. (1989*a*). Host-range related structural features of the acidic extracellular polysaccharides of *Rhizobium trifolii* and *Rhizobium leguminosarum*. *Journal of Biological Chemistry* **264**, 1461–6.

Philip-Hollingsworth, S., Hollingsworth, R. I., Dazzo, F. B., Djordjevic, M. A. & Rolfe, B. G. (1989*b*). The effect of interspecies transfer of *Rhizobium* host-specific nodulation genes on acidic polysaccharide structure and *in situ* binding by host lectin. *Journal of Biological Chemistry* **264**, 5710–14.

Picton, J. M. & Steer, M. W. (1982). A model for the mechanism of tip extension in pollen tubes. *Journal of Theoretical Biology* **98**, 15–20.

Picton, J. M. & Steer, M. W. (1983). Evidence for the role of Ca^{2+} ions in tip extension in pollen tubes. *Protoplasma* **115**, 11–17.

Pühler, A., Arnold, W., Buendia-Claveria, A., Kapp, D., Keller, M., Niehaus, K., Quandt, J., Roxlau, A. & Weng, W. M. (1991). The role of the *Rhizobium meliloti* exopolysaccharides EPSI and EPSII in the infection process of alfalfa nodules. In *Advances in Molecular Genetics of Plant–Microbe Interactions*, vol. 1, ed. H. Hennecke & D. P. S. Verma, pp. 189–94. Dordrecht: Kluwer Academic Press.

Ridge, R. W. (1988). Freeze-substitution improves the ultrastructural preservation of legume root hairs. *Botanical Magazine, Tokyo* **101**, 427–41.

Robertsen, B. K., Aman, P., Darvill, A. G., McNeil, M. & Albersheim, P. (1981). Host–symbiont interactions. V. The structure of acidic extracellular polysaccharides secreted by *Rhizobium leguminosarum* and *Rhizobium trifolii*. *Plant Physiology* **67**, 389–400.

Roche, P., Lerouge, P., Prome, J. C., Faucher, C., Vasse, J., Maillet, F., Camut, S., De Billy, F., Dénarié, J. & Truchet, G. (1991). NodRm-1, a sulphated lipo-oligosaccharide signal of *Rhizobium meliloti* elicits hair deformation, cortical cell division and nodule organogenesis on alfalfa roots. In *Advances in Molecular Genetics of Plant–Microbe Interactions*, vol. 1, ed. H. Hennecke & D. P. S. Verma, pp. 119–26. Dordrecht: Kluwer Academic Press.

Sen, D. & Weaver, R. W. (1984). A basis for different rates of N_2-fixation by the same strain of *Rhizobium* in peanut and cowpea root nodules. *Plant Science Letters* **34**, 239–46.

Sherwood, J. E., Truchet, G. L. & Dazzo, F. B. (1984*a*). Effect of nitrate supply on the in-vivo synthesis and distribution of trifoliin A, a *Rhizobium trifolii*-binding lectin, in *Trifolium repens* seedlings. *Planta* **162**, 540–7.

Sherwood, J. E., Vasse, J. M., Dazzo, F. B. & Truchet, G. L. (1984*b*). Development and trifoliin A-binding ability of the capsule of *Rhizobium trifolii*. *Journal of Bacteriology* **159**, 145–52.

Sievers, A. & Schnepf, E. (1981). Morphogenesis and polarity of tubular cells with tip growth. In *Cytomorphogenesis in Plants*, ed. O. Kiermayer, pp. 263–99. Berlin, Heidelberg, New York: Springer-Verlag.

Sinnott, E. W. & Bloch, R. (1939). Cell polarity and the differentiation of root hairs. *Proceedings of the National Academy of Sciences, USA* **25**, 55–8.

Smit, G., Kijne, J. W. & Lugtenberg, B. J. J. (1987). Both cellulose fibrils and a Ca^{2+}-dependent adhesin are involved in the attachment of *Rhizobium leguminosarum* to pea root hair tips. *Journal of Bacteriology* **169**, 4294–301.

Smit, G., Logman, T. J. J., Boerrigter, M. E. T. I., Kijne, J. W. & Lugtenberg, B. J. J. (1989). Purification and partial characterization of the *Rhizobium leguminosarum* biovar *viciae* Ca^{2+}-dependent adhesin, which mediates the first step in attachment of cells of the family Rhizobiaceae to plant root hair tips. *Journal of Bacteriology* **171**, 4054–62.

Smit, G., Tubbing, D. M. J., Kijne, J. W. & Lugtenberg, B. J. J. (1991). Role of Ca^{2+} in the activity of rhicadhesin from *Rhizobium leguminosarum* biovar *viciae*, which mediates the first step in attachment of Rhizobiaceae cells to plant root hair tips. *Archives of Microbiology* **155**, 278–83.

Spaink, H. P., Geiger, O., Sheeley, D. M., Van Brussel, A. A. N., York, W. S., Reinhold, V. N., Lugtenberg, B. J. J. & Kennedy, E. P. (1991). The biochemical function of the *Rhizobium leguminosarum* proteins involved in the production of host specific signal molecules. In *Advances in Molecular Genetics of Plant–Microbe Interactions*, vol. 1, ed. H. Hennecke & D. P. S. Verma, pp. 142–9. Dordrecht: Kluwer Academic Press.

Spaink, H. P., Weinman, J., Djordjevic, M. A., Wijffelman, C. A., Okker, R. J. H. & Lugtenberg, B. J. J. (1989). Genetic analysis and cellular localization of the *Rhizobium* host-specificity determining NodE protein. *EMBO Journal* **8**, 2811–18.

Spaink, H. P., Wijffelman, C. A., Pees, E., Okker, R. J. H. & Lugtenberg, B. J. J. (1987). *Rhizobium* nodulation gene *nodD* as a determinant of host specificity. *Nature* **328**, 337–40.

Tschermak-Woess, E. & Hasitschka, G. (1953). Über Musterbildung in der Rhizodermis und exodermis bei einigen Angiospermen und einer Polypodiacee. *Oesterreicher Botanischen Zeitung* **100**, 646–51.

Van Batenburg, F. H. D., Jonker, R. & Kijne, J. W. (1986). *Rhizobium* induces marked root hair curling by redirection of tip growth, a computer simulation. *Physiologia Plantarum* **66**, 476–80.

Van Brussel, A. A. N., Pees, E., Spaink, H. P., Tak, T., Wijffelman, C. A., Okker, R. J. H., Truchet, G. & Lugtenberg, B. J. J. (1988). Correlation between *Rhizobium leguminosarum nod* genes and nodulation phenotypes on *Vicia*. In *Nitrogen Fixation, Hundred Years Later*, ed. H. Bothe, F. J. de Bruijn & W. E. Newton, p. 483, Stuttgart. New York: Gustav Fischer Verlag.

Van Brussel, A. A. N., Zaat, S. A. J., Canter Cremers, H. C. J., Wijffelman, C. A., Pees, E., Tak, T. & Lugtenberg, B. J. J. (1986). Role of plant root exudate and Sym plasmid-localized nodulation genes in the synthesis by *Rhizobium leguminosarum* of Tsr factor which causes thick and short roots on common vetch. *Journal of Bacteriology* **165**, 517–22.

Van der Schaal, I. A. M., Kijne, J. W., Diaz, C. L. & Van Iren, F. (1983). Pea lectin binding by *Rhizobium*. In *Lectins, Biology, Biochemistry, Clinical Biochemistry*, vol. 2, ed. T. C. Bøg-Hansen & G. A. Spengler, pp. 531–8. Berlin: Walter de Gruyter & Co.

Van Driessche, E. (1988). Structure and function of Leguminosae lectins. In *Advances in Lectin Research*, vol. 1, ed. H. Franz, pp. 73–134. Berlin, Heidelberg, New York: Springer-Verlag.

Ward, H. M. (1887). On the tubercular swellings on the roots of *Vicia faba*. *Philosophical Transactions of the Royal Society of London, Series B* **178**, 539–62.

Wessels, J. G. H. (1988). A steady-state model for apical wall growth in fungi. *Acta Botanica Neerlandica* **37**, 3–16.

Wolpert, J. S. & Albersheim, P. (1976). Host-symbiont interactions. I. The lectins of legumes interact with the O-antigen-containing lipopolysaccharides of their symbiont rhizobia. *Biochemical and Biophysical Research Communications* **70**, 729–37.

Yao, P. Y. & Vincent, J. M. (1969). Host specificity in the root hair curling factor of *Rhizobium* spp. *Australian Journal of Biological Sciences* **22**, 413–23.

Yao, P. Y. & Vincent, J. M. (1976). Factors responsible for the curling and branching of clover root hairs by Rhizobium. *Plant and Soil* **45**, 1–16.

Zurkowski, W. (1980). Specific adsorption of bacteria to clover root hairs, related to the presence of plasmid pWZ2 in cells of *Rhizobium trifolii. Microbios* **27**, 27–32.

N. J. BREWIN, E. A. RATHBUN,
S. PEROTTO, A. GUNDER, A. L. RAE
AND E. L. KANNENBERG

Structure and function of *Rhizobium* lipopolysaccharide in relation to legume nodule development

Introduction

The legume root nodule, which houses the symbiotic nitrogen-fixing *Rhizobium* bacteria, is an attractive system for studying plant–microbe interactions (Brewin, 1991). It shows a high degree of host-strain specificity. Intimate cell-to-cell contact is sustained over several weeks of progressive development. Furthermore, the relationship seems delicately balanced between symbiosis and pathogenicity.

Flavonoid compounds exuded by legume roots induce transcription of *Rhizobium* nodulation (*nod*) genes. The products of these genes in turn synthesise and secrete a glycolipid signal molecule specific for a particular strain of *Rhizobium*, which stimulates cortical cell divisions in the appropriate host and initiates the development of the nodule structure. Subsequent tissue and cell invasion by *Rhizobium* involves a series of intimate surface interactions between plant and bacterial cells (Fig. 1). A secreted plant glycoprotein provides the intercellular matrix for tissue invasion by *Rhizobium* (VandenBosch *et al.*, 1989a). The intracellular phase of symbiosis results from endocytosis following close membrane interactions between the bacterial surface, which carries a specialised form of lipopolysaccharide, and the plant cell membrane, which carries a specialised form of glycocalyx (composed of glycoprotein and glycolipid elements). Intracellular rhizobia are enclosed in an organelle-like structure termed the 'symbiosome' and differentiate into nitrogen-fixing forms. At the same time, Golgi-derived hydrolytic enzymes are secreted into the lumen of the symbiosome compartment, and these are eventually responsible for the host-induced senescence of endosymbiotic bacteria. Thus, the legume nodule system provides a unique opportunity to analyse

Society for Experimental Biology Seminar Series 48: *Perspectives in Plant Cell Recognition*, ed. J. A. Callow & J. R. Green. © Cambridge University Press 1992, pp. 285–96.

a   **b**   **c**

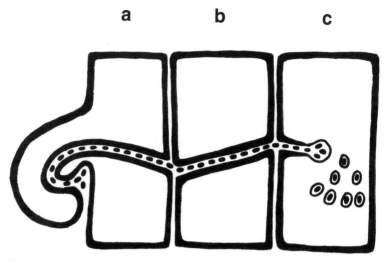

Fig. 1. Diagram of successive stages of tissue and cell invasion by *Rhizobium*. (a) Root hair curling and development of an infection thread as an intracellular tunnel still bounded by plant cell wall and cell membrane. (b) Cell-to-cell spread through trans-cellular infection threads. (c) Endocytosis of rhizobia from thin-walled infection droplets; continued division of the released rhizobia with concomitant division of the plant-derived peribacteroid membrane is followed by differentiation of symbiosomes as a plant-microbe organelle containing nitrogen-fixing bacteroids.

successive stages of bacterial invasion and host control of the course of infection, using bacterial and plant genetics in combination with tissue-specific molecular probes.

This chapter is concerned with the role of lipopolysaccharide (LPS), which is a major component of the surface of *Rhizobium* (Carlson, 1984; Carlson *et al.*, 1988). The lipid part of these macromolecules is anchored in the bacterial outer membrane, while the carbohydrate moiety projects from the outer surface as the major immunogenic determinant (Brewin *et al.*, 1986). Mutants of *Rhizobium* with major defects in the structure of the carbohydrate components of LPS cannot establish a nitrogen-fixing symbiosis in legume root nodules (Carlson *et al.*, 1987; Cava *et al.*, 1989; Priefer, 1989). These observations point to an essential role for LPS in the development of nitrogen-fixing bacteroids and several possible functions have been proposed. Through their effects on membrane stability, LPS macromolecules may contribute to the resistance of bacteria to a variety of physiological stresses encountered within the nodule, for example, high or low pH, salt stress or low oxygen tension (Kannenberg & Brewin, 1989). Furthermore, through their carbohydrate composition, LPS macromolecules may be involved in plant–microbe surface interactions, perhaps involving specific molecular recognition (Karlsson, 1989; Maiti & Podder, 1989) or general 'compatibility' effects dependent on surface charge or hydrophobicity (de Maagd *et al.*, 1989). In order to investigate these possibilities, it will be necessary to characterise more precisely the carbohydrate components of the LPS macromolecules that are essential for symbiosis.

Unfortunately, it has proved very difficult to relate structure to function in LPS because of the complexity and heterogeneity of these molecules, especially in the higher molecular weight forms known as LPS-1. Not only is there pronounced variation from strain to strain (Carlson, 1984), but even within a single strain there is a family of LPS macromolecules with different sizes and compositions (Carlson *et al.*, 1988). Moreover the size distribution and antigenicity of LPS has been shown to vary between free-living and nitrogen-fixing (bacteroid) forms of *Rhizobium* (Brewin *et al.*, 1986; VandenBosch *et al.*, 1989*b*; Sindhu *et al.*, 1990), and between different free-living cultures of the same *Rhizobium* strain as a result of exposure to different physiological conditions such as low oxygen tension, low pH, or different carbon sources (Kannenberg & Brewin, 1989; Sindhu *et al.*, 1990). As a means to investigate LPS structure and function, we have isolated a range of mutants with defects in LPS biosynthesis, some of which were also found to be defective in nodule development and symbiotic nitrogen fixation (Fix$^-$). The goal of the mutagenesis programme was to isolate mutants with relatively minor structural defects in the LPS-1 macromolecule that resulted in the impairment of nodule development and symbiotic nitrogen fixation (Fix$^-$). Mutants isolated after nitrosoguanidine (NTG) or transposon mutagenesis were tested against a panel of monoclonal antibodies in order to identify minor variations in the molecular architecture of LPS. Subsequently, the LPS of particularly interesting mutant derivatives was examined by polyacrylamide gel electrophoresis using taurodeoxycholate as detergent in order to identify the molecular species that were missing or modified as a result of these *lps* mutations. In this way, the molecular structure of LPS could be correlated with its possible role in nodule development and symbiotic nitrogen fixation.

Results

The strain chosen for mutagenesis was *R. leguminosarum* bv. *viciae* 3841 (=300 *str*), because this strain has already been well characterised genetically (Beringer *et al.*, 1980) and physiologically (Glenn *et al.*, 1980; Wang *et al.*, 1982). The starting point for the present study was an LPS antigen of strain 3841 that had been identified using monoclonal antibody MAC 203 and was of particular interest because its expression was developmentally regulated (Bradley *et al.*, 1988). This epitope is a component of the LPS-1 macromolecular species that contains the O-antigen determinants (Diebold & Noel, 1989; Wood *et al.*, 1989). Immunogold labelling studies showed that the epitope is expressed in bacteroids but not in immature parts of the pea root nodule, nor in free living cultures under normal growth conditions (VandenBosch *et al.*, 1989*b*). Physiological studies demonstrated that the expression of MAC 203 antigen could be induced by growth of free-living cultures under low oxygen or low pH conditions (Kannenberg & Brewin, 1989). In order to investigate whether the developmentally regulated LPS epitope recognised by MAC 203 carries an important determinant for nodule development it was necessary to isolate mutants lacking this particular epitope and to identify the biochemical nature of the lesions involved.

Following mutagenesis with transposon Tn5, a derepressed mutant was obtained which gave constitutive expression of the LPS antigen recognised by MAC 203. This transposon-induced mutation, which mapped to a chromosomal location, simultaneously resulted in constitutive expression of MAC 203 antigen and no expression of another LPS-1 antigen recognised by monoclonal antibody MAC 281 (Wood *et al.*, 1989). Moreover, this mutant formed normal nitrogen-fixing nodules on peas, demonstrating that the presence of some LPS components (in this case the MAC 281 antigen), was not essential for a nitrogen-fixing symbiosis, as has been shown for some LPS-defective mutants of *R. meliloti* (Clover *et al.*, 1989). This strain (B574) was further mutagenised with NTG in order to isolate mutants that failed to react with MAC 203 (Table 1). These mutants were screened for reactivity with a panel of monoclonal antibodies following growth on culture medium containing either glucose or succinate: these substrates induce, respectively, a low or a high pH culture medium (Kannenberg & Brewin, 1989) and were used to test standard inducing or non-inducing conditions for MAC 203 antigen (Table 1). Where possible, ex-nodule bacteria were also recovered in sufficient quantities to test for reactivity with monoclonal antibodies.

Most of the mutants obtained (21/25) induced normal nitrogen-fixing (Fix$^+$) nodules on peas, but some induced symbiotically defective nodules that were deficient or delayed in bacterial release (endocytosis) into plant cells: inoculation of pea seedlings with these mutants resulted in nodules with little or no capacity for nitrogen fixation (Fix$^-$). Structural modifications of LPS from mutants were examined using a range of monoclonal

Table 1. *Reactivity of* Rhizobium leguminosarum *strains with monoclonal antibodies, following mutagenesis of strain B574 with nitrosoguanidine*

R. *leguminosarum* strain		Reactivity of MAC antibodies[b]							Symbiotic phenotype
Class	Culture[a]	203	281	300	301	251	57	302	
Wild-type	3841 S	−	+	+	(−)	+	+	+	
	G	+	+	+	−	+	+	+	
	N	+	+	+	+	+	+	+	Fix[+]
Parental	B574 S	+	−	+	(−)	+	+	+	
3841:Tn5	G	+	−	+	−	+	+	+	
	N	+	−	+	+	+	+	+	Fix[+]
Class I	B633 S	−	−	−	−	−	−	−	
	G	−	−	−	−	−	−	−	
	N	−	−	−	−	−	−	−	Fix[−]
Class II	B631 S	−	−	−	−	+	+	(−)	
	G	−	−	−	+	+	+	+	
	N	−	−	(+)	+	+	+	+	Fix[−]
Class III	B632 S	−	−	+	−	−	−	+	
	G	−	−	+	+	−	−	+	
	N	−	−	+	+	−	−	+	Fix[+]
Class IV	B636 S	−	(+)	−	−	−	+	+	
	G	−	−	+	+	−	+	+	
	N	+	−	+	+	+	+	+	Fix[+]

[a]Colony immunoassays were conducted using samples of bacteria previously cultured on succinate (S), glucose (G) or isolated from pea nodule homogenates (N).
[b]Relevant antigens gave: strong expression, +; reduced expression, (+); very weak expression, (−); no expression, −.

antibodies with different epitope specificities, and also by analysis of mobility changes after electrophoresis on polyacrylamide gels. Among the Fix⁻ mutants, two classes of LPS modification could be distinguished: those which completely lacked LPS-1 (the form of LPS that carries O-antigen), and those such as mutant B631 which showed a modified and truncated form of LPS-1 when grown under low oxygen or low pH culture conditions (Fig. 2). Mutants with an LPS structure and symbiotic phenotype similar to that of B631 were also obtained after Tn5 muta-genesis of *R. leguminosarum* 3841. All these mutations mapped to the

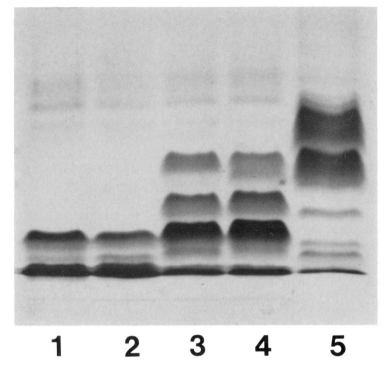

Fig. 2. Analysis of LPS molecules from extracts of *R. leguminosarum* 3841 and mutant derivatives. Bacterial samples were solubilised in taurodeoxycholate, fractionated by polyacrylamide gel electrophoresis, and visualised after periodate treatment and silver-staining. Track 1, B631 (Fix$^-$); track 2, B661 (Fix$^-$); track 3, B574 (Fix$^+$); track 4, 3841 (Fix$^+$); track 5, 3841 bacteroids.

same region of chromosomal DNA and could be suppressed by the same cosmid carrying a segment of DNA derived from the wild-type strain. These results suggest that strain B631 and related Fix$^-$ *lps* mutants are unable to polymerise O-antigen units onto a backbone LPS-1 molecule (Fig. 3). Moreover, because the LPS-1 components isolated from strain 3841 bacteroids are almost exclusively of the slow migrating (high molecular weight) form, it is presumed that these high molecular weight forms of LPS-1 are an essential component of normal nodule development.

Discussion

There are several current theories concerning the role of *Rhizobium* LPS in symbiosis, and they are not mutually exclusive. One possibility is that

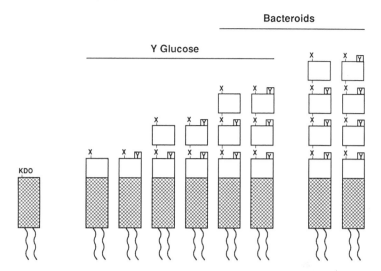

Fig. 3. Schematic model illustrating the successive addition of O-antigen units (square blocks) onto a 'core' oligosaccharide structure (oblong block), attached to lipid A chains embedded in the bacterial outer membrane. We propose that mutants such as B631 and B661 are unable to add the second O-antigen unit to the core structure under physiologically stressed conditions, such as low pH or low oxygen concentration. Hence, this Fix⁻ mutant cannot polymerise the higher molecular weight forms characteristic of the LPS of bacteroids of the parent strain 3841.

LPS stabilises the bacterial outer membrane against physiological stresses, such as extreme acidity, which might be encountered in the endophytic environment (Kannenberg & Brewin, 1989; Bhat *et al.*, 1991). Another possibility is that LPS promotes a physical interaction with the plant plasma membrane (Gharyal *et al.*, 1989), by providing either a surface that has appropriate charge density and hydrophobicity (de Maagd *et al.*, 1989), or a specific carbohydrate group that is recognised by the plant membrane (Ho *et al.*, 1990a,b).

Membrane-enclosed bacteria released into the plant cytoplasm eventually differentiate into nitrogen-fixing endosymbionts. These bacteria, termed bacteroids, are individually enclosed by the peribacteroid membrane, which is antigenically very similar to the plasma membrane (Perotto *et al.*, 1991). Following endocytosis, bacteroids continue to divide in conjunction with division of the peribacteroid membrane, so that in pea or clover nodule cells each bacterium is individually enclosed. (By contrast, in *Glycine* and *Phaseolus*, this synchrony breaks down later in nodule development so that a dozen bacteroids may be enclosed within a single membrane sac.) The mechanism that drives the division and segregation of peribacteroid membrane sacs may be related to the original mechanism of endocytosis or phagocytosis, through close surface interactions between bacterial and plant membranes. Cytological evidence from pea and soybean nodules suggests that the bacteroid membrane is not encapsulated by extracellular polysaccharide, and similarly the peribacteroid membrane is not masked by any plant cell wall material (Roth & Stacey, 1989*a,b*). Some evidence for surface adhesion between these two surfaces has been obtained by analysis of isolated membranes from nodule homogenates (Bradley *et al.*, 1986). However, the molecular basis for this interaction is not understood and could be investigated by experiments designed to test whether isolated LPS components or plant membrane glycoproteins or glycolipids are capable of direct molecular interaction with each other, as has been demonstrated in the case of bacterial pathogens interacting with animal cells (Karlsson, 1989).

Another general class of theory that is now testable by experiment suggests that the significance of developmentally regulated LPS structure may be related to the physiological stresses experienced by intracellular bacteria within the symbiosome compartment. Bacteroids of *Bradyrhizobium japonicum* synthesise stress proteins (H. Hennecke, personal communication), resembling GroEL and GroES, which are induced in *Salmonella* after intracellular infection of macrophages (Bucmeier & Heffron, 1990). Moreover, the LPS of *Rhizobium* also appears to be adapted to survive physiological stress because it carries 27-hydroxyoctacosanoic acid as a membrane-spanning component of lipid A (Bhat *et al.*, 1991), and because epitope modification of the O-antigen, which occurs during nodule development, can be simulated by culturing free-living bacteria under conditions of low oxygen or low pH (Kannenberg & Brewin, 1989). The symbiosome can be considered as a form of lysosomal compartment within the plant endomembrane system (Truchet & Coulomb, 1973; Mellor, 1989), and this may be the basis for the physiological stress encountered by bacteroids. Two characteristic activities of the peribacteroid membrane would tend to acidify the enclosed space: these are dicarboxylic acid transport (Yang *et al.*, 1990) and proton transport by a vectorial membrane ATPase (Udvardi & Day, 1989). Furthermore, Golgi-derived acid hydrolases are present in the peribacteroid fluid (Brewin, 1990). Hence the symbiosome apparently has the potential to develop into a lytic vesicle if the internal pH should drop significantly and this would lead, directly or indirectly, to the degradation of enclosed bacteroids. However, an essential feature of the *Rhizobium*–legume symbiosis may be that bacteroids can counterbalance the plant-induced acidification of the symbiosome compartment (Kannenberg & Brewin, 1989), first by the uptake of dicarboxylic acids for respiration, and second by the excretion of ammonia as the product of nitrogen fixation (Brewin, 1990). One obvious prediction of this model is that the pH of the peribacteroid space should decrease following growth of nitrogen-fixing nodules in an argon/oxygen atmosphere where the nitrogenase-dependent production of ammonia would be impossible. A second prediction would be that mutants of *Rhizobium* that were unable to polymerise O-antigen to make higher molecular weight forms of LPS might be more susceptible to pH or osmotic stress, and this might be the reason for their failure to survive in the intracellular environment.

Future work on the role of LPS in cell and tissue invasion during nodule development will involve a combination of molecular genetics, biochemistry and bacterial physiology. Site-directed transposon mutagenesis will extend the range of LPS-defective phenotypes available. The use of promoter probes coupled to reporter genes such as β-galactosidase or glucuronidase will permit the physiological analysis of *lps* gene expression, both *in vitro* and *in situ* during nodule development. A biochemical analysis of LPS is required in order to understand how the O-antigen groups are coupled together and connected to the 'core' oligosaccharide structure which is itself connected to lipid A: this investigation will be facilitated by the study of *lps* mutants and by the use of monoclonal antibodies as probes for LPS molecular substructure.

Conclusions

The results of our mutational analysis suggest that a part of the region of the LPS-1 macromolecule that is recognised by MAC203 antibody is essential for symbiotic nitrogen fixation. Moreover, our data suggest that a precondition for normal development of rhizobia within pea nodules is the ability to synthesise relatively long-chain LPS-1 macromolecules by polymerisation of the repeating O-antigen subunits under the physiologically stressed conditions encountered within the nodule. Because the LPS-1 components isolated from strain 3841 bacteroids are almost exclusively of the slow migrating (high molecular weight) form, it is presumed that these high molecular weight forms of LPS-1 are an essential component of normal nodule development. It seems probable that the apparent inability of strain B631 to polymerise O-antigen units onto a backbone LPS-1 macromolecule is the essential difference that distinguishes this Fix$^-$ mutant from other *lps* mutants which also carry antigenically modified forms of O-antigen but which still give rise to a Fix$^+$ phenotype (Fig. 3).

Acknowledgements

We are grateful to Dr G. W. Butcher and colleagues from the AFRC Institute of Animal Physiology and Genetics, Babraham, Cambridge, for their support in relation to the production and screening of monoclonal antibodies. E.L.K. was supported by Deutsche Forschungsgemeinschaft, and S.P. by the Trustees of the John Innes Foundation.

References

Beringer, J. E., Brewin, N. J. & Johnston, A. W. B. (1980). The genetics of *Rhizobium* in relation to symbiotic nitrogen fixation. *Heredity* **45**, 161–86.

Bhat, U. R., Mayer, H., Yokota, A., Hollingsworth, R. I. & Carlson, R. W. (1991). Occurrence of lipid A variants with 27-hydroxyocta-cosanoic acid in lipopolysaccharides from the *Rhizobiaceae* group. *Journal of Bacteriology* **173**, 2155–9.

Bradley, D. J., Butcher, G. W., Galfre, G., Wood, E. A. B. & Brewin, N. J. (1986). Physical association between the peribacteroid membrane and lipopolysaccharide from the bacteroid outer membrane in *Rhizobium*-infected pea root nodule cells. *Journal of Cell Science* **85**, 47–61.

Bradley, D. J., Wood, E. A., Larkins, A. P., Galfre, G., Butcher, G. W. & Brewin, N. J. (1988). Isolation of monoclonal antibodies reacting with peribacteroid membranes and other components of pea root nodules containing *Rhizobium leguminosarum*. *Planta* **173**, 149–60.

Brewin, N. J. (1990). The role of the plant plasma membrane in symbiosis. In *The Plant Plasma Membrane*, ed. C. Larsson & I. M. Moller, pp. 351–75. Berlin, Heidelberg, New York: Springer-Verlag.

Brewin, N. J. (1991). Development of the legume root nodule. *Annual Review of Cell Biology* **7**, 191–226.

Brewin, N. J., Wood, E. A., Larkins, A. P., Galfre, G. & Butcher, G. W. (1986). Analysis of lipopolysaccharide from root nodule bacteroids of *Rhizobium leguminosarum* using monoclonal antibodies. *Journal of General Microbiology* **132**, 1959–68.

Buchmeier, N. A. & Heffron, F. (1990). Induction of *Salmonella* stress proteins upon infection of macrophages. *Science* **248**, 730–2.

Carlson, R. W. (1984). The heterogeneity of *Rhizobium* lipopolysaccharides. *Journal of Bacteriology* **158**, 1012–17.

Carlson, R. W., Hollingsworth, R. L. & Dazzo, F. B. (1988). A core oligosaccharide component from the lipopolysaccharide of *Rhizobium trifolii* ANU843. *Carbohydrate Research* **176**, 127–35.

Carlson, R. W., Kalembasa, S., Turowksi, D., Pachori, P. & Noel, K. D. (1987). Characterisation of the lipopolysaccharide from a *Rhizobium phaseoli* mutant that is defective in infection thread development. *Journal of Bacteriology* **169**, 4923–8.

Cava, J. R., Elias, P. M., Turowski, D. A. & Noel, K. D. (1989). *Rhizobium leguminosarum* CFN42 genetic regions encoding lipopolysaccharide structures essential for complete nodule development on bean plants. *Journal of Bacteriology* **171**, 8–15.

Clover, R. H., Kieber, J. & Signer, E. R. (1989). Lipopolysaccharide mutants of *Rhizobium meliloti* are not defective in symbiosis. *Journal of Bacteriology* **171**, 3961–7.

de Maagd, R. A., Rao, A. S., Mulders, I. H. M., Goosen-de Roo, L., van Loosdrecht, M. C. M., Wijffelman, C. A. & Lugtenberg, B. J. J. (1989). Isolation and characterization of mutants of *Rhizobium leguminosarum* bv. *viciae* 248 with altered lipopolysaccharides: possible role of surface charge or hydrophobicity in bacterial release from the infection thread. *Journal of Bacteriology* **171**, 1143–50.

Diebold, R. & Noel, K. D. (1989). *Rhizobium leguminosarum* exopolysaccharide mutants: biochemical and genetic analyses and symbiotic behaviour on three hosts. *Journal of Bacteriology* **171**, 4821–30.

Gharyal, P. K., Ho, S.-C., Wang, J. L. & Schindler, M. (1989). O-antigen from *Bradyrhizobium japonicum* lipopolysaccharide inhibits intercellular (symplast) communication between soybean (*Glycine max*) cells. *Journal of Biological Chemistry* **264**, 12119–21.

Glenn, A. R., Poole, P. S.. & Hudman, J. F. (1980). Succinate uptake by free-living and bacteroid forms of *Rhizobium leguminosarum*. *Journal of General Microbiology* **119**, 267–71.

Ho, S.-C., Schindler, M. & Wang, J. L. (1990). Carbohydrate binding activities of *Bradyrhizobium japonicum*. II. Isolation and characterization of a galactose-specific lectin. *Journal of Cell Biology* **111**, 1639–43.

Ho, S.-C., Wang, J. L. & Schindler, M. (1990). Carbohydrate binding activities of *Bradyrhizobium japonicum*. I. Saccharide-specific inhibition of homotypic and heterotypic adhesion. *Journal of Cell Biology* **111**, 1631–8.

Kannenberg, E. L. & Brewin, N. J. (1989). Expression of a cell surface antigen from *Rhizobium leguminosarum* 3841 is regulated by oxygen and pH. *Journal of Bacteriology* **171**, 4543–8.

Karlsson, K. A. (1989). Animal glycosphingolipids as membrane attachment sites for bacteria. *Annual Review of Biochemistry* **58**, 309–50.

Maiti, T. K. & Podder, S. K. (1989). Differential binding of peanut agglutinin with lipopolysaccharide of homologous and heterologous *Rhizobium*. *FEMS Microbiology Letters* **65**, 279–84.

Mellor, R. B. (1989). Bacteroids in the *Rhizobium*–legume symbiosis inhabit a plant internal lytic compartment: implications for other microbial endosymbioses. *Journal of Experimental Botany* **40**, 831–9.

Perotto, S., VandenBosch, K. A., Butcher, G. W. & Brewin, N. J. (1991). Molecular composition and development of the plant glycocalyx associated with the peribacteroid membrane of pea root nodules. *Development*, **112**, 763–73.

Priefer, U. B. (1989). Genes involved in lipopolysaccharide production and symbiosis are clustered on the chromosome of *Rhizobium leguminosarum* biovar *viciae* VF39. *Journal of Bacteriology* **171**, 6161–8.

Roth, L. E. & Stacey, G. (1989*a*). Bacterium release into host cells of nitrogen-fixing soybean nodules: the symbiosome membrane comes from three sources. *European Journal of Cellular Biology* **49**, 13–23.

Roth, L. E. & Stacey, G. (1989*b*). Cytoplasmic membrane systems involved in bacterium release into soybean nodule cells as studied with two *Bradyrhizobium japonicum* mutant strains. *European Journal of Cellular Biology* **49**, 24–32.

Sindhu, S. S., Brewin, N. J. & Kannenberg, E. L. (1990). Immunochemical analysis of lipopolysaccharides from free-living and endosymbiotic forms of *Rhizobium leguminosarum*. *Journal of Bacteriology* **172**, 1804–13.

Truchet, G. & Coulomb, P. H. (1973). Mise en évidence et évolution du systéme phytolysosomal dans les cellules des différentes zones de nodules radiculaires de *pois* (*Pisum sativum* L.). Notion d'hétérophagie. *Journal of Ultrastructural Research* **43**, 36–57.

Udvardi, M. K. & Day, D. A. (1989). Electrogenic ATPase activity on the peribacteroid membrane of soybean (*Glycine max* L.) root nodules. *Plant Physiology* **90**, 982–7.

VandenBosch, K. A., Bradley, D. J., Knox, J. P., Perotto, S., Butcher, G. W. & Brewin, N. J. (1989*a*). Common components of the infection thread matrix and the intercellular space identified by immunocytochemical analysis of pea nodules and uninfected roots. *EMBO Journal* **8**, 335–42.

VandenBosch, K. A., Brewin, N. J. & Kannenberg, E. L. (1989*b*). Developmental regulation of a *Rhizobium* cell surface antigen during growth of pea root nodules. *Journal of Bacteriology* **171**, 4537–42.

Wang, T. L., Wood, E. A. & Brewin, N. J. (1982). Growth regulators, *Rhizobium* and nodulation in peas. The cytokinin content of a wild-type and Ti-plasmid containing strain of *R. leguminosarum*. *Planta* **155**, 350–5.

Wood, E. A., Butcher, G. W., Brewin, N. J. & Kannenberg, E. L. (1989). Genetic derepression of a developmentally regulated lipopolysaccharide antigen from *Rhizobium leguminosarum* 3841. *Journal of Bacteriology* **171**, 4549–55.

Yang, L.-J. O., Udvardi, M. K. & Day, D. A. (1990). Specificity and regulation of the dicarboxylate carrier on the peribacteroid membrane of soybean nodules. *Planta* **182**, 437–44.

Index